BIRKHÄUSER

Operator Theory: Advances and Applications

Vol. 208

Founded in 1979 by Israel Gohberg

Linearization Models for Complex Dynamical Systems

Topics in Univalent Functions, Functional Equations and Semigroup Theory

Mark Elin
David Shoikhet

L O L S | Linear Operators & Linear Systems

Birkhäuser

Authors:

Mark Elin
David Shoikhet
Department of Mathematics
ORT Braude College
Karmiel 21982
Israel
e-mail: mark_elin@braude.ac.il
davs@braude.ac.il

2010 Mathematics Subject Classification: 37Fxx, 30C45, 47H20, 47B33, 30D05

Library of Congress Control Number: 2010924097

Bibliographic information published by Die Deutsche Bibliothek.
Die Deutsche Bibliothek lists this publication in the Deutsche Nationalbibliografie;
detailed bibliographic data is available in the Internet at http://dnb.ddb.de

ISBN 978-3-0346-0508-3

P.O. Box 133, CH-4010 Basel, Switzerland
Part of Springer Science+Business Media
Printed on acid-free paper produced from chlorine-free pulp. TCF∞

ISBN 978-3-0346-0508-3
ISBN 978-3-0346-0509-0 (eBook)

9 8 7 6 5 4 3 2 1

www.birkhauser.ch

Contents

Preface

Interactions between Complex Analysis, Iteration Theory and Dynamical Systems have fascinated mathematicians for more than a century. Richness and beauty of these topics originated from the transparency of geometric ideas and depth of analytic methods forming a peculiarity of this area. Iteration algorithms and continuous dynamical processes are at the heart of many related fields, e.g., branching stochastic processes, composition operators, control theory and theory of linear operators in indefinite metric spaces. For recent books in these topics see, for example, [27, 104, 127, 130, 2, 40] and [120].

Each iteration procedure generates a discrete dynamical system, which can be considered a semigroup with respect to an integer parameter. The study of the behavior of such semigroups in a domain in the complex plane $\mathbb{C}$ goes back to the classical works of Julia, Fatou, Denjoy and Wolff and has been developed in different directions (including higher dimensional generalizations and dynamics in hyperbolic metric spaces and operator algebras, see [120] and references therein).

In parallel, and even earlier, the study of the local behavior of iterations near a fixed point was based on linearization models provided by the functional equations of Schröder, Abel and Bottcher. In 1884 Kœnigs showed how to solve Schröder's functional equation

$$h\left(F(z)\right) = \lambda h(z)$$

in a neighborhood of an attractive fixed point of F by using a limit scheme bearing his name.

Thus, the iteration properties of a given mapping F can be understood by means of the iteration properties of the linear mapping λw. This approach has been developed over a period of about a century by the combined efforts of many mathematicians (Valiron, Hadamard, Pommerenke, Baker and Cowen, among others). The key tool in these investigations is the ability to construct an intertwining mapping h for different types of holomorphic functions as well as the analysis of its uniqueness and geometric and analytic properties.

Although the theory of discrete dynamical systems was developed extensively, little was known about semigroups with respect to a continuous parameter. At the same time, continuous semigroups have various applications in Markov Stochastic

Processes [79, 126], Control Theory and Optimization [81]. For example, one of the central problems in Markov branching processes is to describe the asymptotic behavior of the process in a neighborhood of the so-called extinction probability which can be understood as the common attractive fixed point of a corresponding semigroup.

In 1978, Berkson and Porta [17] showed that each continuous semigroup $\{F_t\}_{t\geq 0}$ of holomorphic self-mappings on the open unit disk Δ is everywhere differentiable with respect to its parameter. Hence the limit

$$f(z) := \lim_{t\to 0^+} \frac{z - F_t(z)}{t} \tag{0.0.1}$$

exists and defines a holomorphic function f in Δ. This function is called the infinitesimal generator of the semigroup. Furthermore, it can be shown (see [17], [115] and [130]) that given f, the semigroup can be reproduced by the solution of the Cauchy problem:

$$\begin{cases} \dfrac{\partial F_t(z)}{\partial t} + f\left(F_t(z)\right) = 0, & t \geq 0, \\ F_0(z) = z, & z \in \Delta. \end{cases}$$

Moreover, in the same work Berkson and Porta showed that a holomorphic function f is a semigroup generator if and only if it admits the representation

$$f(z) = (z - \tau)(1 - z\overline{\tau})\, p(z), \quad z \in \Delta, \tag{0.0.2}$$

for some point $\tau \in \overline{\Delta}$ and some function $p \in \operatorname{Hol}(\Delta, \mathbb{C})$ with $\operatorname{Re} p(z) \geq 0$, $z \in \Delta$; and this representation is unique.

Based on this fact, Siskakis investigated in [136] continuous analogs of Schröder's and Abel's functional equations:

$$h(F_t(z)) = \lambda^t h(z) \tag{0.0.3}$$

and

$$h(F_t(z)) = h(z) + t. \tag{0.0.4}$$

It seems he was the first to focus on the connection between their solutions and the semigroup generators. In particular, he showed that if a one-parameter continuous semigroup has an interior fixed point, then Schröder's equation (0.0.3) (or Abel's equation (0.0.4) in the opposite case) should have the unique normalized solution.

The linearization models provided by these equations allowed Siskakis to study weighted semigroups of composition operators as well as the Cesàro and Volterra averaging operators on classical Hilbert and Banach spaces of analytic functions (Hardy spaces, Bergman spaces etc.).

Linearization models for continuous semigroups attracted special attention in the last decade. Recently, Contreras, Díaz-Madrigal and Pommerenke in a series of works [32, 33, 34] used these models to study dynamical properties of evolution equation and the boundary behavior of Kœnigs functions.

Also, we would like to emphasize that linearization models for discrete and continuous time dynamical systems are driving forces for the Composition Operator Theory on function spaces in the unit disk of the complex plane. Naturally, the eigenvalue problems for composition operators on Hardy, Bergman and Dirichlet spaces of analytic functions involve Schröder's functional equations. They can be considered linearization models for self-mappings of the underlying domain (see, for example, the books [127, 18] and [40]). In fact, as a part of operator theory, research into composition with a fixed function acting on a space of analytic functions is of recent origin, dating back to the mid-1960s, while investigation of continuous semigroups of composition operators started intensively from the work of Berkson and Porta at the beginning of the 1980s (see for example, the survey [136]).

Recent developments of the Denjoy–Wolff Theory and Julia–Carathéodory boundary versions of the Schwarz–Pick Lemma in the context of the generation theory for semigroups of holomorphic mappings lead to an independent study of linearization models for complex dynamical systems lying at the interface of analytic function theory and operator theory.

In addition, solutions of Schröder's and Abel's equations were applied to solve the Kœnigs embedding problem in [49] and the rigidity problem in the spirit of the Burns–Krantz theorem in [51, 53]. Different aspects of the boundary rigidity problem inspired by Abel's functional equation and its linearization model for holomorphic semigroups and generators can be integrated into the boundary interpolation theory for Schur's functions on the unit disk.

It turns out, that intertwining mappings for continuous semigroups have transparent geometric properties: they should be either starlike, spirallike, or convex in one-direction mappings. Thus these linearization models link to new questions in the classical Geometric Function Theory. Such questions have been investigated in [56, 5], as well as new covering and distortion theorems for spirallike and starlike functions studied in [47, 65, 132].

Finally, the nice geometry of linearization models enable the study of backward flow invariant domains for continuous semigroups [68, 69].

This book is devoted to a systematic and detailed survey and treatment of linearization models for one-parameter continuous semigroups, functional equations, different classes of univalent functions which serve intertwining mappings for these semigroups, and their applications to various problems of complex dynamics.

We believe that the book will be useful to a wide readership with an interest in these areas. Although many results have been previously published in journal papers, we endeavor to facilitate the understanding of their proofs to students and nonspecialists.

Some topics of this book have been given in special courses for students of ORT Braude College, Karmiel. More advanced issues were presented at the Seminars on Nonlinear Analysis at the Technion, Haifa, on Complex Analysis at the Bar–Ilan University, Ramat-Gan, on Operator Theory in the Banach Center of Polish Academy of Sciences, on Complex Analysis at the University Rome II Tor Vergata.

We would like to express our thanks to our colleagues who have read the manuscript, made useful remarks and suggested improvements. We are grateful to Professor Lawrence Zalcman who greatly influenced the conceptual approaches applied here. We also benefited from discussions with Professors Simeon Reich and Dov Aharonov. Many results included in the book were obtained due to collaboration with them.

We are very thankful to Professors Filippo Bracci, Manuel Contreras and Santiago Díaz-Madrigal for their many useful discussions related to these topics. Finally, special thanks to Marina Levenshtein who browsed through the entire manuscript and made helpful corrections.

Chapter 1

Geometric Background

1.1 Some classes of univalent functions

1.1.1 Starlike functions

Let D be a domain (open connected set) in the complex plane $\mathbb{C}$. The set of all holomorphic functions on D will be denoted by $\mathrm{Hol}(D,\mathbb{C})$. The subset of $\mathrm{Hol}(D,\mathbb{C})$, which consists of all univalent functions (with no normalization), we denote by $\mathrm{Univ}(D)$. The set of all holomorphic self-mappings of D we denote by $\mathrm{Hol}(D)$. This set is a semigroup with respect to composition operation.

The concept of univalent starlike functions was introduced by Alexander [7] in 1915 (see also Goodman [75] and Golusin [74]).

Definition 1.1. *A domain $\Omega \subset \mathbb{C}$ is said to be* **starlike** *(with respect to the origin) if for every point $w \in \Omega$, the linear segment joining w to zero:*

$$(0, w] := \{tw : \ t \in (0,1]\}$$

lies entirely in Ω.

It is clear that if a domain Ω is starlike, then $0 \in \overline{\Omega}$.

Let D be a simply connected domain in $\mathbb{C}$.

Definition 1.2. *A univalent function $h \in \mathrm{Hol}(D,\mathbb{C})$ is called a* **starlike function on** D *if $h(D)$ is a starlike domain.*

- *If $0 \in h(D)$, then h is said to be* **starlike with respect to an interior point**.
- *If $0 \in \partial h(D)$, then h is said to be* **starlike with respect to a boundary point**.

In Fig. 1.1 we see images of starlike functions: (i) — with respect to an interior point, and (ii) — with respect to a boundary point. We will mostly consider the case where $D = \Delta = \{z \in \mathbb{C} : \ |z| < 1\}$ is the open unit disk in $\mathbb{C}$.

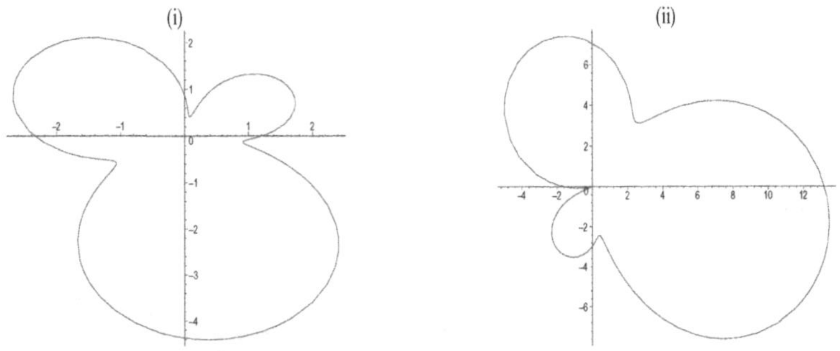

Figure 1.1: Images of starlike functions.

For any starlike function on Δ with respect to an interior point there exists a unique point $\tau \in \Delta$ such that $h(\tau) = 0$.

If h is a starlike function on Δ with respect to a boundary point, then there exists a unique point $\tau \in \partial\Delta$ and a sequence $\{z_n\}_{n=1}^{\infty} \subset \Delta$ converging to τ, such that $h(\tau) := \lim\limits_{n\to\infty} h(z_n) = 0$.

In both cases we use the notation $S^*[\tau]$ for the set of starlike functions (with respect to an interior or a boundary point) satisfying the condition $h(\tau) = 0$, $\tau \in \overline{\Delta}$.

The class $S^* \subset S^*[0]$ of starlike functions normalized by the conditions $h(0) = 0$ and $h'(0) = 1$ has been considered one of the first objects of interest in the classical geometric function theory.

1.1.2 Class $S^*[0]$. Nevanlinna's condition

The following well-known theorem is due to Nevanlinna [106] (see also [74, 75]).

Theorem 1.1. *Let $h \in \mathrm{Univ}(\Delta)$ and $h(0) = 0$. Then $h \in S^*[0]$ if and only if*

$$\operatorname{Re} \frac{zh'(z)}{h(z)} > 0, \quad z \in \Delta. \tag{1.1.1}$$

Remark 1.1. Actually, one can show that if a holomorphic function h satisfies (1.1.1) and is locally univalent at zero, that is, $h'(0) \neq 0$, then $h(0) = 0$ and $h \in \mathrm{Univ}(\Delta)$. Hence, in this case $h \in S^*[0]$.

Using Nevanlinna's condition we can easily check that if $h \in S^*[0]$, then the image $h(\Delta_r)$ of the disk $\Delta_r := \{z : |z| < r\}$ is a starlike domain for each $r \in (0, 1)$.

Stronger inequalities than Nevanlinna's condition (1.1.1) led to the study of some interesting subclasses of $S^*[0]$. In particular, in 1936 Robertson [121] introduced the class S^*_λ of **starlike functions of order** λ:

$$S^*_\lambda = \left\{ h \in S^* : \operatorname{Re} \frac{zh'(z)}{h(z)} > \lambda, \quad z \in \Delta \right\}, \tag{1.1.2}$$

where $\lambda \geq 0$.

It is clear that for $\lambda > 0$, $S^*_\lambda \subset S^*_0 = S^*$ and $S^*_\lambda = \emptyset$ when $\lambda > 1$ because $\lim\limits_{z\to 0} \dfrac{zh'(z)}{h(z)} = 1$ for any function $h \in S^*$.

Remark 1.2. Since $h \in S^*$ is normalized by the condition $h'(0) = 1$, it is easy to check (using Remark 1.1) that $h_1(z) := z \left(\dfrac{h(z)}{z}\right)^{1-\lambda} \in S^*_\lambda$.

Maximum values of many functionals on the class S^* as well as on the class $\operatorname{Univ}(\Delta)$ including Taylor coefficients (Bieberbach problem) are attained for the so-called **Koebe function**:

$$k(z) = \frac{z}{(1-z)^2} = z + 2z^2 + 3z^3 + \ldots + nz^n + \ldots = \sum_{n=1}^{\infty} nz^n. \tag{1.1.3}$$

This function is in S^*. It maps Δ in a one-to-one manner onto the domain that consists of the entire complex plane except for the slit along the negative real axis from $w = -\infty$ to $w = -1/4$.

In link with Remark 1.2, the function

$$k_\lambda(z) = \frac{z}{(1-z)^{2(1-\lambda)}}$$

will be referred to as the λ**-starlike Koebe function**.

1.1.3 Classes $S^*[\tau]$, $\tau \in \Delta$. Hummel's representation

By using the Möbius transformation $M_\tau(z) = \dfrac{\tau - z}{1 - \bar{\tau} z}$, $\tau \in \Delta$, one can easily describe the classes $S^*[\tau]$ of univalent starlike functions such that $h(\tau) = 0$, $\tau \in \Delta$.

Theorem 1.2. *Let $h \in \operatorname{Hol}(\Delta, \mathbb{C})$ be a locally univalent function on Δ. Then h belongs to $S^*[\tau]$, $\tau \in \Delta$, if and only if the function*

$$q(z) = \frac{h'(z)\,(z-\tau)(1-z\bar{\tau})}{h(z)}$$

is a well-defined holomorphic function on Δ with

$$\operatorname{Re} q(z) > 0.$$

A different formulation of this fact was given in 1978 by Wald in [144] (see also [75] and [130]). However, this result cannot be useful when τ approaches the boundary of Δ because of degeneration of the Möbius transformation M_τ.

Note also that if $h \in S^*[\tau]$ and

$$D_r(\tau) := \{z \in \mathbb{C} : |M_\tau(z)| < r\}$$

(the so-called pseudohyperbolic disk centered at τ, see [40, 130]), then the image $h(D_r(\tau))$ is a starlike domain for each $r \in (0,1)$.

Another deep connection between classes $S^*[0]$ and $S^*[\tau]$, $\tau \in \Delta$, was established by Hummel [84, 85]; see also [130].

Theorem 1.3. *Let $h_1 \in \mathrm{Hol}(\Delta, \mathbb{C})$, $h_1(0) = 0$, and $h_2 \in \mathrm{Hol}(\Delta, \mathbb{C})$ be two functions related by the formula:*

$$h_2(z) = \Psi_\tau(z) h_1(z), \qquad z \in \Delta, \tag{1.1.4}$$

where

$$\Psi_\tau(z) = \frac{1}{1 - |\tau|^2} \cdot \frac{(z-\tau)(1-z\bar{\tau})}{z}.$$

Then $h_1 \in S^*[0]$ if and only $h_2 \in S^*[\tau]$. Moreover, the inverse transform can be defined by the formula

$$h_1(z) = \Psi_\tau(M_\tau(z)) h_2(z).$$

Furthermore, Hummel proved that if $h \in S^*[\tau]$, $\tau \in \Delta$, then for each $\varepsilon > 0$, there exists $\rho \in (0,1)$ such that the function h satisfies a 'weak' Nevanlinna's condition:

$$\mathrm{Re}\, \frac{zh'(z)}{h(z)} > -\varepsilon$$

in the annulus $\rho < |z| < 1$; see [130] for details.

1.1.4 Spirallike functions. Špaček's condition

The class of spirallike functions (sometimes called also logarithmically spirallike) is an important generalization of starlike functions.

Definition 1.3. *A domain $\Omega \subset \mathbb{C}$ is said to be* **spirallike** *(with respect to the origin) if there is a number $\mu \in \mathbb{C}$ with $\mathrm{Re}\,\mu > 0$ such that for every point $w \in \Omega$ the spiral curve*

$$\{e^{-t\mu} w : \ t \geq 0\} \tag{1.1.5}$$

lies entirely in Ω.

It is clear that if the number μ is real, then Ω is actually starlike.

Since $\mathrm{Re}\,\mu > 0$ we have $\lim_{t\to\infty} e^{-t\mu} w = 0$ for any point $w \in \mathbb{C}$ and then $0 \in \overline{\Omega}$ for each spirallike domain Ω.

Definition 1.4. *If a univalent function h on a simply connected domain D maps D onto a spirallike domain, we say that h is a* **spirallike function**.

- *If for each point $w \in h(D)$ the image $h(D)$ contains the spiral curve* (1.1.5), *then h is said to be a* **μ-spirallike function**.
- *If $0 \in h(D)$, then the function h is said to be* **spirallike with respect to an interior point**.
- *If $0 \in \partial h(D)$, then the function h is said to be* **spirallike with respect to a boundary point**.

In Fig. 1.2 we see images of spirallike functions ((i) — with respect to an interior point, and (ii) — with respect to a boundary point) and spiral lines of the form $we^{-(1+i)t}$ which are contained in the images.

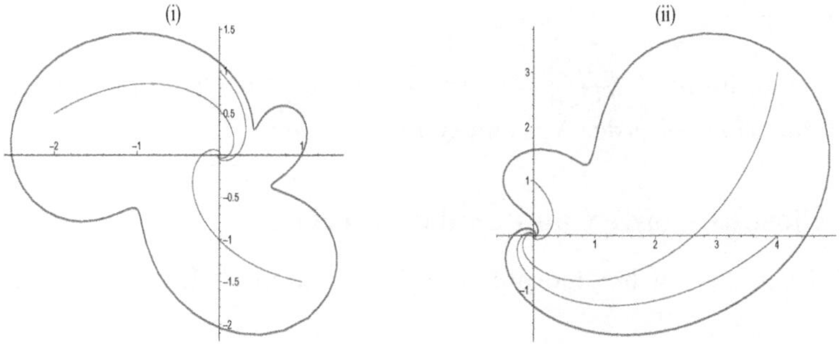

Figure 1.2: Images of spirallike functions.

If a spirallike function h on the open unit disk Δ has no interior null point (i.e., h is spirallike with respect to a boundary point), there exists a unique point $\tau \in \partial\Delta$ and a sequence $\{z_n\}_{n=1}^{\infty} \in \Delta$ converging to τ, such that $h(\tau) := \lim\limits_{n\to\infty} h(z_n) = 0$. For a point $\tau \in \overline{\Delta}$, the set of all spirallike functions h that satisfy $h(\tau) = 0$ will be denoted by $\mathrm{Spiral}[\tau]$.

It seems that Montel [105] was the first mathematician who suggested the study of spirallike functions. A generalization of Nevanlinna's condition was given by Špaček [137]. His result can be reformulated as follows.

Theorem 1.4. *Let $h \in \mathrm{Univ}(\Delta)$ have the form*

$$h(z) = z + \sum_{k=2}^{\infty} a_k z^k, \tag{1.1.6}$$

(i.e., $h(0) = 0$ and $h'(0) = 1$). Then $h \in \mathrm{Spiral}[0]$ if and only if

$$\operatorname{Re}\left[e^{-i\theta}\frac{zh'(z)}{h(z)}\right] > 0, \quad z \in \Delta, \tag{1.1.7}$$

for some $\theta \in (-\frac{\pi}{2}, \frac{\pi}{2})$. Moreover, h is μ-spirallike, where $\operatorname{Re}\mu > 0$, if and only if condition (1.1.7) holds for $\theta = \arg\mu$.

Similarly as for starlike functions, one can consider classes of functions satisfying a stronger condition than (1.1.7). To do so, we use the following notion.

Definition 1.5. *A function $h \in \mathrm{Spiral}[0]$ is said to be* **spirallike of order** λ *if for some $\theta \in \mathbb{R}$, it satisfies the following inequality:*

$$\operatorname{Re}\left[e^{-i\theta}\frac{zh'(z)}{h(z)}\right] > \lambda \geq 0, \quad z \in \Delta, \tag{1.1.8}$$

and $h'(0) = 1$.

Since $\lim\limits_{z\to 0} \operatorname{Re}\left[e^{-i\theta}\frac{zh'(z)}{h(z)}\right] = \cos\theta$ for any function $h \in \mathrm{Spiral}[0]$, the class of spirallike functions of order λ is empty for $\lambda > \cos\theta$.

1.1.5 Close-to-convex and φ-like functions

The following class of holomorphic functions was introduced by Kaplan in 1952 [89].

Definition 1.6. *A function $h \in \mathrm{Hol}(\Delta, \mathbb{C})$ is called* **close-to-convex** *if there is a function $h_1 \in S^*[0]$ such that*

$$\operatorname{Re}\frac{zh'(z)}{h_1(z)} > 0.$$

Obviously, any starlike function $h \in S^*[0]$ is close-to-convex. At the same time, one can show (see, for example, [75]) that any close-to-convex function belongs to $\mathrm{Univ}(\Delta)$, but not necessarily to $S^*[0]$.

Brickman [22] introduced another intensely studied generalization of classes $S^*[0]$ and $\mathrm{Spiral}[0]$. The generalization, in fact, includes all univalent functions preserving the origin.

Definition 1.7. *Let $h \in \mathrm{Hol}(\Delta, \mathbb{C})$ satisfy $h(0) = 0$ and $h'(0) \neq 0$. Let $\varphi \in \mathrm{Hol}(h(\Delta), \mathbb{C})$ satisfy $\varphi(0) = 0$ and $\operatorname{Re}\varphi'(0) > 0$. Then h is a* **φ-like function** *if*

$$\operatorname{Re}\frac{zh'(z)}{\varphi(h(z))} > 0.$$

If we set in this definition $\varphi(w) = aw$ with a real positive number a, then the class of φ-like functions with this choice of φ coincides with the class $S^*[0]$. Setting $\varphi(w) = \mu w$ with $\operatorname{Re}\mu > 0$, we get the subclass of Spiral[0] which consists of μ-spirallike functions. Furthermore, any close-to-convex function $h \in \mathrm{Hol}(\Delta, \mathbb{C})$ relative to a starlike function $h_1 \in S^*[0]$ is also φ-like with $\varphi(w) = h_1\left(h^{-1}(w)\right)$. As a matter of fact, each univalent function that preserves the origin can be represented as a φ-like function with a suitable choice of φ.

Theorem 1.5 ([22]). *Let $h \in \mathrm{Hol}(\Delta, \mathbb{C})$ satisfy $h(0) = 0$. Then $h \in \mathrm{Univ}(\Delta)$ if and only if h is φ-like for some φ.*

Below we discuss a dynamic approach to the study of starlike, spirallike and, more generally, φ-like functions on Δ. This approach will clarify Theorem 1.5.

1.2 Boundary behavior of holomorphic functions

Let f be a continuous function on the open unit disk Δ, and ζ be a boundary point of Δ. The limit

$$\lim_{r\to 1^-} f(r\zeta)$$

(if it exists) is called the **radial limit** of f at ζ. We say that f has the **unrestricted limit** A at the point $\zeta \in \partial\Delta$ if the limit

$$\lim_{z\to\zeta,\, z\in\Delta} f(z) = A$$

exists finitely. In this case one can define $f(\zeta) = A$ and then f becomes continuous in $\Delta \cup \{\zeta\}$.

A different notion of the limit at a boundary point is based on the so-called non-tangential approach regions at this point.

Definition 1.8. *For a point ζ on the unit circle $\partial\Delta$ and $\kappa > 1$, a* **non-tangential approach region** *at ζ is the set*

$$\Gamma(\zeta, \kappa) = \{z \in \Delta : \ |z - \zeta| < \kappa(1 - |z|)\}. \tag{1.2.1}$$

The term "non-tangential" refers to the fact that $\Gamma(\zeta, \kappa)$ lies in a sector W in Δ (with vertex at ζ), which is symmetric about the radius to ζ; i.e., the boundary curve of $\Gamma(\zeta, \kappa)$ has a corner at ζ of angle less than π.

Definition 1.9. *We will say that a complex-valued function f on Δ has a* **non-tangential (or angular) limit** *L at a point $\zeta \in \partial\Delta$ if $f(z) \to L$ as $z \to \zeta$, $z \in \Gamma(\zeta, \kappa)$ for each $\kappa > 1$. In this case we write*

$$L = \angle\lim_{z\to\zeta} f(z).$$

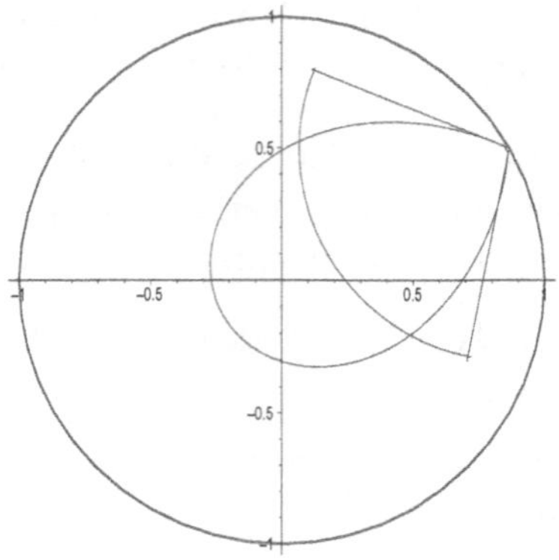

Figure 1.3: A non-tangential approach region and a Stolz angle at the point $\zeta = e^{i\pi/6}$.

For $0 < a < \frac{\pi}{2}$ and $r < 2\cos a$, a **Stolz angle** at $\zeta \in \partial\Delta$ is the set

$$D_{a,r}(\zeta) = \{z \in \Delta : \ |\arg(1 - z\bar{\zeta})| < a, \ |z - \zeta| < r\}$$

(see Fig 1.3).

It is clear that $f \in \mathrm{Hol}(\Delta, \mathbb{C})$ has a non-tangential limit L at a point ζ if and only if $f(z) \to L$ as $z \to \zeta$, for each Stolz angle $D_{a,r}(\zeta)$ at ζ.

Once again one can set $f(\zeta) := \angle \lim_{z\to\zeta} f(z)$.

Theorem 1.6 (Lindelöf's Theorem, [101]). *Let $\zeta \in \partial\Delta$ and let $f \in \mathrm{Hol}(\Delta, \mathbb{C})$ be bounded on each non-tangential approach region at ζ. If for some continuous curve $\gamma \in \Delta$ ending at ζ, there exists the limit*

$$L = \lim_{z\to\zeta} f(z), \quad z \in \gamma,$$

then the angular limit

$$\angle \lim_{z\to\zeta} f(z) = L$$

also exists.

Remark 1.3. For univalent functions the boundedness condition in this theorem can be omitted (see Corollary 2.17 [113]). That is, *a univalent function f has an angular limit at the point ζ if and only if it has a limit along some curve ending at ζ. Moreover, both limits coincide.*

Definition 1.10. *A holomorphic function f is said to have an* **angular derivative** *at a boundary point $\zeta \in \partial\Delta$, if the angular limit $f(\zeta)$ of f at ζ exists finitely and* *the angular limit*

$$\angle \lim_{z\to\zeta} \frac{f(z)-f(\zeta)}{z-\zeta} =: f'(\zeta) \tag{1.2.2}$$

exists.

Theorem 1.7 (see [113], p. 79). *Let $f \in \mathrm{Hol}(\Delta, \mathbb{C})$ and $\zeta \in \partial\Delta$. Then the function f has a finite angular derivative $f'(\zeta)$ at the point ζ if and only if f' has the finite angular limit $f'(\zeta)$ at ζ. In this case*

$$f'(\zeta) = \angle \lim_{z\to\zeta} f'(z).$$

Theorem 1.8 (see [113], p. 80). *Let f be a univalent function ($f \in \mathrm{Univ}(\Delta)$), and let f have a finite angular limit $f(\zeta)$ at $\zeta \in \partial\Delta$. If*

$$\lim_{z\to\zeta,\, z\in\Gamma} \frac{f(z)-f(\zeta)}{z-\zeta} = A \tag{1.2.3}$$

for some curve $\Gamma \subset \Delta$ ending at ζ, or if

$$\lim_{z\to\zeta,\, z\in\Gamma} f'(z) = A,$$

then $f'(\zeta) = A$ exists.

Definition 1.11. *Let $f \in \mathrm{Hol}(\Delta, \mathbb{C})$ and $\zeta \in \partial\Delta$.*

- *The function f is called* **conformal at the boundary point** *ζ if the angular derivative $f'(\zeta)$ exists finitely and $f'(\zeta) \neq 0$.*
- *The function f is called* **isogonal at the point** *ζ if there is a finite angular limit $f(\zeta)$ and if*

$$\angle \lim_{z\to\zeta} \arg \frac{f(z)-f(\zeta)}{z-\zeta} \tag{1.2.4}$$

exists finitely.

It is obvious that if f is conformal at the point $\zeta \in \partial\Delta$, then it is isogonal at ζ.

Theorem 1.9 (see [113], p. 81). *Let $f \in \mathrm{Univ}(\Delta)$ and $\zeta \in \partial\Delta$. The function f is isogonal at ζ if and only if f and $\arg f'$ have finite angular limits at ζ. In this case*

$$\angle \lim_{z\to\zeta} \frac{(z-\zeta)f'(z)}{f(z)-f(\zeta)} = 1. \tag{1.2.5}$$

The quotient

$$Q_f(\zeta, z) := \frac{(z-\zeta)f'(z)}{f(z)-f(\zeta)} \tag{1.2.6}$$

is usually called the **Visser–Ostrowski quotient**. Condition (1.2.5) is called the **Visser–Ostrowski condition.**

Note, however, that $f \in \mathrm{Hol}(\Delta, \mathbb{C})$ satisfying the Visser–Ostrowski condition is not necessarily isogonal (see [113]).

Definition 1.12. *We say that f satisfies a* **generalized Visser–Ostrowski condition** *if $Q_f(\zeta) := \angle \lim\limits_{z\to\zeta} Q_f(\zeta, z)$ exists finitely and is different from zero.*

1.3 The Julia–Wolff–Carathéodory and Denjoy–Wolff Theorems

In this section we continue to consider the boundary behavior of bounded holomorphic functions that are not necessarily univalent. Without loss of generality one can assume that $F \in \mathrm{Hol}(\Delta)$, i.e., the lower bound for $|F(z)|$ is the unity.

First let F be an arbitrary continuous self-mapping of the open unit disk Δ. Recall that a point $\zeta \in \Delta$ is said to be a **fixed point** of F if $F(\zeta) = \zeta$. Since we have not assumed that the function F is continuous on the boundary of Δ, we need to extend the notion of fixed point to include the case of a fixed point on the unit circle. Namely, a point $\zeta \in \overline{\Delta}$ will be called a **fixed point** of F if

$$\lim_{r\to 1^-} F(r\zeta) = \zeta. \tag{1.3.1}$$

Obviously, an interior fixed point of a self-mapping F of Δ is a common fixed point of all iterates $\{F^n\}_{n=1}^{\infty}$, $F^1 = F$, $F^{n+1} = F \circ F^n$.

In addition, *if a holomorphic self-mapping F has an interior fixed point ζ,* then the classical **Schwarz–Pick Lemma** asserts that *F leaves invariant each closed pseudohyperbolic disk $D_r(\zeta) := \left\{ z : \left| \frac{z-\zeta}{1-z\overline{\zeta}} \right| \le r \right\}$, $r < 1$, centered at this point* (see, for example, [40, 130]), i.e.,

$$\left| \frac{F(z)-\zeta}{1-F(z)\overline{\zeta}} \right| \le \left| \frac{z-\zeta}{1-z\overline{\zeta}} \right| .$$

Moreover, *if F is not the identity, then equality holds if and only if F is an elliptic automorphism of Δ.* Hence, *if F is not the identity mapping of Δ, then it has at most one fixed point inside the disk.*

At the same time, holomorphic functions may have many fixed points on the circle. It seems that the first descriptions of the boundary behavior of holomorphic self-mappings of the open unit disk were given by Julia (Julia's Lemma, [88]), Wolff (Boundary Schwarz–Wolff's Lemma, [145]–[147]) and Carathéodory (see [26]).

Combining their results one obtains the following characterization of a fixed-point-free holomorphic self-mapping of the unit disk Δ.

Theorem 1.10 (Julia–Wolff–Carathéodory Theorem). *Let $F \in \mathrm{Hol}(\Delta)$. Then the following are equivalent.*

(i) *F has no fixed points in Δ;*

(ii) *there is a unique unimodular point $\tau \in \partial\Delta$ such that*

$$\alpha := \angle \lim_{z \to \tau} \frac{F(z) - \tau}{z - \tau}$$

exists with $0 < \alpha \leq 1$;

(iii) *there is a unique unimodular point $\tau \in \partial\Delta$ such that*

$$\liminf_{z \to \tau} \frac{1 - |F(z)|}{1 - |z|} = \alpha \leq 1;$$

(iv) *there is a unique unimodular point $\tau \in \partial\Delta$ such that*

$$\sup_{z \in \Delta} \frac{\varphi_\tau(F(z))}{\varphi_\tau(z)} = \alpha \leq 1,$$

where

$$\varphi_\tau(z) := \frac{|1 - z\bar{\tau}|^2}{1 - |z|^2} = \frac{|z - \tau|^2}{1 - |z|^2}, \quad z \in \Delta. \tag{1.3.2}$$

Moreover,

(a) *the boundary points τ and the numbers α in* (i)–(iv) *are the same;*

(b) *the non-tangential limit in (ii) can be replaced by the radial limit.*

Remark 1.4. This theorem can be considered a boundary version of the Schwarz–Pick Lemma. Geometrically it means that each horocycle

$$\Delta_r(\tau) = \left\{ z \in \Delta : \frac{|z - \tau|^2}{1 - |z|^2} \leq r \right\}$$

internally tangent to $\partial\Delta$ at τ is F-invariant (see Fig 1.4).

As we already mentioned, a holomorphic self-mapping of Δ may have many boundary fixed points. However, for a fixed-point-free mapping F, condition (iv) means that there is a unique fixed point τ such that all the disks internally tangent to $\partial\Delta$ at τ are invariant.

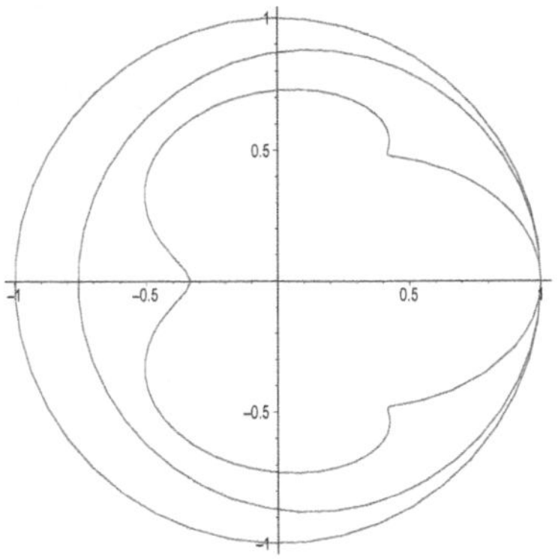

Figure 1.4: A horocycle and its image

Remark 1.5. The number α $(= F'(\tau))$ in this theorem is sometimes called the **Julia number of** F. Note that condition (iv) implies the inequality:

$$\frac{|F^n(z) - \tau|^2}{1 - |F^n(z)|^2} \leq \alpha^n \frac{|z - \tau|^2}{1 - |z|^2} \,. \tag{1.3.3}$$

So, if $\alpha < 1$, then $\tau \in \partial\Delta$ *is an attractive boundary fixed point of* F, *i.e., the semigroup* $\{F^n\}_{n=0}^{\infty}$ *of iterates of* F *converges to* τ *uniformly on each compact subset of* Δ *with the power rate of convergence in the sense of the non-euclidean "distance"* $\varphi_\tau(z)$ *defined by* (1.3.2).

The question is:

◊ *Whether this point is also attractive when* $\alpha = 1$*?*

The affirmative answer to this question is given in the next assertion, following Wolff [145, 146, 147] and Denjoy [41]; see also [40, 127, 117, 130].

Theorem 1.11 (Denjoy–Wolff Theorem). *If* $F \in \mathrm{Hol}(\Delta)$ *is not the identity and not an elliptic automorphism of* Δ, *then there is a unique point* $\tau \in \overline{\Delta}$ *such that the one-parameter discrete semigroup* $\mathcal{S} = \{F^n\}_{n=0}^{\infty}$ *of iterates of* F *converges to* τ *uniformly on compact subsets of* Δ.

The point $\tau \in \overline{\Delta}$ mentioned in Theorems 1.10 and 1.11 is called the **Denjoy–Wolff point** for a self-mapping F. If $\tau \in \partial\Delta$, it is also variously called **Wolff's point** or the **sink point** of F.

It is clear that if F has no fixed point in Δ then the limit point $\tau \in \partial\Delta$ of its iterates must be Wolff's point of F.

1.4 Functions of positive real part

We have already seen that the classes $S^*[\tau]$ and Spiral[0], as well as the classes of close-to-convex and φ-like functions, are characterized by a lower bound of the real part of some analytic expressions. Now let us recall some known properties of the class $\mathcal{P}$ of holomorphic functions in the open unit disk Δ of positive real part. Namely,

$$\mathcal{P} := \{p \in \mathrm{Hol}(\Delta, \mathbb{C}) : \ \mathrm{Re}\, p(z) \geq 0 \quad \text{for all } z \in \Delta\}$$

and

$$\mathcal{P}_0 := \{p \in \mathcal{P} : \ p(0) = 1\}.$$

The class $\mathcal{P}_0$ is usually called the **class of Carathéodory**.

It should be noted that $p \in \mathcal{P}$ is not required to be univalent.

In the same way that the Koebe function is essential for the class $S^*[0]$, the linear fractional function

$$C(z) := \frac{1+z}{1-z} = 1 + 2z + 2z^2 + 2z^3 + \ldots = 1 + 2\sum_{n=1}^{\infty} z^n$$

is crucial for the class $\mathcal{P}$. Sometimes this function is called the **Cayley transform**. It is univalent in Δ and maps Δ onto the right half-plane $\Pi_+ := \{w \in \mathbb{C} : \ \mathrm{Re}\, w > 0\}$.

The following remarkable fact was established by Herglotz in 1911 [83].

Theorem 1.12. *Let $p \in \mathrm{Hol}(\Delta, \mathbb{C})$. Then $p \in \mathcal{P}_0$ if and only if*

$$p(z) = \oint_{\partial\Delta} C(z\bar{\zeta})\, d\sigma_p(\zeta), \quad z \in \Delta, \tag{1.4.1}$$

for some probability measure σ_p on the unit circle $\partial\Delta$.

It follows by (1.4.1) that each element of class $\mathcal{P}$ can be represented in the form

$$p(z) = \int\limits_{\partial\Delta} \frac{1+z\bar{\zeta}}{1-z\bar{\zeta}}\, d\sigma_p(\zeta) + i\,\mathrm{Im}\, p(0) \tag{1.4.2}$$

with a positive measure σ_p on the unit circle $\partial\Delta$.

Formulas (1.4.1) and (1.4.2) are variously called the **Herglotz or Riesz–Herglotz integral representation** for classes $\mathcal{P}_0$ and $\mathcal{P}$, respectively.

Theorem 1.13 (Harnack's inequality). *If $p \in \mathcal{P}$, then it satisfies the estimates:*

$$\mathrm{Re}\, p(0)\, \frac{1-|z|}{1+|z|} \leq \mathrm{Re}\, p(z) \leq \mathrm{Re}\, p(0)\, \frac{1+|z|}{1-|z|}, \quad z \in \Delta. \tag{1.4.3}$$

Other more advanced properties of the class $\mathcal{P}$ are given below.

It follows by (1.4.2) that for each $\tau \in \partial\Delta$, the angular limit

$$\delta = (\delta_p(\tau)) := \angle \lim_{z \to \tau} (1 - z\bar{\tau}) p(z)$$

exists finitely and is a real nonnegative number.

We call this number the **charge** of p at the boundary point $\tau \in \partial\Delta$.

Consider the class

$$\mathcal{P}^+[\tau] = \{p \in \mathcal{P} : \ \delta_p(\tau) > 0\}$$

of functions with the strictly positive charge at a point $\tau \in \partial\Delta$.

By using the Julia–Wolff–Carathéodory Theorem (see Theorem 1.10) one can establish a somewhat more precise lower estimate for functions of the classes $\mathcal{P}$ and $\mathcal{P}^+[\tau]$ than the right-hand side of Harnack's inequality (1.4.3).

Without loss of generality we assume that $\tau = 1$.

Theorem 1.14 (see [132]). *A function p belongs to the class $\mathcal{P}$ if and only if the following inequality holds:*

$$\operatorname{Re} p(z) \geq a \frac{1-|z|^2}{|1-z|^2}, \quad z \in \Delta$$

for some $a \geq 0$. Moreover, in this case the number a can be set as $a = \frac{1}{2}\delta_p(1)$. Thus, p belongs to $\mathcal{P}^+[1]$ if and only if a can be found greater than zero.

In particular, for $z = x \in (-1, 1)$ we have

$$|\sigma_p| \frac{1+x}{1-x} \geq \operatorname{Re} p(x) \geq \frac{\delta_p(1)}{2} \cdot \frac{1+x}{1-x},$$

where σ_p is the Riesz–Herglotz measure for p and $|\sigma_p| = \int_{\partial\Delta} d\sigma_p(\zeta)$.

Now consider the class

$$\mathcal{P}^+[1,-1] = \{p \in \mathcal{P} : \ \delta_+ > 0 \ \text{ and } \ \delta_- > 0\},$$

where

$$\delta_+ = \delta_p(1) = \angle \lim_{z\to 1}(1-z)p(z), \quad \delta_- = \delta_{\frac{1}{p}}(-1) = \angle \lim_{z\to -1} \frac{z+1}{p(z)}.$$

First we note that it follows again from (1.4.2) that

$$\delta_+ = \delta_p(1) \leq 2\,|\sigma_p| = 2 \operatorname{Re} p(0)$$

and

$$\delta_- = \delta_{\frac{1}{p}}(-1) \leq 2 \operatorname{Re} \frac{1}{p(0)}.$$

Since $\operatorname{Re} p(0) \cdot \operatorname{Re} \dfrac{1}{p(0)} \leq 1$, we find that $\delta_+ \cdot \delta_- \leq 4$.

In addition, as in the above lemma, we have

$$\left|\frac{1}{p(z)}\right| \geq \operatorname{Re} \frac{1}{p(z)} \geq \frac{\delta_-}{2} \cdot \frac{1-|z|^2}{|1+z|}.$$

Again, since $\left|\frac{1}{p(z)}\right| \cdot \operatorname{Re} p(z) \le 1$ we obtain

$$|p(z)| \le \frac{2}{\delta_-} \cdot \frac{|1+z|^2}{1-|z|^2}.$$

Thus we can make the following assertion.

Theorem 1.15 (see [132]). *A function $p \in \mathcal{P}$ belongs to the class $\mathcal{P}^+[1,-1]$ if and only if*

$$b\,\frac{|1+z|^2}{1-|z|^2} \ge |p(z)| \ge \operatorname{Re} p(z) \ge a\,\frac{1-|z|^2}{|1-z|^2}$$

for some positive a and b. In this case the numbers a and b can be chosen to be $a = \frac{\delta_+}{2}$ and $b = \frac{2}{\delta_-}$. Moreover,

$$\delta_+ \cdot \delta_- \le 4,$$

and equality holds if and only if $p(z) = \alpha C(z) = \alpha \frac{1+z}{1-z}$ for some $\alpha > 0$.

In particular, if $z = x \in (-1,1)$ we have for $p \in \mathcal{P}^+[1,-1]$:

$$\frac{2}{\delta_-}\frac{1+x}{1-x} \ge \operatorname{Re} p(x) \ge \frac{\delta_+}{2}\frac{1+x}{1-x}.$$

Chapter 2

Dynamic Approach

Dynamic approach to the study of starlike and spirallike functions is based on the following observation. Let $f \in \mathrm{Univ}(\Delta)$, then f is spirallike if and only if there is $\mu \in \mathbb{C}$ with $\operatorname{Re} \mu > 0$ such that

$$e^{-\mu t} f(\Delta) \subset \Delta$$

for all $t \geq 0$. Then, for each $t \geq 0$, the function

$$F_t(z) = f^{-1}\left(e^{-\mu t} f(z)\right) \tag{2.0.1}$$

is a well-defined holomorphic self-mapping of the open unit disk Δ.

It is easy to verify that the family of these functions $\{F_t\}_{t\geq 0}$ satisfies the so-called semigroup property (see Definition 2.1 below) and if $f(\tau) = 0$, $\tau \in \Delta$, then $\lim\limits_{t\to\infty} F_t(z) = \tau$ for all $z \in \Delta$. Moreover, as we will see, $\lim\limits_{t\to\infty} F_t(z) = \tau \in \partial\Delta$ if and only if the angular limit $\angle \lim\limits_{z\to\tau} f(z)$ exists and equals zero. In the latter case f is a spirallike (or starlike) function with respect to a boundary point.

Therefore, to describe various properties of starlike and spirallike functions (with respect to an interior or to a boundary point), it should be natural to study the asymptotic behavior of one-parameter semigroups of holomorphic self-mappings of Δ.

2.1 Semigroups and generators

Let A be either $\mathbb{N}^+ := \{0, 1, 2, \dots\}$ or $\mathbb{R}^+ := [0, \infty)$ and let D be a domain in $\mathbb{C}$.

Definition 2.1. *A family $\mathcal{S} = \{F_t\}_{t\in A} \subset \mathrm{Hol}(D)$ of holomorphic self-mappings of D is called a* **one-parameter semigroup** *if*

(i) $F_{t+s} = F_t \circ F_s$ *for all s and t in A;*

(ii) $F_0(z) = z$ *for all $z \in \Delta$; that is, F_0 is the identity mapping on D.*

- *In the case when* $A = \mathbb{N}^+$, *the family* $\mathcal{S} = \{F_0, F_1, F_2, \dots, F_n, \dots\}$, $F_n \in \mathrm{Hol}(D)$, *is said to be a* **one-parameter discrete semigroup**.
- *In the case when* $A = \mathbb{R}^+$, *the family* $\mathcal{S} = \{F_t\}_{t\geq 0}$, $F_t \in \mathrm{Hol}(D)$, *is said to be a* **one-parameter continuous semigroup** *if*

$$\lim_{t\to s} F_t(z) = F_s(z) \quad \textit{for all} \quad s \geq 0 \quad \textit{and} \quad z \in D.$$

If $A = \mathbb{N}_+$, then a one-parameter discrete semigroup $\mathcal{S} = \{F_n\}_{n=0}^{\infty}$ actually consists of iterates of $F = F_1$, because of conditions (i) and (ii), i.e., $F_0 = I$, $F_n = F \circ F_{n-1}$, $n = 1, 2, \dots$.

If $A = \mathbb{R}_+$, then the only right continuity at zero of a semigroup, in fact, implies its continuity (right and left) on all of $\mathbb{R}^+ = [0, \infty)$. Moreover, in this case the semigroup is differentiable on $\mathbb{R}^+$ with respect to the parameter $t \geq 0$. For the one-dimensional case this result is due to Berkson and Porta [17] (see also [130]); for the higher-dimensional case see Abate[3] (see also Reich and Shoikhet [115, 118] for Banach spaces).

Theorem 2.1 (see [130], pp. 63, 66). *Let* D *be a simply connected domain in* $\mathbb{C}$. *Let* $\mathcal{S} = \{F_t\}_{t\geq 0}$ *be a one-parameter semigroup of holomorphic self-mappings of* D, *such that for each* $z \in D$,

$$\lim_{t\to 0^+} F_t(z) = z. \tag{2.1.1}$$

Then for each $z \in D$ *there exists the limit*

$$\lim_{t\to 0^+} \frac{z - F_t(z)}{t} = f(z), \tag{2.1.2}$$

which is a holomorphic function on D. *The convergence in* (2.1.2) *is uniform on each subset strictly inside* D. *Moreover, the semigroup* $\mathcal{S}$ *can be defined as a (unique) solution of the Cauchy problem:*

$$\begin{cases} \dfrac{\partial F_t(z)}{\partial t} + f(F_t(z)) = 0, \quad t \geq 0, \\ F_0(z) = z, \quad z \in D. \end{cases} \tag{2.1.3}$$

So, a semigroup satisfying (2.1.1) is, in fact, differentiable in parameter, hence, continuous.

Definition 2.2. *The function* $f \in \mathrm{Hol}(D, \mathbb{C})$ *defined by the limit* (2.1.2) *is called the* **(infinitesimal) generator** *of the one-parameter continuous semigroup* $\mathcal{S} = \{F_t\}_{t\geq 0}$.

Definition 2.3. *A function* $f \in \mathrm{Hol}(\Delta, \mathbb{C})$ *is called a* **semi-complete** *(respectively,* **complete***)* **vector field** *if for any initial point* $z \in \Delta$, *the Cauchy problem* (2.1.3) *has the unique solution* $\{F_t(z)\} \subset \Delta$ *defined for all* $t \geq 0$ *(respectively, for all* $-\infty < t < \infty$*).*

Thus, Theorem 2.1 asserts that *a holomorphic function f on Δ is a semi-complete vector field if and only if it is a generator of a one-parameter continuous semigroup of holomorphic self-mappings of Δ.*

The family of all holomorphic generators on D will be denoted by $\mathcal{G}(D)$. If D is convex, this set is a real cone in $\mathrm{Hol}(D,\mathbb{C})$ [115, 130]. Different descriptions of the class $\mathcal{G} = \mathcal{G}(\Delta)$ (or semi-complete vector fields on Δ) can be found in [4, 6, 17, 76]. We will formulate some of them in the next section.

Meanwhile observe that since a discrete semigroup is completely defined by an arbitrary self-mapping of Δ, its elements are not necessarily univalent functions. At the same time, it follows by the uniqueness of the solution of the Cauchy problem (2.1.3) that *each element of a continuous semigroup is a univalent function on Δ.*

Furthermore, an element of a semigroup $\mathcal{S}$ maps Δ onto Δ if and only if the generator f of $\mathcal{S}$ (defined by (2.1.2)) is a complete vector field. In this case the semigroup $\mathcal{S}$ can be extended to a one-parameter group that automatically consists of automorphisms of Δ, so that

$$(F_t)^{-1} = F_{-t}, \quad -\infty < t < \infty.$$

The set of complete vector fields on D (that is, the set of group generators) is usually denoted by $\mathrm{aut}(D)$ (see, for example, [87], [142], [130]).

2.2 Flow invariance conditions and parametric representations of semigroup generators

In this section we make references to some criteria that guarantee that a holomorphic function $f \in \mathrm{Hol}(\Delta,\mathbb{C})$ is a semi-complete vector field, or, which is the same, a semigroup generator on Δ.

First, we suppose that a function $f \in \mathrm{Hol}(\Delta,\mathbb{C})$ is continuous on the closed disk $\overline{\Delta}$. The following fact was established in [115] (see also [130] for details).

Theorem 2.2 (Boundary flow invariance condition). *Let $f \in \mathrm{Hol}(\Delta,\mathbb{C})$ have a continuous extension to $\overline{\Delta}$. Then $f \in \mathcal{G}$ if and only if*

$$\mathrm{Re}\Big(f(z)\bar{z}\Big) \geq 0 \quad \textit{for all} \quad z \in \partial\Delta.$$

This theorem gives a simple criterion that makes a holomorphic function, which is continuous on the closed disk, a semigroup generator. However, there are holomorphic functions on the open disk Δ that have no continuous extension to $\overline{\Delta}$. To describe such generators we need interior flow invariance conditions. Such a criterion was given in Aharonov *et al* [6].

Theorem 2.3. *Let $f \in \mathrm{Hol}(\Delta,\mathbb{C})$. Then $f \in \mathcal{G}$ if and only if*

$$\mathrm{Re}\Big(f(z)\bar{z}\Big) \geq \mathrm{Re}\Big(f(0)\bar{z}\Big)(1-|z|^2), \quad z \in \Delta. \tag{2.2.1}$$

Moreover, the equality in (2.2.1) *holds if and only if f is a complete vector field, i.e., f is the generator of a group of automorphisms.*

Using this theorem and Harnack's inequality (1.4.3), one can establish a more qualified flow invariance condition, which can be considered a distortion theorem for the class $\mathcal{G}$.

Theorem 2.4 (see [6] and Proposition 3.5.3 [130]). *A function $f \in \mathrm{Hol}(\Delta, \mathbb{C})$ is a semi-complete vector field if and only if $\mathrm{Re}\, f'(0) \geq 0$ and the following inequality holds:*

$$\mathrm{Re}\Big(f(0)\bar{z}\Big)(1-|z|^2) + \mathrm{Re}\, f'(0)\,\frac{1+|z|}{1-|z|}\,|z|^2 \geq \mathrm{Re}\Big(f(z)\bar{z}\Big)$$
$$\geq \mathrm{Re}\Big(f(0)\bar{z}\Big)(1-|z|^2) + \mathrm{Re}\, f'(0)\,\frac{1-|z|}{1+|z|}\,|z|^2 \tag{2.2.2}$$

for all $z \in \Delta$. Moreover, the equality in (2.2.2) *holds if and only if* $\mathrm{Re}\, f'(0) = 0$. *In this case f is complete.*

Theorem 2.5 below is an infinitesimal version of the Schwarz–Pick Lemma. To formulate it recall that the (hyperbolic) Poincaré metric on Δ is a function $\rho\colon \Delta \times \Delta \mapsto \mathbb{R}^+$ defined by

$$\rho(z,w) = \frac{1}{2}\log\frac{1+|M_z(w)|}{1-|M_z(w)|}, \qquad \text{where } M_z(w) = \frac{z-w}{1-\bar{z}w}. \tag{2.2.3}$$

Definition 2.4. *A complex-valued continuous function f on Δ is called ρ-***monotone** *if for each pair $z, w \in \Delta$ and for $r \geq 0$ the following condition holds:*

$$\rho(z + rf(z), w + rf(w)) \geq \rho(z,w), \tag{2.2.4}$$

whenever $z + rf(z)$ and $w + rf(w)$ belong to Δ.

Theorem 2.5 (see [116]). *Let $f \in \mathrm{Hol}(\Delta, \mathbb{C})$. The following assertions are equivalent:*

(i) $f \in \mathcal{G}$;

(ii) *f is a ρ-monotone function;*

(iii) *the function f satisfies the inequality*

$$\mathrm{Re}\left[\frac{f(z)\bar{z}}{1-|z|^2} + \frac{f(w)\bar{w}}{1-|w|^2}\right] \geq \mathrm{Re}\,\frac{\bar{z}f(w) + w\overline{f(z)}}{1-\bar{z}w} \tag{2.2.5}$$

for all $z, w \in \Delta$.

We have already mentioned that the class $\mathcal{G}$ is a real cone. It turns out that this cone is isomorphic to the cone $\mathbb{C} \times \mathcal{P}$.

Theorem 2.6 (see [4]). *A function $f \in \mathrm{Hol}(\Delta, \mathbb{C})$ belongs to $\mathcal{G}$ if and only if it admits the following parametric representation*

$$f(z) = a - \bar{a}z^2 + zq(z), \tag{2.2.6}$$

where $a \in \mathbb{C}$ and $q \in \mathcal{P}$. Moreover, f is complete if and only if $\operatorname{Re} q(z) = 0$, *i.e., $q(z) \equiv ib$, $b \in \mathbb{R}$.*

Thus all complete vector fields are, in fact, polynomials of at most order 2 of the form

$$f(z) = a - \bar{a}z^2 + ibz, \ a \in \mathbb{C}, \ b \in \mathbb{R}.$$

Another useful parametric representation of the class $\mathcal{G}$ is due to Berkson and Porta [17] (see, also [4] and [130]).

Theorem 2.7. *A function $f \in \mathrm{Hol}(\Delta, \mathbb{C})$ is of the class $\mathcal{G}$ if and only if there are a point $\tau \in \overline{\Delta}$ and a function $p \in \mathcal{P}$ such that*

$$f(z) = (z - \tau)(1 - z\bar{\tau})p(z). \tag{2.2.7}$$

Moreover, this representation is unique.

It is clear that if τ in (2.2.7) is of modulus less than one, then it is a null point of $f \in \mathcal{G}$. Moreover, the Berkson–Porta formula (2.2.7) shows that any function $f \in \mathcal{G}$ must have at most one null point inside Δ. In this case formula (2.2.7) can be considered a characterization of a subcone of $\mathcal{G}$ of generators that vanish at the point $\tau \in \Delta$:

$$\{f \in \mathcal{G} : \ f(\tau) = 0\} = \left\{f \in \mathcal{G} : \ f(z) = (z - \tau)(1 - z\bar{\tau})p(z), \ p \in \mathcal{P}\right\}. \tag{2.2.8}$$

If $\tau \in \partial\Delta$, then f has no interior null point in Δ. However, it turns out that

$$f(\tau) := \angle \lim_{z \to \tau} f(z) = 0,$$

i.e., f has at least one boundary null point $\tau \in \partial\Delta$. It can be shown, by using the Riesz–Herglotz integral representation (1.4.1) and the Berkson–Porta formula (2.2.7), that the angular derivative

$$f'(\tau) = \angle \lim_{z \to \tau} \frac{f(z)}{z - \tau} \ \left(= \angle \lim_{z \to \tau} f'(z)\right)$$

exists finitely and is a real nonnegative number. (The precise formulation of this fact is given below in Theorem 2.10.)

Note once again that a function $f \in \mathcal{G}$ may have more than one boundary null point. For our purposes, we are especially interested in the class $\mathcal{G}[\zeta]$, $\zeta \in \overline{\Delta}$, defined by

$$\mathcal{G}[\zeta] := \left\{f \in \mathcal{G} : \ f(\zeta) = 0 \ \text{ and } \ f'(\zeta) \ \text{ exists finitely}\right\}. \tag{2.2.9}$$

Observe that if $\zeta \in \Delta$, then it must be the unique interior null point of f and $f'(\zeta)$ exists. Therefore, in this case $\mathcal{G}[\zeta]$ coincides with the cone defined in (2.2.8) with $\tau = \zeta$. In the case when $\zeta \in \partial\Delta$, one can show that for each $f \in \mathcal{G}[\zeta]$, the angular derivative $f'(\zeta)$ is a real number.

In general, a point $\zeta \in \partial\Delta$ is called a **boundary regular null point** of $f \in \mathrm{Hol}(\Delta, \mathbb{C})$ if $\angle \lim_{z\to\zeta} f(z) = 0$ and $\angle \lim_{z\to\zeta} f'(z)$ exists finitely.

Example 2.1. Consider two holomorphic functions f_1 and f_2 defined by $f_1(z) = z(1-z^n)$ and $f_2(z) = z\sqrt{1-z^n}$. By Theorem 2.7, f_1 and f_2 belongs to $\mathcal{G}$. Obviously, they have the same null point set $\left\{0, z_k = e^{\frac{2\pi ik}{n}},\ k = 0, 1, \dots, n-1\right\}$. The points $\{z_k\}_{k=0}^{n-1}$ are boundary regular null points for f_1 while they are irregular for f_2.

The following assertion is a continuous analog of the classical Julia–Carathéodory Theorem (see, for example, [40, 127]). Without loss of generality one can assume that $\zeta = 1$.

Theorem 2.8 (see [132], cf. also [62, 64, 34, 59]). *Let $f \in \mathcal{G}$. The following assertions are equivalent:*

(i) $f \in \mathcal{G}[1]$, *i.e.,* $\eta = 1$ *is a boundary regular null point of* f;

(ii) $\limsup_{z\to 1} \mathrm{Re}\, \dfrac{f(z)}{z-1} =: \beta > -\infty$;

(iii) *the function f admits the following representation*

$$f(z) = -(1-z)^2 q(z) + \frac{\beta}{2}\left(z^2 - 1\right), \tag{2.2.10}$$

where $\beta \in \mathbb{R}$, $q \in \mathcal{P}$ *and* $\angle \lim_{z\to 1}(z-1)q(z) = 0$.

Remark 2.1. The above theorem with a stronger condition (see Section 2.4 below):

(ii′) $\liminf_{z\to 1} \mathrm{Re}\, \dfrac{f(z)}{z-1} =: \beta > -\infty$

instead of (ii) was proved in [132]. It was shown in [59] that the lower limit can be replaced by the upper limit.

Observe also that the equivalence of conditions (ii) and (iii) implies that $f'(1)$ must be a real number that equals β.

Now comparing formulas (2.2.7) and (2.2.10) we see that τ in (2.2.7) can be equal to 1 if and only if β in (2.2.10) is nonnegative. In this case formula (2.2.7) becomes

$$f(z) = -(1-z)^2 p(z), \tag{2.2.11}$$

and, consequently, representations (2.2.11) with $p \in \mathcal{P}$ of the form $p(z) = q(z) + \dfrac{\beta}{2}\dfrac{1+z}{1-z}$ and (2.2.10) are equivalent.

The subclass of $\mathcal{G}$ of the generators for which

$$f(\zeta) := \angle \lim_{z\to\zeta} f(z) = 0 \quad \text{and} \quad \beta = f'(\zeta) := \angle \lim_{z\to\zeta} f'(z) \tag{2.2.12}$$

with $\operatorname{Re}\beta \geq 0$ will be denoted by $\mathcal{G}^+[\zeta] \subset \mathcal{G}[\zeta]$ (obviously, if $\zeta \in \Delta$, then the cones $\mathcal{G}^+[\zeta]$ and $\mathcal{G}[\zeta]$ coincide).

Finally, following [76] and [132] we introduce the subcone $\mathcal{G}[1,-1]$ by the formula

$$\mathcal{G}[1,-1] := \mathcal{G}^+[1] \bigcap \mathcal{G}[-1]. \tag{2.2.13}$$

Corollary 2.1 (see [132]). *A function $f \in \operatorname{Hol}(\Delta, \mathbb{C})$ is of the class $\mathcal{G}[1,-1]$ if and only if it admits the representation (2.2.10) with $\beta \geq 0$ and $q \in \mathcal{P}$ such that*

$$\angle \lim_{z\to -1} \frac{q(z)}{z+1}$$

exists finitely and

$$\angle \lim_{z\to 1} (z-1)q(z) = 0.$$

Another representation of the class $\mathcal{G}[1,-1]$ was given by Goryainov in [76] (see Section 2.4 below).

Since the classes $\mathcal{G}[1]$ and $\mathcal{G}^+[1]$ play a crucial role in our study of the asymptotic behavior of one-parameter continuous semigroups, as well as in the study of starlike and spirallike functions with respect to a boundary point, we prove Theorem 2.8 simultaneously in Section 2.4. In addition, the class $\mathcal{G}[1,-1]$ is useful to describe unbounded starlike and spirallike functions.

2.3 The Denjoy–Wolff and Julia–Wolff–Carathéodory Theorems for semigroups

Recall that for a continuous semigroup $\mathcal{S} = \{F_t\}_{t\geq 0}$, $F_t \in \operatorname{Hol}(\Delta)$, a common fixed point $\zeta \in \Delta$ of its elements:

$$F_t(\zeta) = \zeta \quad \text{for all } t \geq 0$$

is called a **stationary point** of the semigroup. Thus *a continuous semigroup has at most one stationary point in Δ*. Moreover, it follows by the semigroup property that $\zeta \in \Delta$ is a stationary point of $\mathcal{S} = \{F_t\}_{t\geq 0}$ if and only if $F_{t_0}(\zeta) = \zeta$ at least for one $t_0 > 0$.

In addition, the uniqueness of the solution of the Cauchy problem (2.1.3) implies that *a point $\zeta \in \Delta$ is a stationary point of a continuous semigroup $\mathcal{S} = \{F_t\}_{t\geq 0}$ if and only if it is a null point of the semigroup generator $f \in \mathcal{G}$*. In this case $f \in \mathcal{G}[\zeta]$, i.e., it has the form (2.2.7) with $\tau = \zeta$.

Since our main object of interest is starlike and spirallike functions with respect to a boundary point, we turn now to the asymptotic behavior of semigroups with no stationary point in Δ.

For continuous semigroups an analog of the Denjoy–Wolff Theorem can be found in [17, 117, 119, 130].

Theorem 2.9 (Denjoy–Wolff Theorem for semigroups). *Let $S = \{F_t\}_{t\geq 0} \subset \mathrm{Hol}(\Delta)$ be a continuous semigroup of holomorphic self-mappings on Δ. If for at least one t_0 the function F_{t_0} is not the identity and is not an elliptic automorphism of Δ, then there is a unique point $\tau \in \overline{\Delta}$ such that the semigroup $\{F_t\}_{t\geq 0}$ converges to τ as $t \to \infty$ uniformly on compact subsets of Δ.*

The point τ in Theorem 2.9 is called the **Denjoy–Wolff point** of a continuous one-parameter semigroup $\mathcal{S} = \{F_t\}_{t\geq 0}$. If $\tau \in \partial\Delta$, it is also variously called **Wolff's point** or the **sink point** of a semigroup.

The asymptotic behavior of continuous semigroups can be also described in terms of their generators. A **continuous version of the Julia–Wolff–Carathéodory Theorem** was established in [62].

Theorem 2.10. *Let $\mathcal{S} = \{F_t\}_{t\geq 0}$ be a one-parameter continuous semigroup generated by $f \in \mathcal{G}$. The following are equivalent:*

(i) *f has no null point in Δ;*

(ii) *f admits the Berkson–Porta representation*

$$f(z) = (1 - z\bar{\tau})(z - \tau)p(z)$$

for some $\tau \in \partial\Delta$ and $\operatorname{Re} p(z) \geq 0$ everywhere;

(iii) *there is a point $\tau \in \partial\Delta$ such that $f \in \mathcal{G}^+[\tau]$;*

(iv) *there is a point $\tau \in \partial\Delta$ such that*

$$\angle \lim_{z\to\tau} \frac{f(z)}{z-\tau} = \beta$$

exists and $\operatorname{Re} \beta \geq 0$;

(v) *there are a point $\tau \in \partial\Delta$ and a real positive number γ, such that*

$$\frac{|F_t(z) - \tau|^2}{1 - |F_t(z)|^2} \leq e^{-t\gamma} \frac{|z - \tau|^2}{1 - |z|^2}. \tag{2.3.1}$$

Moreover,

(a) *the points $\tau \in \partial\Delta$ in* (ii)–(v) *are the same;*

(b) *the number β in (iv) is, in fact, a nonnegative real number, which is the maximum of all $\gamma \geq 0$ that satisfy* (2.3.1).

Remark 2.2. Note that if $f \in \mathcal{G}[\zeta]$, $\zeta \in \partial\Delta$, with $f'(\zeta) = \beta \in \mathbb{R}$ (not necessarily nonnegative), then the point ζ is also a common fixed point for the semigroup $\mathcal{S} = \{F_t\}_{t\geq 0}$ generated by f. In addition, $\dfrac{\partial F_t}{\partial z}(\zeta) = e^{-t\beta}$ and inequality (2.3.1) holds even if $\beta < 0$ (see [132, 64]). In particular, for the class $\mathcal{G}[1,-1] = \mathcal{G}^+[1] \cap \mathcal{G}[-1]$ we have the following estimates

$$\frac{|F_t(z)-1|^2}{1-|F_t(z)|^2} \leq e^{-t\beta_+}\frac{|z-1|^2}{1-|z|^2}$$

and

$$\frac{|F_t(z)+1|^2}{1-|F_t(z)|^2} \leq e^{-t\beta_-}\frac{|z+1|^2}{1-|z|^2},$$

where

$$\beta_+ = f'(1) \geq 0 \quad \text{and} \quad \beta_- = f'(-1) < 0.$$

We prove these facts (including Theorem 2.10), which are key points in our discussion in the next section.

Here we just observe that in contrast to the case of interior null points, a boundary null point ζ of a generator f may be not a fixed point of the generated semigroup if $f'(\zeta)$ does not exist finitely (see Example 1, p. 104 in [130]).

2.4 Generators with boundary null points

Our purpose in this section is an advanced study of the classes $\mathcal{G}[\zeta]$, $\mathcal{G}^+[\zeta]$ and $\mathcal{G}[1,-1]$ defined in Section 2.2.

- *By $\mathcal{F}[\zeta]$ we denote the class of holomorphic self-mappings of Δ with a* **boundary regular fixed point** *$\zeta \in \partial\Delta$, i.e., for each F of this class $F(\zeta) := \lim\limits_{r\to 1^-} F(r\zeta) = \zeta$ and the angular derivative $F'(\zeta) := \angle\lim\limits_{z\to\zeta} \frac{F(z)-\zeta}{z-\zeta}$ is finite.*

The class of holomorphic self-mappings of Δ with the boundary Denjoy–Wolff point $\tau \in \partial\Delta$ (see Section 2.3) will be denoted by $\mathcal{F}^+[\tau]$. So, by Theorem 2.10 $f \in \mathcal{G}^+[\tau]$ if and only if the semigroup $\mathcal{S} = \{F_t\}_{t\geq 0}$ generated by f belongs to the class $\mathcal{F}^+[\tau]$, i.e., fixes $\tau \in \partial\Delta$ and this point is the Denjoy–Wolff point for $\{F_t\}_{t\geq 0}$. In this case f has no null point in Δ, but may have other null points on the boundary $\partial\Delta$.

Without loss of generality we can set $\tau = 1$. Note that the class $\mathcal{G}^+[1]$ can be also described using the Berkson–Porta representation (2.2.7). Namely, $f \in \mathcal{G}^+[1]$ *if and only if it has the form:*

$$f(z) = -(1-z)^2 p(z), \tag{2.4.1}$$

with $p \in \mathcal{P}$.

If for $f \in \mathcal{G}^+[1]$ the angular derivative $f'(1) = \angle \lim_{z\to 1} \frac{f(z)}{z-1} = 0$ (or what is the same due to (2.4.1), $\angle \lim_{z\to 1}(1-z)p(z) = 0$), then f is said to be a **parabolic type generator**. The subcone of $\mathcal{G}^+[1]$ of parabolic type generators will be denoted by $\mathcal{G}_p[1]$, i.e.,

$$\mathcal{G}_p[1] = \left\{ f \in \mathcal{G} : \ f'(1) = \angle \lim_{z\to 1} \frac{f(z)}{z-1} = 0 \right\}.$$

If for $f \in \mathcal{G}^+[1]$ the angular derivative $f'(1)$ is positive, then f is said to be a **hyperbolic type generator**. The subcone of $\mathcal{G}^+[1]$ of hyperbolic type generators will be denoted by $\mathcal{G}_h[1]$, i.e.,

$$\mathcal{G}_h[1] = \left\{ f \in \mathcal{G} : \ f'(1) = \angle \lim_{z\to 1} \frac{f(z)}{z-1} > 0 \right\}.$$

So, $\mathcal{G}^+[1] = \mathcal{G}_p[1] \cup \mathcal{G}_h[1]$. A detailed study of semigroups generated by elements of classes $\mathcal{G}_p[1]$ and $\mathcal{G}_h[1]$ can be found in Sections 5 and 7 below.

Goryainov in [76] considered an important class $\mathcal{F}[1,-1] = \mathcal{F}^+[1] \cap \mathcal{F}[-1]$ of self-mappings of Δ, which fixes two boundary points 1 and -1 on $\partial\Delta$ in the sense that

$$\angle \lim_{z\to \pm 1} F(z) = \pm 1, \tag{2.4.2}$$

and such that $\tau = 1$ is the Denjoy–Wolff point for F.

The corresponding class of all generators of semigroups in $\mathcal{F}[1,-1]$ is denoted by $\mathcal{G}_{\mathcal{F}}[1,-1]$.

A result in [76] asserts that $f \in \mathcal{G}_{\mathcal{F}}[1,-1]$ *if and only if it admits the representation*

$$f(z) = \alpha(1+z)(1-z)^2 \frac{1+h(z)}{zh(z)-1}, \tag{2.4.3}$$

in which $\alpha \geq 0$ *and* $h \in \mathrm{Hol}(\Delta)$ *or* h *is a constant of modulus less or equal to* 1.

However, it is not clear whether the classes $\mathcal{G}_{\mathcal{F}}[1,-1]$ and $\mathcal{G}[1,-1] = \mathcal{G}^+[1] \cap \mathcal{G}[-1]$ (see (2.2.13)) coincide. The answer should be affirmative if we show that $f \in \mathcal{G}[\zeta]$ *if and only if the semigroup* $\mathcal{S} = \{F_t\}_{t\geq 0}$ *generated by* f *belongs to* $\mathcal{F}[\zeta]$ *for a boundary point* $\zeta \in \partial\Delta$.

Obviously, it is possible to use conjugation of a semigroup $\mathcal{S} \subset \mathcal{F}[1,-1]$ by a Möbius transformation of the unit disk Δ onto itself to get a characterization of the class $\mathcal{G}_{\mathcal{F}}[-1,1]$ of all generators f on Δ, such that the semigroup $\mathcal{S}$ generated by f belongs to the class $\mathcal{F}[-1,1]$. In other words, it fixes -1 and 1 and such that $\tau = -1$ is the Denjoy–Wolff point for $\mathcal{S}$.

It is clear that $\mathcal{G}_{\mathcal{F}}[1,-1] \cap \mathcal{G}_{\mathcal{F}}[-1,1] = \{0\}$ and for each pair $f_+ \in \mathcal{G}_{\mathcal{F}}[1,-1]$ and $f_- \in \mathcal{G}_{\mathcal{F}}[-1,1]$ the sum $f = f_+ + f_- \in \mathcal{G}_{\mathcal{F}}[1,-1] \cup \mathcal{G}_{\mathcal{F}}[-1,1]$. Then the question is how to recognize to which class $\mathcal{G}_{\mathcal{F}}[1,-1]$ or $\mathcal{G}_{\mathcal{F}}[-1,1]$ this element f belongs. Also note that this union contains the subclass

$$\mathcal{G}_{aut}[\pm 1] = \left\{ f(z) = \alpha\left(z^2 - 1\right), \quad \alpha \in \mathbb{R} \right\}$$

of all (quadratic) generators that vanish at the points 1 and -1. In fact, an element $f \in \mathcal{G}_{aut}[\pm 1]$ if and only if it generates a group of *hyperbolic automorphisms* which fixes these points.

It is clear that all mentioned above classes belong to $\mathcal{G}[1]$.

So, the question is:

◊ *Whether each element $f \in \mathcal{G}[1]$ generates a semigroup of the class $\mathcal{F}[1]$?*

Note that, in general, the mere fact that $f \in \mathcal{G}$ vanishes at a boundary point $\zeta \in \partial\Delta$ does not imply that ζ is necessarily a fixed point of the semigroup $\mathcal{S} = \{F_t\}_{t\geq 0} \subset \mathrm{Hol}(\Delta)$ generated by f. Indeed, Example 5.1 in [115] shows that the function $f(z) = z - 1 + \sqrt{1-z}$ is of the class $\mathcal{G}$ and clearly vanishes at the boundary point $\zeta = 1$. However, all elements of the semigroup $\mathcal{S} = \{F_t\}_{t\geq 0}$ generated by f are strictly less than 1 at this point, i.e., $F_t(1) < 1$ for all $t \geq 0$. The point is that the angular derivative of this generator at $\tau = 1$ does not exist finitely, so f does not belong to $\mathcal{G}[1]$.

We first present a theorem for decomposition of the class $\mathcal{G}[1]$ by its subclasses $\mathcal{G}_{aut}[\pm 1]$ and $\mathcal{G}_p[1]$ (cf., Theorem 2.8). Namely, we show that

$$\mathcal{G}[1] = \mathcal{G}_p[1] \oplus \mathcal{G}_{aut}[\pm 1].$$

As a consequence we derive at the fact that $f \in \mathcal{G}[1]$ if and only if $\mathcal{S} = \{F_t\}_{t\geq 0}$ generated by f belongs to $\mathcal{F}[1]$.

Theorem 2.11. *A function $f \in \mathcal{G}[1]$ if and only if it admits the representation:*

$$f(z) = -(1-z)^2 p(z) + \frac{\beta}{2}(z^2 - 1), \tag{2.4.4}$$

where $p \in \mathcal{P}$ and

$$\angle \lim_{z\to 1}(1-z)p(z) = 0, \tag{2.4.5}$$

and $\beta \in \mathbb{R} = (-\infty, \infty)$.

Moreover, $\tau = 1$ is the Denjoy–Wolff point of the semigroup $\mathcal{S} = \{F_t\}_{t\geq 0}$ generated by f if and only if $\beta \geq 0$. In this case f has no null point in Δ.

Proof. Sufficiency. Let $f \in \mathrm{Hol}(\Delta, \mathbb{C})$ admit representation (2.4.4). Then, the first term

$$f_1(z) = -(1-z)^2 p(z)$$

is an element of $\mathcal{G}[1]$ due to the Berkson–Porta representation (see formulas (2.2.7) and (2.4.1)). The second term

$$f_2(z) = \frac{\beta}{2}(z^2 - 1)$$

is the generator of a group of automorphisms on Δ (see Theorem 2.6).

Since $\mathcal{G}$, hence $\mathcal{G}[1]$, is a real cone in $\mathrm{Hol}(\Delta, \mathbb{C})$, the function $f = f_1 + f_2$ belongs to $\mathcal{G}[1]$.

Note that due to (2.4.4) and (2.4.5) we have

$$\angle \lim_{z\to 1} \frac{f(z)}{z-1} \left(= \angle \lim_{z\to 1} f'(z)\right) = \beta.$$

Hence, $z = 1$ is the Denjoy–Wolff point of $\mathcal{S} = \{F_t\}_{t\geq 0}$ if and only if $\beta \geq 0$ by Theorem 2.10.

Necessity. Let $f \in \mathcal{G}[1]$. Observe, that by Theorem 2.5 each element of the class $\mathcal{G}$ satisfies the inequality

$$\operatorname{Re}\left[\frac{f(z)\bar{z}}{1-|z|^2} + \frac{f(w)\bar{w}}{1-|w|^2}\right] \geq \operatorname{Re}\frac{\bar{z}f(w) + w\overline{f(z)}}{1-\bar{z}w} \tag{2.4.6}$$

for all $z, w \in \Delta$.

Now rewriting (2.4.6) in the form

$$\operatorname{Re} f(z)\left(\frac{\bar{z}}{1-|z|^2} - \frac{\bar{w}}{1-z\bar{w}}\right) \geq \operatorname{Re} f(w)\left[\frac{\bar{z}}{1-\bar{z}w} - \frac{\bar{w}}{1-|w|^2}\right]$$

and setting $w = r \in (0,1)$ we have

$$\operatorname{Re} f(z)\left(\frac{\bar{z}}{1-|z|^2} - \frac{r}{1-zr}\right) \geq \operatorname{Re}\frac{f(r)}{r-1}\cdot\frac{r-\bar{z}}{(1-r\bar{z})(1+r)}\,.$$

Now letting r to 1^- and taking into account that $\lim\limits_{r\to 1^-} f(r) = 0$ and

$$\lim_{r\to 1^-}\frac{f(r)}{r-1} = f'(1),$$

we get

$$\operatorname{Re} f(z)\left(\frac{\bar{z}}{1-|z|^2} - \frac{1}{1-z}\right) \geq \operatorname{Re}\frac{f'(1)}{2}\,, \tag{2.4.7}$$

or (after some manipulations):

$$-\operatorname{Re}\frac{f(z)}{(z-1)^2} \geq \frac{1}{2}\operatorname{Re} f'(1)\frac{1-|z|^2}{|1-z|^2}\,. \tag{2.4.8}$$

Setting

$$q(z) = -\frac{f(z)}{(1-z)^2} \tag{2.4.9}$$

and

$$p(z) = q(z) - \frac{1+z}{2(1-z)}\operatorname{Re} f'(1), \tag{2.4.10}$$

we obtain from (2.4.8)–(2.4.10)

$$f(z) = -(1-z)^2\left[p(z) + \frac{1}{2}\frac{1+z}{1-z}\operatorname{Re} f'(1)\right] \tag{2.4.11}$$

with

$$\begin{aligned}\operatorname{Re} p(z) &= \operatorname{Re} q(z) - \frac{1}{2}\operatorname{Re} f'(1)\cdot \operatorname{Re}\frac{1+z}{1-z} \geq \\ &\geq \frac{1}{2}\operatorname{Re} f'(1)\left[\frac{1-|z|^2}{|1-z|^2} - \operatorname{Re}\frac{1+z}{1-z}\right] = 0.\end{aligned}$$

It is clear that (2.4.11) is equivalent to (2.4.4) with $\beta = \operatorname{Re} f'(1)$.

Now we want to show that $f'(1)$ is, in fact, a real number, or what is the same, that $f'(1) = \beta$. To this end, we observe that by the Riesz–Herglotz formula (1.4.2) for each $p \in \mathcal{P}$ the limit

$$\delta_p(1) = \angle \lim_{z\to 1}(1-z)p(z)$$

exists and is a nonnegative real number (see, for example, [130]). Therefore,

$$f'(1) = \angle \lim_{z\to 1}\frac{f(z)}{z-1} = \beta\,(= \operatorname{Re} f'(1))$$

by (2.4.4). This immediately implies (2.4.5). The theorem is proved. □

Remark 2.3. It can be seen easily from the proof of Theorem 2.11 that actually the the existence of the limit $\beta = \angle \lim_{z\to 1}\frac{f(z)}{z-1}$ can be replaced by a formally weaker condition:

$$\liminf_{r\to 1^-} \operatorname{Re}\frac{f(r)}{r-1} > -\infty. \tag{2.4.12}$$

Indeed, repeating the necessary part of the proof of Theorem 2.11, we see that (2.4.12) implies representation (2.4.4) with $\beta = \liminf_{r\to 1^-} \operatorname{Re}\frac{f(r)}{r-1}$. In turn, this representation and the Riesz–Herglotz formula (1.4.2) applied to $p(z)$ show that, in fact,

$$\lim_{r\to 1^-}(1-r)p(r)\left(= \lim_{r\to 1^-}\left[\frac{f(r)}{r-1} - \frac{\beta}{2}(r+1)\right]\right)$$

exists and equals zero (cf., [62]). Finally, using Harnack's inequality (1.4.3) for the function p, we get that the function

$$h(z) = (1-z)p(z)$$

is bounded on each non-tangential approach region at the point $z = 1$. Thus, condition (2.4.5) follows from Lindelöf's Theorem 1.6, and we are done.

This observation leads to the following infinitesimal version of the Julia–Carathéodory Theorem.

Theorem 2.12. *Let $f \in \mathcal{G}$ and let $\mathcal{S} = \{F_t\}_{t\geq 0}$ be the semigroup generated by f. Then $f \in \mathcal{G}[1]$ if and only if $\mathcal{S} \in \mathcal{F}[1]$. Moreover, if*

$$\beta = f'(1) := \angle \lim_{z\to 1} f'(z), \tag{2.4.13}$$

then for each $t \geq 0$

$$(F_t)'(1) := \lim_{z\to 1} (F_t)'(z) = e^{-t\beta}. \tag{2.4.14}$$

Proof. Sufficiency. Let $\mathcal{S} = \{F_t\}_{t\geq 0}$ be the semigroup generated by $f \in \mathcal{G}$. Suppose that $\mathcal{S} \in \mathcal{F}[1]$, that is, for each $t \geq 0$ there exists the angular limit

$$\alpha_t := (F_t)'(1) \left(= \angle \lim_{t\to 1} \frac{F_t(z) - 1}{z - 1} \right) < \infty. \tag{2.4.15}$$

Then it follows again by the classical Julia–Carathéodory Theorem (see [40, 127]) that

$$\frac{|1 - F_t(z)|^2}{1 - |F_t(z)|^2} \leq \alpha_t \frac{|1 - z|^2}{1 - |z|^2}.$$

Differentiating this inequality at $t = 0^+$ and using the semigroup property we obtain

$$\liminf_{t\to 0^+} \frac{1 - \alpha_t}{t} = \beta > -\infty,$$

and

$$\operatorname{Re} f(z) \left(\frac{\bar{z}}{1 - |z|^2} - \frac{1}{1 - z} \right) \geq \beta \geq 2.$$

Letting $z = r$ and r tend to 1^-, we get the inequality

$$\liminf_{r\to 1^-} \frac{f(r)}{r - 1} \geq \frac{\beta}{2},$$

which is equivalent (2.4.12). Remark 2.3 now implies the required inclusion: $f \in \mathcal{G}[1]$.

Necessity. Suppose that $f \in \mathcal{G}[1]$. First we show that $z = 1$ is a common fixed point for the semigroup $\mathcal{S} = \{F_t\}_{t\geq 0}$.

Indeed, if $\beta = f'(1)$, then

$$f = f_1 + f_2,$$

where $f_1(z) = -(1 - z)^2 p(z)$ and $f_2(z) = \frac{\beta}{2}(z^2 - 1)$ are elements of $\mathcal{G}[1]$.

In addition, as we mentioned above, if $\mathcal{S}_1 = \{F_t^{(1)}\}_{t\geq 0}$ is the semigroup generated by f_1, then $z = 1$ is the Denjoy–Wolff point of this semigroup; hence,

$$\varphi(F_t^{(1)}(z)) \leq \varphi(z), \tag{2.4.16}$$

where

$$\varphi(z) = \frac{|1-z|^2}{1-|z|^2}.$$

Also, if $\mathcal{S}_2 = \{F_t^{(2)}\}_{t\geq 0}$ is the semigroup (actually, group) generated by f_2, it follows by direct calculations that

$$\varphi\left(F_t^{(2)}(z)\right) \leq e^{-\beta t}\varphi(z). \tag{2.4.17}$$

Now it follows by the semigroup product formula (see, for example, [118]) that

$$F_t = \lim_{n\to\infty}\left(F_{\frac{t}{n}}^{(2)} \circ F_{\frac{t}{n}}^{(1)}\right)^n, \tag{2.4.18}$$

where $(F)^n$ denotes the n-fold iterate of a mapping $F : \Delta \mapsto \Delta$. Then, applying (2.4.16), (2.4.17) and (2.4.18), by induction we get:

$$\varphi(F_t(z)) \leq e^{-t\beta}\varphi(z),$$

or, explicitly,

$$\frac{|1-F_t(z)|^2}{1-|F_t(z)|^2} \leq e^{-t\beta}\frac{|1-z|^2}{1-|z|^2}.$$

The latter inequality immediately implies that for each $t \geq 0$

$$\lim_{r\to 1^-} F_t(r) = 1,$$

or, what is the same, because of the boundedness of F_t,

$$\angle \lim_{z\to 1} F_t(z) = 1. \tag{2.4.19}$$

It remains to show that

$$(F_t)'(1) = \angle \lim_{z\to 1}\frac{F_t(z)-1}{z-1} = e^{-t\beta}. \tag{2.4.20}$$

To this end we note that it follows by representation (2.4.4) and the Cauchy problem (2.1.3) that

$$\frac{\partial F_t(z)}{\partial t} = (1-F_t(z))^2 p(F_t(z)) - \frac{\beta}{2}((F_t(z))^2 - 1).$$

This implies that

$$\log\frac{F_t(z)-1}{z-1} = -\int_0^t (1-F_s(z)p(F_s(z))ds - \frac{\beta}{2}\int_0^t (F_s(z)+1)ds.$$

Taking into account (2.4.5) we obtain (2.4.20). The theorem is proved. □

The classes $\mathcal{G}[1]$, $\mathcal{G}_h[1]$ and $\mathcal{G}[1,-1]$ play a crucial role in the study of spirallike and starlike functions with respect to a boundary point.

Another useful characterization of these classes can be given as follows.

Theorem 2.13 (cf. [48]). *Let $f \in \mathcal{G}_h[1]$, i.e., $f \in \mathcal{G}[1]$ with $f'(1) = \angle \lim_{z\to 1} f'(z) = \beta_+ > 0$. Then for each $c \in \left(0, \frac{\beta_+}{2}\right]$ there is a unique $F = F_c \in \mathcal{F}^+[1]$ with*

$$\angle \lim_{z\to 1} F'(z) =: F'(1) = \frac{2c}{\beta_+} \le 1,$$

and such that

$$f(z) = -c(1-z)^2 \frac{1+F(z)}{1-F(z)}. \tag{2.4.21}$$

Moreover, if $f \in \mathcal{G}[1,-1] \cap \mathcal{G}_h[1]$ with $\beta_- = \angle \lim_{z\to 1} f'(z)$, then $F \in \mathcal{F}[1,-1]$, with

$$\angle \lim_{z\to -1} F'(z) = -\frac{\beta_-}{2c}.$$

Conversely, if $F \in \mathcal{F}^+[1]$ (respectively, $F \in \mathcal{F}[1,-1]$), then for each $c > 0$ the function f defined by (2.4.21) *belongs to $\mathcal{G}_h[1]$ (respectively, f belongs to $\mathcal{G}[1,-1] \cap \mathcal{G}_h[1]$, i.e., $\angle \lim_{z\to \pm 1} f(z) = 0$ with $f'(1) > 0$ and $f'(-1) < 0$).*

Proof. Let $f \in \mathcal{G}_h[1] \subset \mathcal{G}^+[1]$. Then one can present f in the form

$$f(z) = -(1-z)^2 p(z)$$

with $\operatorname{Re} p(z) \ge 0$. For each $c > 0$ there is a unique holomorphic function $F = F_c$ on Δ such that $|F(z)| < 1,\ z \in \Delta$, and

$$p(z) = c\,\frac{1+F(z)}{1-F(z)}. \tag{2.4.22}$$

Furthermore, since

$$\beta_+ = f'(1) = \angle \lim_{z\to 1} \frac{f(z)}{z-1} > 0,$$

we get

$$\beta_+ = c \lim_{r\to 1^-} \frac{1-r}{1-F(r)}(1+F(r)). \tag{2.4.23}$$

This relation shows that

$$\lim_{r\to 1^-} F(r) = 1$$

and

$$\lim_{r\to 1^-} \frac{1-r}{1-F(r)} = \frac{\beta_+}{2c} > 0.$$

Now it follows by the Julia–Carathéodory Theorem that

$$\angle \lim_{z\to 1} F'(z) = \angle \lim_{z\to 1} \frac{1-F(z)}{1-z} = \frac{2c}{\beta_+} =: \alpha_+ > 0.$$

If $c \in \left(0, \frac{\beta_+}{2}\right]$, then $\alpha_+ \le 1$ and it follows by Wolff's Lemma (see, for example, [40] and [127]) that $\zeta = 1$ is the Denjoy–Wolff point for F.

Now let $f \in \mathcal{G}[1,-1]$ and set

$$\beta_- = f'(-1) = \angle \lim_{z\to -1} f'(z).$$

Then

$$\begin{aligned} \angle \lim_{z\to -1} \frac{f(z)}{z+1} = \beta_- &= \angle \lim_{z\to -1} -(1-z)^2 c \frac{1+F(z)}{1-F(z)} \frac{1}{1+z} \\ &= -4c \angle \lim_{z\to -1} \frac{1+F(z)}{1+z} \cdot \angle \lim_{z\to -1} \frac{1}{1-F(z)}. \end{aligned} \tag{2.4.24}$$

Again it follows by the Julia–Carathéodory Theorem and (2.4.24) that

$$\angle \lim_{z\to -1} \frac{1+F(z)}{1+z} =: \alpha_-$$

exists finitely and

$$\alpha_- = -\frac{1}{2c}\beta_-.$$

Reverse considerations complete our proof. □

Corollary 2.2. *Let $f \in \mathcal{G}[1,-1]$ with $f'(1) = \beta_+ \ge 0$ and $f'(-1) = \beta_-$. Then $-\beta_- \ge \beta_+$. Moreover, the equality $-\beta_- = \beta_+$ is possible if and only if f is a generator of a group of hyperbolic automorphisms of Δ, i.e.,*

$$f(z) = \frac{\beta_+}{2}(z^2 - 1).$$

Proof. The assertion is obvious if $\beta_+ = 0$ since $\beta_- < 0$. Let us assume that $\beta_+ > 0$ and $-\beta_- \in (0, \beta_+)$. Then the function f_1 defined by

$$f_1(z) = f(z) + \frac{\beta_-}{2}(z^2 - 1)$$

belongs to the class $\mathcal{G}[1,-1]$, because this class is a real cone. In addition, we have

$$f_1'(1) = \beta_+ + \beta_- > 0$$

while

$$f_1'(-1) = \beta_- - \beta_- = 0.$$

Then $f_1 \neq 0$ and both points 1 and -1 are Denjoy–Wolff points of the semigroup generated by f. This contradiction implies that $\beta_+ \leq -\beta_-$. Moreover, the same considerations show that $\beta_+ = -\beta_-$ if and only if $f_1(z) \equiv 0$. Hence, $f(z)$ has the form

$$f(z) = \frac{\beta_-}{2}(1 - z^2) = \frac{\beta_+}{2}(z^2 - 1). \tag{2.4.25}$$

Thus, f belongs to $\mathcal{G}_{aut}[\pm 1]$. This complete our proof. □

Corollary 2.3 (cf., [76]). *If $F \in \mathcal{F}[1, -1]$ then*

$$F'(1) \cdot F'(-1) \geq 1. \tag{2.4.26}$$

Furthermore, the equality $F'(1) \cdot F'(-1) = 1$ is possible if and only if F is a hyperbolic automorphism of Δ.

Proof. Indeed, if $F \in \mathcal{F}[1, -1]$, then it follows by Theorem 2.13 that f defined by (2.4.21) belongs to $\mathcal{G}[1, -1]$. In addition, (2.4.26) and (2.4.21) imply that

$$F'(1) \cdot F'(-1) = \alpha_+ \cdot \alpha_- = -\frac{2c}{\beta_+} \cdot \frac{\beta_-}{2c} = -\frac{\beta_-}{\beta_+} \geq 1$$

(see the previous corollary).

If $\alpha_+ \cdot \alpha_- = 1$, then $-\beta_- = \beta_+$ and by the same corollary we have

$$f(z) = -c(1 - z)^2 \frac{1 + F(z)}{1 - F(z)} = \frac{\beta_+}{2}(z^2 - 1).$$

Hereby, F must be of the form

$$F(z) = \frac{(z + 1) - \alpha_+(1 - z)}{(z + 1) + \alpha_+(1 - z)}, \qquad \text{where } \alpha_+ = \frac{2c}{\beta_+}.$$

The assertion is proved. □

2.5 Univalent functions and semi-complete vector fields

To complete the auxiliary part of the work we now state some relations between starlike (spirallike) functions and semigroup generators.

First we observe that each μ-spirallike (starlike, if μ is real) function h on D determines a family $\{F_t\}_{t \geq 0}$ of holomorphic self-mappings of D defined by

$$F_t(z) = h^{-1}\left(e^{-\mu t} h(z)\right), \quad \operatorname{Re} \mu > 0.$$

Obviously this family forms a one-parameter continuous semigroup (see Fig. 2.1). Differentiating this equality at $t = 0^+$ we obtain that

$$\mu h(z) = h'(z) f(z), \tag{2.5.1}$$

where f is the generator of the semigroup $\mathcal{S} = \{F_t\}_{t \geq 0}$.

As a matter of fact, it turns out that equation (2.5.1) is a criterion for a univalent function f to be spirallike. More precisely:

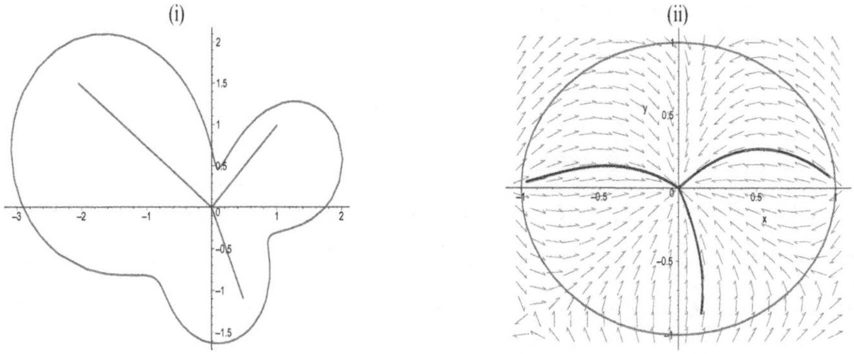

Figure 2.1: A semigroup defined by a starlike function

Theorem 2.14 (see [56, 54, 55]). *Let a holomorphic function h be univalent on a simply connected domain D. Then h is a spirallike function if and only if there exist a semigroup generator $f \in \mathcal{G}(D)$ and a complex number μ with $\operatorname{Re}\mu > 0$ such that*

$$\mu h(z) = h'(z)f(z), \quad z \in \Delta. \tag{2.5.2}$$

In the case where μ is a real positive number, h is, actually, starlike.

Corollary 2.4. *Let $D = \Delta = \{z \in \mathbb{C} : |z| < 1\}$ and $f \in \mathcal{G}$. Assume that for some μ, $\operatorname{Re}\mu > 0$, equation (2.5.2) is fulfilled with some $h \in \operatorname{Univ}(\Delta)$. The following assertions hold.*

(a) *If f has an interior null point τ, then h is a spirallike (starlike) function with respect to an interior point with $h(\tau) = 0$. Moreover, h is starlike (spirallike) on each pseudohyperbolic disk $D_r(\tau) \subset \Delta$ centered at τ.*

(b) *If f has no interior null point, then there exists a boundary point τ such that*

$$\angle \lim_{z\to\tau} h(z) = \angle \lim_{z\to\tau} f(z) = 0.$$

In this case h is a spirallike (starlike) function with respect to a boundary point. Moreover, h is starlike (spirallike) on each horodisk $\Delta_r(\tau) \subset \Delta$ internally tangent to $\partial\Delta$ at τ.

Using these facts and the Berkson–Porta representation (2.2.7) of generators, one can easily obtain the classical Nevanlinna condition (1.1.1), as well as Wald's Theorem 1.2 and Špaček's condition (1.1.7) (see details in [58] and [130]).

In general, a connection between univalent functions and semigroup generators is based on the notion of φ-like domains introduced by Brickman in [22] (see Section 1.1.5).

Definition 2.5. *Let $\Omega \subset \mathbb{C}$ be a domain containing $w = 0$, and let $\varphi \in \mathrm{Hol}(\Omega, \mathbb{C})$ with $\varphi(0) = 0$ and $\mathrm{Re}\,\varphi'(0) > 0$. Then Ω is a* **φ-like domain** *if for any $w \in \Omega$ the Cauchy problem*

$$\begin{cases} \dfrac{\partial v(t,w)}{\partial t} + \varphi(v(t,w)) = 0, \\ v(0,w) = w \end{cases} \tag{2.5.3}$$

has a solution $v(t,w)$ defined for all $t \geq 0$ such that $v(t,w) \in \Omega$ for all $t \geq 0$.

Theorem 2.15 ([22]). *Let $h \in \mathrm{Univ}(\Delta)$ satisfy $h(0) = 0$. Then the image $h(\Delta)$ is a φ-like domain if and only if h is φ-like, i.e., $\mathrm{Re}\,\dfrac{zh'(z)}{\varphi(h(z))} > 0$ for all $z \in \Delta$.*

Remark 2.4. Note that the solution $v = v(t, \cdot)$ of the Cauchy problem (2.5.3) is a semigroup of holomorphic self-mappings of Ω. Condition $\mathrm{Re}\,\varphi'(0) > 0$, in fact, means that $w = 0$ is an attractive common fixed point of this semigroup.

Furthermore, if $\Omega = h(\Delta)$ for some $h \in \mathrm{Univ}(\Delta)$, then as in Lemma 3.7.1 in [58], it easy to verify that the function

$$f(z) = \frac{\varphi(h(z))}{h'(z)}$$

is the generator of the semigroup $\{u(t,\cdot)\}_{t\geq 0} \subset \mathrm{Hol}(\Delta)$, defined by

$$u(t,z) = h^{-1}\left(v(t,h(z))\right).$$

Therefore, Definition 2.5 with $\Omega = f(\Delta)$ can service a generalized notion of a φ-like function h even if $h(0) \neq 0$. To generalize the notion of starlike (or spirallike) function with respect to a boundary point, it would be preferable to omit the restriction $0 \in \Omega = h(\Delta)$.

Namely,

- *Let $h \in \mathrm{Univ}(\Delta)$ and let $\varphi \in \mathrm{Hol}(h(\Delta), \mathbb{C})$ satisfy either*

  (i) *$\varphi(0) = 0$ and $\mathrm{Re}\,\varphi'(0) > 0$ when $\tau \in \Delta$,*

  or

  (ii) $\angle \lim\limits_{z\to\tau} \varphi(h(z)) = 0$ and $\angle \lim\limits_{z\to\tau} \mathrm{Re}\,\varphi'(h(z)) > 0$ when $\tau \in \partial\Delta$.

  A function h with $h(\tau) = 0$ for $\tau \in \overline{\Delta}$, is called a **generalized φ-like function** *if for any $z \in \Delta$ the Cauchy problem*

$$\begin{cases} \dfrac{\partial u(t,z)}{\partial t} + \dfrac{\varphi\left(h(u(t,z))\right)}{h'(u(t,z))} = 0, \\ u(0,z) = z \end{cases} \tag{2.5.4}$$

  has a solution $u(t,z)$ defined for all $t \geq 0$ such that $\{u(t,z),\ t \geq 0\} \subset \Delta$ for all initial values $z \in \Delta$.

Substituting in this definition $\varphi(w) = w$ (or, more generally, $\varphi(w) = \mu w$ with $\operatorname{Re}\mu > 0$), we get the class of starlike (or spirallike) functions on Δ.

Note that for the case $h(0) = 0$ our definition of generalized φ-like functions is equivalent to those given in [22]. At the same time, if h has no null point in Δ, setting $\varphi(w) = w$ we obtain another class of univalent functions that are starlike with respect to a boundary point. We study this class in the next chapter.

Chapter 3

Starlike Functions with Respect to a Boundary Point

Although starlike functions normalized by the condition $h(0) = 0$ have been studied intensively during the last century, until 1981 a few works only dealt with starlike functions with respect to a boundary point.

The first observation related to such functions was most likely that of Egervàry [46] who studied the mapping properties of the Cesàro means of the partial sums of the geometric series $z/(1-z) = z + z^2 + z^3 + \ldots + z^n + \ldots$. He showed that the Cesàro means of the first order are starlike with respect to a boundary point.

An intensive study of the class of starlike functions with respect to a boundary point was initiated by Robertson in his paper [123].

3.1 Robertson's classes. Robertson's conjecture

Robertson [123] introduced the following classes of functions:

Definition 3.1. *Let the class G^* consist of the constant function $h(z) \equiv 1$ and all functions $h \in \operatorname{Univ}(\Delta)$ starlike with respect to a boundary point and normalized by $h(0) = 1$ and $h(1) = \lim\limits_{r \to 1^-} h(r) = 0$. In addition, assume that for any $h \in G^*$ there is a real α such that*

$$\operatorname{Re} e^{i\alpha} h(z) > 0, \quad z \in \Delta.$$

Thus, $G^* \setminus \{h(z) \equiv 1\}$ is the subclass of $S^*[1]$ consisting of the functions h that take values $h(z)$, $z \in \Delta$, in a half-plane that contains $h(0) = 1$ (see Fig. 3.1).

Definition 3.2. *Let G denote the class of functions*

$$h(z) = 1 + d_1 z + d_2 z^2 + \ldots + d_n z^n + \ldots,$$

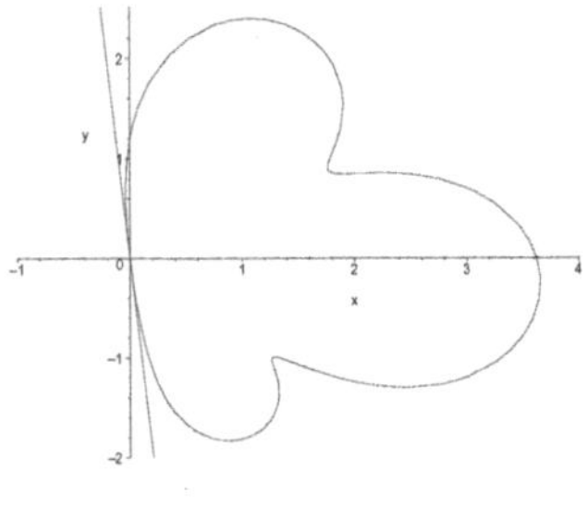

Figure 3.1: A function of Robertson's class.

holomorphic and non-vanishing in Δ*, which satisfy*

$$\mathrm{Re}\left[\frac{2zh'(z)}{h(z)}+\frac{1+z}{1-z}\right]>0 \tag{3.1.1}$$

for all $z\in\Delta$.

Robertson [123] conjectured that

$$G=G^{*}. \tag{3.1.2}$$

Actually, he partially proved this relation. Namely, Robertson showed that

$$G_{*}\subset G\subset G^{*},$$

where G_* is the subset of G^* consisting of functions holomorphic on the closed disk $\overline{\Delta}$. In [102] Lyzzaik completed the proof of Robertson's conjecture (see also Section 3.3) and showed that (3.1.2) holds without any additional restriction.

Note also that Robertson–Lyzzaik's result means that inequality (3.1.1) is a characterization of those functions from $S^*[1]$ that are equal to 1 at zero and have images in a wedge of angle π with its vertex at the origin. Furthermore, we will see below that the smallest wedge containing $h(\Delta)$ is exactly of angle π if and only if h satisfies the Visser–Ostrowski condition (see Section 1.2).

In the same paper Robertson proved that if $h\in G,\ h\not\equiv 1$, then $\log h$ is a close-to-convex function relative to the starlike function $-k(z)$, where k is the Koebe function $k(z)=\frac{z}{(1-z)^2}$. In other words, $h\in G,\ h\not\equiv 1$, satisfies the inequality:

$$\mathrm{Re}\left[(1-z)^2\frac{h'(z)}{h(z)}\right]<0,\quad z\in\Delta. \tag{3.1.3}$$

As a matter of fact, we will show below that a function $h\in\mathrm{Univ}(\Delta)$ with $h(1)=0$ is starlike with respect to a boundary point if and only if condition

(3.1.3) holds. Moreover, if, in addition, h satisfies a generalized Visser–Ostrowski condition

$$\angle \lim_{z\to 1} \frac{(z-1)h'(z)}{h(z)} = \mu, \quad 0 < \mu \leq 1,$$

then inequality (3.1.1) is equivalent to (3.1.3).

To prove these statements as well as Robertson's conjecture (3.1.2), we need some auxiliary assertions.

3.2 Auxiliary lemmas

Following Silverman and Silvia [135] we define the following classes of functions.

Definition 3.3. *Let $\lambda \in (0,2]$. Denote by $G(\lambda)$*[1] *the class of functions*

$$h(z) = 1 + d_1 z + d_2 z^2 + \ldots + d_n z^n + \ldots,$$

holomorphic and non-vanishing in Δ, which satisfy

$$\operatorname{Re}\left[\frac{2zh'(z)}{\lambda h(z)} + \frac{1+z}{1-z}\right] > 0 \tag{3.2.1}$$

for all $z \in \Delta$.

In particular, the class $G(1)$ is Robertson's class G (see Definition 3.2).

The first result gives an integral representation for the class $G(\lambda)$, which itself is interesting.

Lemma 3.1 (see [56], cf. [135]). *Let $h \in \operatorname{Hol}(\Delta, \mathbb{C})$. Then $h \in G(\lambda)$ if and only if it admits the following representation:*

$$h(z) = (1-z)^{\lambda} \exp\left[-\lambda \oint_{|\zeta|=1} \log(1 - z\bar{\zeta}) d\sigma(\zeta)\right] \tag{3.2.2}$$

with some probability measure σ on the unit circle $|\zeta| = 1$.

Proof. First, suppose that a function $h \in \operatorname{Hol}(\Delta, \mathbb{C})$ is represented by (3.2.2). We see at once that $h(0) = 1$. By simple calculation we get

$$\frac{2zh'(z)}{\lambda h(z)} + \frac{1+z}{1-z} = 1 + \oint_{|\zeta|=1} \frac{2z\bar{\zeta}}{1 - z\bar{\zeta}} d\sigma(\zeta).$$

Since $\operatorname{Re} \dfrac{z\bar{\zeta}}{1 - z\bar{\zeta}} > -\dfrac{1}{2}$, we obtain inequality (3.2.1).

[1]The classes $\mathcal{G}_\alpha$, $\alpha \in [0,1)$, originally introduced in [135], coincide with the classes $G(\lambda)$ for $\lambda = 2(1-\alpha)$.

Suppose now that $h \in G(\lambda)$. Consider the function $p(z) = \dfrac{2zh'(z)}{\lambda h(z)} + \dfrac{1+z}{1-z}$, which evidently belongs to the Carathéodory class $\mathcal{P}_0$ and thus can be represented by the Riesz–Herglotz formula (1.4.1):

$$p(z) = \oint_{|\zeta|=1} \frac{1+z\bar{\zeta}}{1-z\bar{\zeta}}\, d\sigma(\zeta)$$

with a probability measure σ. This implies that

$$\frac{h'(z)}{\lambda h(z)} = \oint_{|\zeta|=1} \left[\frac{\bar{\zeta}}{1-z\bar{\zeta}} - \frac{1}{1-z}\right] d\sigma(\zeta). \tag{3.2.3}$$

Integrating both sides of this equality we obtain (3.2.2) □

Corollary 3.1. *Let $h \in \mathrm{Hol}(\Delta, \mathbb{C})$ and $\lambda_1, \lambda_2 \in (0, 2]$. Then $h \in G(\lambda_1)$ if and only if the function $h^{\frac{\lambda_2}{\lambda_1}}$ belongs to $G(\lambda_2)$.*

Lemma 3.2. *Let $h \in \mathrm{Hol}(\Delta, \mathbb{C})$ with $h(0) = 1$. Then $h \in G(\lambda)$ if and only if there is a function $h_1 \in S^*_\kappa$, $h_1(0) = 0$, starlike of order $\kappa = 1 - \frac{\lambda}{2}$, such that*

$$h(z) = \frac{(1-z)^\lambda h_1(z)}{z}. \tag{3.2.4}$$

In particular, $h \in G$ if and only if the function

$$h_1(z) := \frac{zh(z)}{1-z}$$

belongs to $S^*_{1/2}$.

Proof. Let $h_1 \in S^*_\kappa$ be starlike of order $\kappa \in [0, 1)$. It follows by the definition of the class S^*_κ (see formula (1.1.2)) that the function $\dfrac{\frac{zh_1'(z)}{h_1(z)} - \lambda}{1-\lambda}$ is of Carathéodory's class $\mathcal{P}_0$. Using the Riesz–Herglotz formula (1.4.1) we get the following representation for h_1:

$$h_1(z) = z \exp\left[-\lambda \oint_{|\zeta|=1} \log(1-z\bar{\zeta}) d\sigma(\zeta)\right]. \tag{3.2.5}$$

Let h and h_1 be connected by (3.2.4) and set $\lambda = 2 - 2\kappa$. Then by (3.2.2) and (3.2.5) we obtain our assertion. □

Lemma 3.3. *Let $0 < \lambda_1 < \lambda_2 \leq 2$. Then $G(\lambda_1) \subset G(\lambda_2)$.*

Proof. Let $h \in G(\lambda_1)$. We have

$$h(z) = (1-z)^{\lambda_1} \exp\left[-\lambda_1 \oint_{|\zeta|=1} \log(1-z\bar{\zeta})d\sigma(\zeta)\right]$$
$$= (1-z)^{\lambda_2} \exp\left[-\lambda_2 \oint_{|\zeta|=1} \log(1-z\bar{\zeta})d\widetilde{\sigma}(\zeta)\right],$$

where the probability measure $\widetilde{\sigma}$ on the unit circle is defined by

$$\widetilde{\sigma} = \frac{\lambda_1}{\lambda_2}\sigma + \frac{\lambda_2-\lambda_1}{\lambda_2}\delta$$

(δ is the Dirac measure at the point $\zeta = 1$). Then the assertion follows by the integral representation (3.2.2). □

Lemma 3.4. *The set*

$$\bigcup_{n=1}^{\infty} \left\{ \prod_{j=1}^{n} \left(\frac{1-z}{1-z\bar{\zeta}_j}\right)^{\lambda_j} : \sum_{j=1}^{n} \lambda_j = \lambda,\ |\zeta_j| = 1 \right\}$$

is dense in $G(\lambda)$ in the topology of uniform convergence on compact subsets of Δ.

Proof. Replacing the integral in (3.2.2) by integral sums we have for any $h \in G(\lambda)$:

$$h(z) = \lim_{n\to\infty} (1-z)^{\lambda} \exp\left[-\lambda \sum_{j=1}^{n} \log(1-z\bar{\zeta}_j)\sigma_j\right]$$
$$= \lim_{n\to\infty} (1-z)^{\lambda} \prod_{j=1}^{n} (1-z\bar{\zeta}_j)^{-\lambda\sigma_j},$$

with $\sum\limits_{j=1}^{n} \sigma_j = 1$. Writing $\lambda_j = \lambda\sigma_j$ we complete the proof. □

Lemma 3.5. *Let $h \in G(\lambda)$, $h \not\equiv 1$. Then the function $F(z) := -\log h(z)$, $F(0) = 0$, is close-to-convex relative to the Koebe function $k(z) = \frac{z}{(1-z)^2}$, i.e.,*

$$\operatorname{Re}\left[(1-z)^2 \frac{h'(z)}{h(z)}\right] \left(= -\operatorname{Re} \frac{z(-\log h(z))'}{k(z)}\right) < 0, \quad z \in \Delta.$$

Proof. Using Lemma 3.1 (or formula (3.2.3) from its proof) we see that

$$-(1-z)^2 \frac{h'(z)}{h(z)} = \lambda \oint_{\partial\Delta} (1-z)\frac{1-\bar{\zeta}}{1-z\bar{\zeta}} d\sigma(\zeta).$$

Since the integrand takes values in the right half-plane, we have the required assertion. □

The following result was originally formulated in [135] (see also [56]).

Lemma 3.6. *Let $h \in G(\lambda)$ and $h(z) \not\equiv 1$. Then the function $1-h$ is close-to-convex, hence h is univalent in Δ.*

Proof. Let $h \in G(\lambda)$ and $h(z) \not\equiv 1$. Then $h \in G(2)$ by Lemma 3.3. Now Lemma 3.2 implies that the function $h_1(z) = \frac{zh(z)}{(1-z)^2}$ belongs to the class $S_0^* = S^*$. In turn, by Lemma 3.5 we have

$$\operatorname{Re} \frac{z(1-h(z))'}{h_1(z)} = -\operatorname{Re}(1-z)^2 \frac{h'(z)}{h(z)} > 0$$

for all $z \in \Delta$. This completes our proof. □

3.3 A generalization of Robertson's conjecture

In terms of our notation, Robertson's conjecture is that the class $G(1)$ coincides with the class G^* of starlike functions with respect to a boundary point h whose images lie in a half-plane with $h(0) = 1$ and $h(1) = 0$.

Following Silverman and Silvia [135] we show now that $h \in \operatorname{Univ}(\Delta)$ with $h(0) = 1$ and $h(1) = 0$ is a starlike function with respect to a boundary point if and only if $h \in G(\lambda)$ for some $\lambda \in (0, 2]$. Moreover, in this case the image $h(\Delta)$ lies in a wedge of angle $\lambda\pi$. In particular, if $\lambda \in (0, 2]$ this proves Robertson's conjecture.

Theorem 3.1. *If $h \in G(\lambda)$, $h \not\equiv 1$, then $h \in S^*[1]$, i.e., h is a starlike function with respect to a boundary point with $h(1) = 0$, and its image $h(\Delta)$ is contained in a wedge of angle $\lambda\pi$.*

Proof. Since $h \in G(\lambda)$ and $h \not\equiv 1$, it follows from Lemma 3.5 that there exists a function $p \in \mathcal{P}$ of positive real part such that

$$(1-z)^2 \frac{h'(z)}{h(z)} = -\frac{1}{p(z)},$$

or, equivalently,

$$h(z) = h'(z) \cdot \left(-(1-z)^2 p(z)\right).$$

It is clear that the function f defined by $f(z) = -(1-z)^2 p(z)$ satisfies the Berkson–Porta representation (2.2.7) with $\tau = 1 \in \partial\Delta$. Consequently, by Theorem 2.7, f is a semi-complete vector field. Moreover, $f \in \mathcal{G}^+[1]$. Therefore, Theorem 2.14 implies that h is a starlike function with respect to a boundary point and $h(1) = 0$. In other words, $h \in S^*[1]$.

Further, by Lemma 3.4 one can approximate $h \in G(\lambda)$ by functions of the form

$$h_n(z) := \prod_{j=1}^{n} \left(\omega_j(z)\right)^{\lambda_j}, \quad \text{where } \omega_j(z) = \frac{1-z}{1-z\bar{\zeta}_j}, \ \sum_{j=1}^{n} \lambda_j = \lambda, \ |\zeta_j| = 1.$$

Each function $\omega_j(z)$ maps the open unit disk Δ onto a half-plane. Thus, for some $\beta_j \in \mathbb{R}$ we have $\arg\left(e^{i\beta_j}\omega_j(z)\right) < \frac{\pi}{2}$, $j = 1, \dots, n$.

Let $\beta = \sum\limits_{j=1}^{n} \lambda_j \beta_j$. Then for each $z \in \Delta$,

$$\left|\arg\left(e^{i\beta}h_n(z)\right)\right| = \left|\arg\left(e^{i\beta}\prod_{j=1}^{n}\left(\omega_j(z)\right)^{\lambda_j}\right)\right|$$
$$= \left|\sum_{j=1}^{n}\lambda_j \arg\left(e^{i\beta_j}\omega_j(z)\right)\right| < \sum_{j=1}^{n}\lambda_j\frac{\pi}{2} = \frac{\lambda\pi}{2}.$$

Hence $h(\Delta)$ is contained in a wedge of angle $\lambda\pi$. □

Theorem 3.2. *If $h \in S^*[1]$ and the image $h(\Delta)$ lies in a wedge of angle $\lambda\pi$, then $h \in G(\lambda)$.*

We will prove this theorem by using a geometrical idea of Lyzzaik [102]. This idea is based on the existence of an approximation sequence consisting of starlike functions with respect to interior points and the Carathéodory Theorem on kernel convergence. Other analytic approaches, which have a more constructive character for approximation problems, are given in Section 9.1.

Proof. Since the image $h(\Delta)$ lies in a wedge of angle $\lambda\pi$, the function $h_0(z) = h(z)^{\frac{2}{\lambda}}$, $h_0(0) = 1$, is a well-defined univalent function on Δ. Furthermore, $h_0(1) = 0$ and h_0 is starlike with respect to a boundary point, i.e., $h_0 \in S^*[1]$. For each $n = 1, 2, \dots$, let D_n be a starlike domain defined by

$$D_n = h_0(\Delta) \cup \left\{w \in \mathbb{C} : \ |w| < \frac{1}{n}\right\},$$

and let $h_n : \Delta \mapsto D_n$ be the univalent (starlike) mapping of Δ onto D_n such that $h_n(0) = 1$ and $\arg h_n'(0) = \arg h_0'(0)$. It follows by Carathéodory's theorem on kernel convergence that h_n converges to h_0 uniformly on compact subsets of Δ as n tends to infinity. Because each $h_n(\Delta) = D_n$ is a starlike domain that contains the origin, there is a point $\tau_n \in \Delta$ such that $h_n(\tau_n) = 0$, i.e., $h_n \in S^*[\tau_n]$. Then by Theorem 1.3 there are starlike functions $\psi_n \in S^*[0]$ that satisfy

$$h_n(z) = \frac{\psi_n(z)}{z}(z - \tau_n)(1 - \bar{\tau}_n z), \quad z \in \Delta. \tag{3.3.1}$$

Since $1 = h_n(0) = -\tau_n\psi_n'(0)$, we get

$$h_n'(0) = \frac{\psi_n''(0)}{2\psi_n'(0)} + \psi_n'(0)\left(1 + |\tau_n|^2\right).$$

Thus the convergence $h_n'(0) \to h_0'(0)$ and the estimate $|\psi_n''(0)| \le 4|\psi_n'(0)|$ imply that the sequence $\{\psi_n'(0)\}$ is uniformly bounded. Therefore, there is a subsequence of $\{\psi_n\}$ that converges to a starlike function $-\widetilde{\psi}$, $\widetilde{\psi} \in S^*[0]$. One can assume that the corresponding subsequence of $\{\tau_n\}$ converges to a point $\tau \in \overline{\Delta}$. Now it follows by (3.3.1) that

$$h_0(z) = \frac{-\widetilde{\psi}(z)}{z}(z-\tau)(1-\bar{\tau}z), \quad z \in \Delta.$$

Letting z approach 1 we conclude that $\tau = 1$. Hence

$$h_0(z) = \frac{\widetilde{\psi}(z)}{z} \cdot (1-z)^2.$$

This equality implies that $\widetilde{\psi}'(0) = 1$, and then $\widetilde{\psi} \in S^*$. Raising to the power $\lambda/2$, we obtain

$$h(z) = \left(\frac{\widetilde{\psi}(z)}{z}\right)^{\lambda/2} (1-z)^{\lambda}.$$

Note that the function $z\left(\frac{\widetilde{\psi}(z)}{z}\right)^{\lambda/2}$ is starlike of order $\kappa = 1 - \frac{\lambda}{2}$. Using Lemma 3.2 we conclude that $h \in G(\lambda)$. This completes the proof. □

3.4 Angle distortion theorems

3.4.1 Smallest exterior wedge

We have already shown that a function $h \in S^*[1]$, $h(0) = 1$, belongs to $G(\lambda)$ if and only if its image $h(\Delta)$ lies in a wedge of angle $\lambda\pi$. This fact is a geometric explanation of Lemma 3.3. Now the question is:

◊ *Given* $h \in S^*[1]$, *find the minimal* $\lambda \in (0, 2]$ *such that* $h \in G(\lambda)$.

In other words, one should determine the angle of the minimal wedge $W^*(h)$ that contains $h(\Delta)$. Obviously, $W^*(h)$ contains the point $w = 1$ because of the normalization. An additional question that arises in this context is the precise location of $W^*(h)$, say, with respect to the real axis.

Theorem 3.3 (cf. [56, 63]). *Let* $h \in S^*[1]$, $h(0) = 1$. *Then*

(i) *h satisfies a generalized Visser–Ostrowski condition (see Definition 1.12)*

$$\angle \lim_{z \to 1} \frac{(z-1)h'(z)}{h(z)} = \kappa,$$

with $0 < \kappa \leq 2$;

(ii) *the radial limit*

$$\lim_{r \to 1^-} \arg h(r) = \theta^*$$

exists and $|\theta^*| < \pi$;

(iii) *the minimal wedge $W^*(h)$, which contains the image $h(\Delta)$, is*

$$W^*(h) = \left\{ w \in \mathbb{C} : \; -\frac{\kappa\pi}{2} + \theta^* < \arg w < \frac{\kappa\pi}{2} + \theta^* \right\}. \qquad (3.4.1)$$

Consequently, $h \in G(\kappa)$, and for any $\lambda < \kappa$, $h \notin G(\lambda)$.

Proof. Suppose that $W^*(h)$ is of angle $\kappa\pi$ for some κ, $0 < \kappa \leq 2$. Then $h \in G(\kappa)$ by Theorem 3.2. So, by Lemma 3.1, it can be represented as follows:

$$h(z) = (1-z)^{\kappa} \exp\left[-\kappa \oint_{|\zeta|=1} \log(1 - z\bar{\zeta}) d\sigma(\zeta) \right] \qquad (3.4.2)$$

for some probability measure σ on the unit circle $|\zeta| = 1$.

We claim that the measure σ in this formula is mutually singular with the Dirac measure δ at the point $\zeta = 1 \in \partial\Delta$. Indeed, decomposing σ relative to δ, one can write $\sigma = t\widetilde{\sigma} + (1-t)\delta$, where $0 \leq t \leq 1$, and $\widetilde{\sigma}$ and δ are mutually singular probability measures. Thus

$$h(z) = (1-z)^{t\kappa} \exp\left[-t\kappa \oint_{|\zeta=1} \log(1 - z\bar{\zeta}) d\widetilde{\sigma}(\zeta) \right];$$

that is, $h \in G(t\kappa)$. By Theorem 3.1, $h(\Delta)$ is contained in a wedge of angle $t\kappa\pi$. If $t < 1$, our supposition is contradicted. Therefore, $t = 1$, and the measures $\sigma = \widetilde{\sigma}$ and δ are mutually singular.

Further, using (3.4.2) we see that

$$\frac{(z-1)h'(z)}{h(z)} = \kappa \oint_{|\zeta|=1} \frac{1-\bar{\zeta}}{1-z\bar{\zeta}} d\sigma(\zeta), \quad z \in \Delta. \qquad (3.4.3)$$

Let $\{z_n = x_n + iy_n\}_{n=1}^{\infty}$ be any sequence that tends to 1 non-tangentially, i.e., $\{z_n\}_{n=1}^{\infty}$ lies in some Stolz angle

$$D_{a,r}(1) = \{ z \in \Delta : \; |\arg(1-z)| < a, \; |z-1| < r \}.$$

Consider the functions $\phi_n : \partial\Delta \to \mathbb{C}$, $n = 1, 2, \ldots$, defined by

$$\phi_n(\zeta) := \frac{1-\bar{\zeta}}{1-z_n\bar{\zeta}}, \quad \zeta \in \partial\Delta.$$

Each function ϕ_n maps the unit circle $\partial\Delta$ onto the circle $|\xi - c_n| = |c_n|$, where $c_n = (1-\bar{z}_n)/(1-|z_n|^2)$. Hence, there is a natural number N independent of $\zeta \in \partial\Delta$ such that

$$|\phi_n(\zeta)|^2 \leq 4|c_n|^2 = 4\frac{1 + \left(\frac{y_n}{1-x_n}\right)^2}{\left[1 + x_n - \frac{y_n^2}{1-x_n}\right]^2} \leq 2(1+a^2)$$

for all $n \geq N$. Using (3.4.3) and applying Lebesgue's Bounded Convergence Theorem we obtain

$$\angle \lim_{z\to 1} \frac{(z-1)h'(z)}{h(z)} = \kappa \lim_{n\to\infty} \oint_{|\zeta|=1} \phi_n(\zeta) d\sigma(\zeta) = \kappa,$$

i.e., h satisfies a generalized Visser–Ostrowski condition.

To prove (ii) we note that by Lemma 3.1

$$\arg h(z) = \kappa \left[\arg(1-z) - \oint_{|\zeta|=1} \arg\left(1 - z\bar{\zeta}\right) d\sigma(\zeta) \right], \tag{3.4.4}$$

and, consequently, for any $r \in (0,1)$,

$$\arg h(r) = -\kappa \oint_{|\zeta|=1} \arg\left(1 - r\bar{\zeta}\right) d\sigma(\zeta).$$

Since $|\arg(1 - r\bar{\zeta})| < \frac{\pi}{2}$, Lebesgue's Bounded Convergence Theorem implies that the radial limit

$$\theta^* := \lim_{r\to 1^-} \arg h(r)$$

exists and $|\theta^*| < \dfrac{\kappa\pi}{2}$. Moreover,

$$\theta^* = -\kappa \oint_{|\zeta|=1} \arg\left(1 - \bar{\zeta}\right) d\sigma(\zeta).$$

Combining the last equality with (3.4.4) we see that

$$\begin{aligned} &|\arg h(z) - \theta^*| \\ &= \kappa \left| \arg(1-z) - \oint_{|\zeta|=1} \arg\left(1 - z\bar{\zeta}\right) d\sigma(\zeta) + \oint_{|\zeta|=1} \arg\left(1 - \bar{\zeta}\right) d\sigma(\zeta) \right| \\ &\leq \kappa \oint_{|\zeta|=1} \left| \arg \frac{(1-z)(1-\bar{\zeta})}{1 - z\bar{\zeta}} \right| d\sigma(\zeta). \end{aligned}$$

Since the function $\dfrac{(1-z)(1-\bar{\zeta})}{1-z\bar{\zeta}}$ takes values in the right half-plane, we have

$$|\arg h(z) - \theta^*| < \frac{\kappa\pi}{2}.$$

Taking into account that the minimal wedge $W^*(h)$ is of angle $\kappa\pi$, we complete the proof. □

A consequence of Theorems 2.13, 2.14 and 3.3 is the following assertion.

Corollary 3.2 (cf. [97, 99]). *Let* $h \in \mathrm{Hol}(\Delta, \mathbb{C})$, $h(0) = 1$. *Then* $h \in S^*[1]$ *if and only if it satisfies the condition*

$$h(z) = -\frac{(1-z)^2}{4}\, h'(z)\, \frac{1+F(z)}{1-F(z)},$$

where $F \in \mathrm{Hol}(\Delta)$ *is such that* $F(1) = 1$ *and* $F'(1) = \alpha \in (0, 1]$.

Moreover, the smallest wedge which contains $h(\Delta)$ *is exactly of angle* $2\alpha\pi$.

Now it is natural to distinguish the classes of starlike functions that satisfy a generalized Visser–Ostrowski condition. Namely, for each $\mu \in (0, 2]$ we say that

- *a univalent function* h *belongs to the class* $\mathrm{Star}_\mu[1]$ *if it is of* $S^*[1]$, *is normalized by* $h(0) = 1$, *and satisfies*

$$Q_h(1) := \angle \lim_{z \to 1} \frac{(z-1)h'(z)}{h(z)} = \mu.$$

Corollary 3.3. *The class* $\mathrm{Star}_\mu[1]$ *consists of those univalent functions in* $G(\mu)$ *that do not belong to* $G(\lambda)$ *for* $\lambda < \mu$; *and*

$$G(\mu) \setminus \{h \equiv 1\} = \bigcup_{\lambda \le \mu} \mathrm{Star}_\lambda[1].$$

3.4.2 Biggest interior wedge

Given a starlike function $h \in S^*[\tau]$ (with respect to an interior or boundary point $\tau \in \overline{\Delta}$), the following additional questions are seen to be natural:

◊ find conditions which ensure that the image $h(\Delta)$ contains a wedge;

◊ find the size and the location of a maximal wedge $W_*(h)$, which is contained in $h(\Delta)$.

To answer these questions we need the following definition.

Definition 3.4. *We say that a function* $h \in S^*[\tau]$ *belongs to the subclass* $S^*[\tau, \eta]$, *where* $\eta \in \partial\Delta$, $\eta \ne \tau$, *if the angular limit*

$$Q_h(\eta) := \angle \lim_{z \to \eta} \frac{(z-\eta)h'(z)}{h(z)} \tag{3.4.5}$$

exists finitely and is different from zero. In the case where $\tau = 1$, *we say that* $h \in \mathrm{Star}_{\mu,\nu}[1, \eta]$ *if* $h \in \mathrm{Star}_\mu[1]$ *and the angular limit* (3.4.5) *at the boundary point* $\eta \ne 1$ *equals* ν.

Thus each function $h \in S^*[\tau, \eta]$ satisfies a generalized Visser–Ostrowski condition at the boundary point η (see Definition 1.12).

To proceed, we note that the inequality $\eta\neq\tau$ implies that for each $h\in S^*[\tau,\eta]$,

$$\angle\lim_{z\to\eta} h(z)=\infty.$$

We require representation formulas for the classes of starlike functions $S^*[\tau]$ and $S^*[\tau,\eta]$, respectively $\mathrm{Star}_\mu[1]$ and $\mathrm{Star}_{\mu,\nu}[1,\eta]$. For a boundary point w, denote by δ_w the Dirac measure (δ-function) at this point.

Lemma 3.7 (cf. [75] and [64]). *Let $\tau\in\overline{\Delta}$ and $\eta\in\partial\Delta$, $\eta\neq\tau$. Let $h\in\mathrm{Hol}(\Delta,\mathbb{C})$ satisfy $h(\tau)=0$. Then*

(i) *$h\in S^*[\tau]$ if and only if it has the form*

$$h(z)=C(z-\tau)(1-z\bar{\tau})\cdot\exp\left[-2\oint_{\partial\Delta}\log(1-z\bar{\zeta})d\widetilde{\sigma}(\zeta)\right],\tag{3.4.6}$$

where $\widetilde{\sigma}$ is an arbitrary probability measure on the unit circle and $C\neq 0$.

(ii) *Moreover, $h\in S^*[\tau,\eta]$ if and only if it has the form*

$$\begin{aligned}h(z)=\ & C(z-\tau)(1-z\bar{\tau})(1-z\bar{\eta})^{-2a}\cdot\\ &\cdot\exp\left[-2(1-a)\oint_{\partial\Delta}\log(1-z\bar{\zeta})d\sigma(\zeta)\right],\end{aligned}\tag{3.4.7}$$

where σ is a probability measure on the unit circle singular relative to δ_η, $C\neq 0$ and $a\in(0,1]$. In this case, $Q_h(\eta)=-2a$.

Remark 3.1. The constant C in (3.4.6) can be chosen starting from a normalization of the functions under consideration. Because functions of the classes $\mathrm{Star}_\mu[1]$ are normalized by $h(0)=1$, one sets $C=1$. On the other hand, since a starlike function h is a solution of a linear homogeneous equation (see Theorem 2.14)), C arises in the integration process of this equation.

Remark 3.2. In the case where $\tau=0$, formula (3.4.6) is the well-known representation of functions of the class $S^*[0]$; see, for example, [8]. For instance, the Koebe function $k(z)=\frac{z}{(1-z)^2}$ is obtained by choosing the measure $\widetilde{\sigma}$ to be the Dirac δ-function at the point 1. On the other hand, for $\tau=1$, formula (3.4.6) follows by Lemma 3.1. In turn, formula (3.4.6) is more general than formula (3.2.2).

Proof. (i) First, suppose that $\tau=0$, and let h be normalized by $h(0)=0$ and $h'(0)=1$. Recall that Nevanlinna's criterion (Theorem 1.1) asserts that $h\in S^*[0]$ if and only if

$$q(z):=\frac{zh'(z)}{h(z)}$$

has a positive real part. (Note that the same fact follows by Theorem 2.14, because of the Berkson–Porta representation of generators vanishing at the origin: $g(z) = zp(z)$ with $\operatorname{Re} p(z) \geq 0$).

Representing q by the Riesz–Herglotz formula, we write

$$\frac{zh'(z)}{h(z)} = \oint_{\partial\Delta} \frac{1 + z\bar{\zeta}}{1 - z\bar{\zeta}} \, d\widetilde{\sigma}(\zeta)$$

with some probability measure $\widetilde{\sigma}$. Integrating this equality, we get

$$h(z) = z \exp\left[-2 \oint_{\partial\Delta} \log(1 - z\bar{\zeta}) d\widetilde{\sigma}(\zeta)\right]. \tag{3.4.8}$$

This proves (3.4.6) for the case $\tau = 0$.

Now let $h(\tau) = 0$ for $\tau \in \Delta$, $\tau \neq 0$. It follows from Hummel's Theorem 1.3 that $h \in S^*[\tau]$ if and only if

$$\frac{z}{(z - \tau)(1 - z\bar{\tau})} \, h(z) \in S^*[0].$$

Thus, (3.4.8) implies (3.4.6) for the interior location of τ. The reverse consideration and Hummel's criterion show that if h satisfies (3.4.6) with $\tau \in \Delta$, it must be starlike.

Finally, let $\tau \in \partial\Delta$. Following Lyzzaik [102] (see also the proof of Theorem 3.2) one can approximate $h \in S^*[\tau]$ by a sequence $\{h_n\}$ of functions starlike with respect to interior points τ_n that converge to τ. Also, one can assume that $h_n(0) = h(0)$. Representing each function h_n by (3.4.6),

$$h_n(z) = C_n(z - \tau_n)(1 - z\bar{\tau}_n) \cdot \exp\left[-2 \oint_{\partial\Delta} \log(1 - z\bar{\zeta}) d\widetilde{\sigma_n}(\zeta)\right],$$

we see that

$$h(0) = h_n(0) = -C_n\tau_n.$$

Thus $C_n \to -\frac{h(0)}{\tau}$. Since the set of all probability measures is compact, $\{\widetilde{\sigma_n}\}$ has a subsequence converging to some probability measure $\widetilde{\sigma}$. Therefore, any function $h \in S^*[\tau]$ has the form (3.4.6).

To prove the converse assertion, we suppose that h has the form (3.4.6) with $\tau \in \partial\Delta$. Note that h is starlike if and only if the function $ah(cz)$, $a \neq 0$, $|c| = 1$, is also starlike. Therefore, without loss of generality, one can assume that $\tau = 1$ and h is normalized by $h(0) = 1$, i.e.,

$$h(z) = (1 - z)^2 \cdot \exp\left[-2 \oint_{\partial\Delta} \log(1 - z\bar{\zeta}) d\widetilde{\sigma}(\zeta)\right].$$

Then $h \in G(2)$ by Lemma 3.1. Now Theorem 3.1 implies that h is a starlike function with respect to a boundary point with $h(1) = 0$, i.e., $h \in S^*[1]$. This proves the first assertion.

(ii) Let

$$\widetilde{\sigma} = a\delta_\eta + (1-a)\sigma, \quad 0 \le a \le 1,$$

be the Lebesgue decomposition of $\widetilde{\sigma}$ relative to the Dirac measure δ_η, where the probability measures σ and δ_η are mutually singular. Using this decomposition, we rewrite (3.4.6) in the form of (3.4.7).

Now we calculate

$$\begin{aligned} Q_h(\eta) &= \angle \lim_{z\to\eta} \frac{h'(z)(z-\eta)}{h(z)} \\ &= \angle \lim_{z\to\eta} (z-\eta) \left[\frac{((z-\tau)(1-z\bar{\tau}))'}{(z-\tau)(1-z\bar{\tau})} + \frac{2a\bar{\eta}}{1-z\bar{\eta}} \right. && (3.4.9) \\ &\left. +2(1-a) \oint_{\partial\Delta} \frac{\bar{\zeta}}{1-z\bar{\zeta}}\, d\sigma(\zeta) \right] \\ &= -2a + 2(1-a) \angle \lim_{z\to\eta} \oint_{\partial\Delta} \frac{\bar{\zeta}(z-\eta)}{1-z\bar{\zeta}}\, d\sigma(\zeta). && (3.4.10) \end{aligned}$$

Noting that

$$\left| \frac{\bar{\zeta}(z-\eta)}{1-z\bar{\zeta}} \right| \le \frac{|z-\eta|}{1-|z|},$$

we see that the integrand in the last expression of (3.4.9) is bounded on each non-tangential approach region

$$\Gamma(\eta,\kappa) = \{z : |z-\eta| < \kappa(1-|z|)\}, \ \kappa \ge 1,$$

at the point η. Since the measures σ and δ_η are mutually singular, we conclude by the Lebesgue convergence theorem that the last integral in (3.4.9) is equal to zero, so

$$Q_h(\eta) = -2a.$$

Our proof is complete. □

Remark 3.3. Note, in passing, that for starlike functions with respect to a boundary point of the class $\mathrm{Star}_{\mu,\nu}[1,\eta]$, formula (3.4.7) can be refined in the following way: *A function h belongs to $\mathrm{Star}_{\mu,\nu}[1,\eta]$ if and only if it admits the following representation:*

$$h(z) = (1-z)^\mu (1-z\bar{\eta})^\nu \exp\left[-(\mu+\nu) \oint_{\partial\Delta} \log(1-z\bar{\zeta})d\sigma(\zeta) \right] \qquad (3.4.11)$$

with some probability measure σ on the unit circle, which is singular relative to both δ_1 and δ_η. Furthermore, $-\mu \le \nu \le 0$.

Theorem 3.4 (cf. [128]). *Let $h \in S^*[\tau,\eta]$ with $Q_h(\eta)=\nu$. Let*

$$\theta_* := \lim_{r\to 1^-} \arg h(r\eta).$$

Then the image $h(\Delta)$ contains the wedge

$$W_* = \left\{ w \in \mathbb{C} : \ |\arg w - \theta_*| < \frac{|\nu|\pi}{2} \right\} \tag{3.4.12}$$

and contains no larger wedge with the same bisector.

Proof. By Lemma 3.7, the function h has the form (3.4.7) with $a = -\frac{\nu}{2}$. First, we show that the image $h(\Delta)$ contains the wedge W_* defined by (3.4.12).

Since (as mentioned above) $\angle \lim\limits_{z\to\eta} h(z) = \infty$, for each $\alpha \in \left(0, \frac{\pi}{2}\right)$ and each $R>0$, there exists $r>0$ such that

$$|h(z)| > R, \tag{3.4.13}$$

whenever

$$z \in D_{\alpha,r}(\eta) := \left\{ z \in \Delta : \ |1 - z\bar{\eta}| \le r, \ |\arg(1 - z\bar{\eta})| \le \alpha \right\}.$$

Now the Lebesgue Bounded Convergence Theorem implies the existence of

$$\lim_{z\to\eta} \arg \frac{h(z)}{(1-z\bar{\eta})^{\nu}} = \arg\Big(C(\eta-\tau)(1-\eta\bar{\tau}) \Big) - 2(1-a) \lim_{z\to\eta} \oint\limits_{\partial\Delta} \arg(1 - z\bar{\zeta}) d\sigma(\zeta).$$

Alternatively, by formula (3.4.7), we have

$$\theta_* = \lim_{r\to 1^-} \arg h(r\eta) = \arg\Big(C(\eta-\tau)(1-\eta\bar{\tau}) \Big) - 2(1-a) \lim_{r\to 1^-} \oint_{\partial\Delta} \arg(1 - r\eta\bar{\zeta}) d\sigma(\zeta).$$

Therefore,

$$\lim_{z\to\eta} \arg \frac{h(z)}{(1-z\bar{\eta})^{\nu}} = \theta_*.$$

Fix $\varepsilon>0$. Decreasing r (if necessary), we have

$$\theta_* - \varepsilon < \arg \frac{h(z)}{(1-z\bar{\eta})^{\nu}} < \theta_* + \varepsilon$$

for all $z \in D_{\delta,r}(\eta)$. So, for each point z belonging to the arc

$$\Gamma := \left\{ z \in \Delta : \ |1 - z\bar{\eta}| = r, \ |\arg(1 - z\bar{\eta})| \le \delta \right\} \subset D_{\delta,r}(\eta),$$

i.e., $z = \eta(1 - re^{it})$, $|t| \le \delta$, we get

$$\theta_* - \varepsilon - t|\nu| < \arg h(z) < \theta_* + \varepsilon - t|\nu|.$$

In particular,

$$\arg h(\eta(1 - re^{i\delta})) < \theta_* + \varepsilon - \delta|\nu|$$

and

$$\arg h(\eta(1 - re^{-i\delta})) > \theta_* - \varepsilon + \delta|\nu|.$$

Thus, the curve $h(\Gamma)$ lies outside the disk $|z| \leq R$ and joins two points having arguments less than $\theta_* + \varepsilon - \delta|\nu|$ and greater than $\theta_* - \varepsilon + \delta|\nu|$, respectively. Since h is starlike, we see that $h(\Delta)$ contains the sector

$$\{w \in \mathbb{C} : \ |w| < R, \ |\arg w - \theta_*| < \delta|\nu| - \varepsilon\}.$$

Because R and ε are arbitrary, one concludes

$$\{w \in \mathbb{C} : \ |\arg w - \theta_*| < \delta|\nu|\} \subset h(\Delta).$$

Letting δ tend to $\frac{\pi}{2}$, we obtain

$$W_* = \left\{w \in \mathbb{C} : \ |\arg w - \theta_*| < \frac{|\nu|\pi}{2}\right\} \subset h(\Delta).$$

Further, since h is a starlike function, $\arg h(e^{i\varphi})$ is an increasing function in $\varphi \in (\arg \eta - \pi, \arg \eta + \pi)$. Hence the limits

$$\lim_{\varphi \to (\arg \eta)^\pm} \arg h(e^{i\varphi})$$

exist. Let $\varphi_{n,+} \to (\arg \eta)^+$ and $\varphi_{n,-} \to (\arg \eta)^-$ be two sequences such that the values $h(e^{i\varphi_{n,\pm}})$ are finite. Then, once again by Lemma 3.7,

$$\begin{aligned}
&\lim_{n\to\infty} \left(\arg h(e^{i\varphi_{n,+}}) - \arg h(e^{i\varphi_{n,-}})\right) \\
&= \lim_{n\to\infty} \left((\arg(1 - e^{i\varphi_{n,+}}\bar{\eta}))^\nu - (\arg(1 - e^{i\varphi_{n,-}}\bar{\eta}))^\nu\right) = |\nu|\pi.
\end{aligned}$$

Therefore, the image contains no wedge of angle larger than $|\nu|\pi$. Thus, the wedge W_* defined by (3.4.12) is the largest one contained in $h(\Delta)$.

The proof is complete. □

It is also of interest to establish a converse assertion.

Theorem 3.5. *Let $h \in S^*[\tau]$. Suppose that the image $h(\Delta)$ contains an open wedge*

$$W = \left\{w \in \mathbb{C} : \ |\arg w - \theta| < \frac{\kappa\pi}{2}\right\} \subset h(\Delta),$$

which is maximal in the sense that there is no wedge $W_1 \neq W$ such that $W \subset W_1 \subset h(\Delta)$. Then for some boundary point η, $\eta \neq \tau$, the function h belongs to $S^*[\tau, \eta]$ and satisfies the generalized Visser–Ostrowski condition

$$\angle \lim_{z\to\eta} \frac{(z - \eta)h'(z)}{h(z)} = -\kappa.$$

Proof. By our assumption, the curve $\ell := h^{-1}\Big(\{w:\ \arg w = \theta\}\Big)$ lies in Δ and joins the point τ with a boundary point η. Since W cannot be extended to a larger wedge lying in $h(\Delta)$, for each $\varepsilon > 0$ there are boundary points of the image $h(\Delta)$ belonging to the sectors

$$\left\{w:\ \theta + \frac{\kappa\pi}{2} \le \arg w \le \theta + \frac{\kappa\pi}{2} + \varepsilon\right\}$$

and

$$\left\{w:\ \theta - \frac{\kappa\pi}{2} - \varepsilon < \arg w \le \theta - \frac{\kappa\pi}{2}\right\}.$$

Let $\{\phi_n^+\}$ be a decreasing sequence such that $\phi_n^+ \to \arg\eta$, the values $h\left(e^{i\phi_n^+}\right)$ exist and $\theta + \frac{\kappa\pi}{2} \le \arg h\left(e^{i\phi_n^+}\right) < \theta + \frac{\kappa\pi}{2} + \frac{1}{n}$. Similarly, let $\{\phi_n^-\}$ be an increasing sequence such that $\phi_n^- \to \arg\eta$, the values $h\left(e^{i\phi_n^-}\right)$ exist and $\theta - \frac{\kappa\pi}{2} - \frac{1}{n} < \arg h\left(e^{i\phi_n^-}\right) \le \theta - \frac{\kappa\pi}{2}$.

Since $\eta \ne \tau$, it is possible to define on a neighborhood of η a one-valued continuous branch of the function $\arg(z-\tau)$. Without loss of generality we suppose that all of the points $\left\{e^{i\phi_n^+}, e^{i\phi_n^-}\right\}$ lie in this neighborhood.

By Lemma 3.7 we have

$$\begin{aligned}\arg h(z) &= \arg C + \arg\left((z-\tau)(1-z\bar\tau)\right) - 2a\arg(1-z\bar\eta)\\ &\quad - 2(1-a)\oint_{\partial\Delta} \arg(1-z\bar\zeta)d\sigma(\zeta),\end{aligned}$$

where the probability measure σ is singular relative to δ_η, $C \ne 0$ and $a \in [0,1]$. Note that $h \in S^*[\tau,\eta]$ if and only if $a \ne 0$.

Consider the expression $\arg h\left(e^{i\phi_n^+}\right) - \arg h\left(e^{i\phi_n^-}\right)$, which tends to $\kappa\pi$:

$$\begin{aligned}&\arg h\left(e^{i\phi_n^+}\right) - \arg h\left(e^{i\phi_n^-}\right)\\ &= \left[\arg\left((e^{i\phi_n^+} - \tau)(1 - e^{i\phi_n^+}\bar\tau)\right) - \arg\left((e^{i\phi_n^-} - \tau)(1 - e^{i\phi_n^-}\bar\tau)\right)\right]\\ &\quad - 2a\left[\arg\left(1 - e^{i\phi_n^+}\bar\eta\right) - \arg\left(1 - e^{i\phi_n^-}\bar\eta\right)\right]\\ &\quad - 2(1-a)\oint_{\partial\Delta} \arg\frac{1 - e^{i\phi_n^+}\bar\zeta}{1 - e^{i\phi_n^-}\bar\zeta}\,d\sigma(\zeta).\end{aligned} \tag{3.4.14}$$

The first summand tends to zero while the second one tends to $2a\pi$. The third summand also tends to zero because the integrand is a bounded function tending to zero for each $\zeta \in \partial\Delta$, $\zeta \ne \eta$. Hence by Lebesgue's bounded convergence theorem the integral in (3.4.14) goes to 0.

Then we obtain $\kappa\pi = 2a\pi$. Again by Lemma 3.7 we conclude that $h \in S^*[\tau,\eta]$ with $Q_h(\eta) = -\kappa$. The proof is complete. □

Corollary 3.4. *Let $h \in \mathrm{Star}_\mu[1]$. Then there is a one-to-one correspondence between wedges*

$$W = \left\{ w \in \mathbb{C} : \ |\arg w - \theta| < \frac{-\nu\pi}{2} \right\} \subset h(\Delta),$$

that cannot be extended to a bigger one lying in $h(\Delta)$ and boundary points η such that $\lim_{r\to 1^-} h(r\eta) = \infty$, $\lim_{r\to 1^-} \arg h(r\eta) = \theta$ and h belongs to $\mathrm{Star}_{\mu,\nu}[1,\eta]$ (see Fig. 3.2).

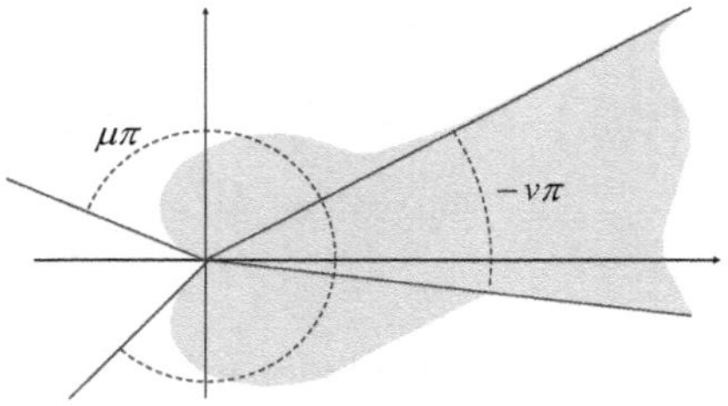

Figure 3.2: Angle distortion.

3.5 Functions convex in one direction

The class of functions we consider here has been studied by several mathematicians (see, for example, Hengartner and Schober [82], Ciozda [30, 31], Lecko [98]) as a subclass of functions defined by Robertson in [122].

Definition 3.5. *We say that a univalent function $g \in \mathrm{Hol}(\Delta, \mathbb{C})$ normalized by*

$$g(0) = 0 \tag{3.5.1}$$

is convex in the positive direction of the real axis if, for each $z \in \Delta$ and $t > 0$,

$$g(z) + t \in g(\Delta) \quad \text{and} \quad \lim_{t\to\infty} g^{-1}(g(z) + t) = \tau \in \partial\Delta. \tag{3.5.2}$$

The class of these functions is denoted by $\Sigma[\tau]$.

We first study the following question:

◊ Given a function of the class $\Sigma[\tau]$, find the minimal horizontal strip that contains its image.

The next question is also natural but more complicated:

◊ Characterize those functions convex in the positive direction of the real axis whose images contain a whole (two-sided) strip and find the size (width) of this strip.

We solve the latter problem for functions having minimal horizontal strips of finite size. The problem is still open for the general case.

To proceed we need the following lemmata.

Lemma 3.8 (cf. [82, 30, 31]). *Let g be a univalent function normalized by* (3.5.1). *Then $g \in \Sigma[\tau]$ if and only if*

$$\operatorname{Re}\left((\tau - z)(1 - z\bar{\tau})g'(z)\right) > 0 \quad \textit{for all } z \in \Delta. \tag{3.5.3}$$

Proof. Let $g \in \Sigma[\tau]$. By Definition 3.5, for each $t \geq 0$ the holomorphic function F_t defined by

$$F_t(z) = g^{-1}\left(g(z) + t\right)$$

maps the unit disk into itself. It is easy to verify that the family $\mathcal{S} = \{F_t\}_{t\geq 0}$ forms a continuous semigroup of holomorphic self-mappings of the unit disk.

Differentiating this semigroup at $t = 0^+$ we get

$$f(z) := -\left.\frac{\partial F_t(z)}{\partial t}\right|_{t=0^+} = -\frac{1}{g'(z)}, \tag{3.5.4}$$

where f is the infinitesimal generator of $\mathcal{S}$ (see, for example, Theorem 2.1). By (3.5.2) the point τ is the Denjoy–Wolff point of $\mathcal{S}$. Therefore, its generator can be represented by the Berkson–Porta formula (see Theorem 2.7):

$$f(z) = (z - \tau)(1 - z\bar{\tau})p(z), \quad \text{where } \operatorname{Re} p(z) \geq 0. \tag{3.5.5}$$

Comparing (3.5.4) and (3.5.5) proves inequality (3.5.3).

Conversely, suppose that g satisfies (3.5.3). Then the function

$$p(z) = \frac{1}{(\tau - z)(1 - z\bar{\tau})g'(z)}$$

has a non-negative real part.

If $p(z) = ib$, $b \in \mathbb{R}$, then $g(z) = \dfrac{1}{ib} \cdot \dfrac{z}{\tau - z}$ belongs to the class $\Sigma[\tau]$.

If $\operatorname{Re} p(z) > 0$, $z \in \Delta$, then by a result of Berkson and Porta (see Theorem 2.7), the function f defined by (3.5.5) is the generator of a semigroup $\mathcal{S} = \{F_t\}_{t\geq 0}$ of holomorphic self-mappings of the unit disk. This semigroup can be defined by the Cauchy problem:

$$\begin{cases} \dfrac{\partial F_t(z)}{\partial t} + f(F_t(z)) = 0, \\ F_0(z) = z, \quad z \in D. \end{cases}$$

Substituting here $f(z)\Big(= (z-\tau)(1-z\bar{\tau})p(z)\Big) = -\dfrac{1}{g'(z)}$ we get

$$g'(F_t(z))\frac{\partial F_t(z)}{\partial t} = 1.$$

Integrating the latter expression on the interval $[0,t]$ we get

$$g(F_t(z)) = g(z) + t, \quad \text{that is,} \;\; g(z) + t \in g(\Delta).$$

Since $\mathcal{S}$ has a Denjoy–Wolff point at τ, it follows that

$$\lim_{t\to\infty} g^{-1}\left(g(z)+t\right)\Big(= \lim_{t\to\infty} F_t(z)\Big) = \tau.$$

This completes our proof. □

Later on we concentrate on the case $\tau = 1$ and consider the class $\Sigma[1]$.

Lemma 3.9. *Let $g \in \Sigma[1]$. Then the limit*

$$\angle \lim_{z\to 1}(1-z)g'(z) = \mu \tag{3.5.6}$$

is either a positive real number or infinity.

Proof. By Lemma 3.8 the function

$$p(z) = \frac{1}{(1-z)^2 g'(z)}$$

is either a non-constant holomorphic function of positive real part or an imaginary constant. In the latter case, $p(z) = ib,\ b \in \mathbb{R}$, and the assertion is evident. Otherwise, one can write

$$\frac{1}{(1-z)g'(z)} = (1-z)p(z).$$

It was shown in [62] that, for any function p of the class $\mathcal{P}$, the angular limit $\angle \lim\limits_{z\to 1}(1-z)p(z)$ exists and is a non-negative real number. This proves our assertion. □

Definition 3.6. *We say that a univalent function g belongs to the class $\Sigma_\mu[1]$ with $0 < \mu \le \infty$, if it is of the class $\Sigma[1]$ and the limit* (3.5.6) *is equal to μ.*

Thus by Lemma 3.9 we have $\Sigma[1] = \bigcup\limits_{0<\mu\le\infty} \Sigma_\mu[1]$.

Example 3.1. The function

$$g_1(z) = \log\frac{1+z}{1-z}$$

belongs to $\Sigma_1[1]$, while the function

$$g_2(z) = \frac{z}{1-z}$$

belongs to $\Sigma_\infty[1]$.

Definition 3.7. *Let $\eta \in \partial\Delta$, $\eta \neq 1$. We say that a univalent function g belongs to the class $\Sigma_{\mu,\nu}[1,\eta]$ if it is of the class $\Sigma_\mu[1]$ and satisfies the conditions*

$$\angle \lim_{z\to\eta} \operatorname{Re} g(z) = -\infty$$

and

$$\angle \lim_{z\to\eta} (z-\eta)g'(z) = \nu.$$

Theorem 3.6. (i) *If $g \in \Sigma_\mu[1]$, $\mu \in (0,\infty)$, then the horizontal strip*

$$T^* = \left\{z : \ |\operatorname{Im} z - a^*| < \frac{\pi\mu}{2}\right\} \quad \text{with} \quad a^* = \lim_{r\to 1^-} \operatorname{Im} g(r) \tag{3.5.7}$$

is the smallest one that contains the image $g(\Delta)$. If $g \in \Sigma_\infty[1]$, then there is no horizontal strip of finite width containing $g(\Delta)$;

(ii) *If $g \in \Sigma_{\mu,\nu}[1,\eta]$, $\mu \in (0,\infty)$, then the horizontal strip*

$$T_* = \left\{z : \ |\operatorname{Im} z - a_*| < \frac{\pi\nu}{2}\right\} \quad \text{with} \quad a_* = \lim_{r\to 1^-} \operatorname{Im} g(r\eta) \tag{3.5.8}$$

is contained in $g(\Delta)$, and there is no strip $T \neq T_*$ with $T_* \subset T \subset g(\Delta)$. Consequently, $\nu \in (0,\mu]$.

Conversely, if $g \in \Sigma_\mu[1]$, $\mu \in (0,\infty)$, and the image $g(\Delta)$ contains some open horizontal strip, then there is ν $(0 < \nu \leq \mu)$ and a point $\eta \in \partial\Delta$ such that g belongs to $\Sigma_{\mu,\nu}[1,\eta]$ (see Fig. 3.3).

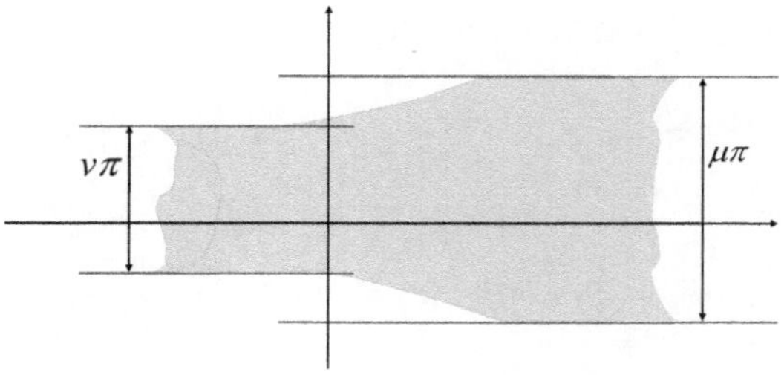

Figure 3.3: Width distortion.

Proof. (i) Suppose that $g \in \Sigma_\mu[1]$, $\mu \in (0,\infty)$, i.e., the limit (3.5.6) is finite. By Lemma 3.8, the function $p(z) = (1-z)^2 g'(z)$ is of nonnegative real part. Hence, there exists a self-mapping $\omega(z)$ such that

$$(1-z)^2 g'(z) = 4\mu \frac{1-\omega(z)}{1+\omega(z)}.$$

Thus

$$\frac{1-\omega(z)}{1-z} = \frac{1+\omega(z)}{4}\,\frac{(1-z)g'(z)}{\mu}.$$

Since the right-hand side of this equality is bounded on each non-tangential approach region at the point $z=1$, one concludes that $\angle\lim\limits_{z\to 1}\omega(z) = 1$. Now we calculate

$$\angle\lim_{z\to 1}\frac{1-\omega(z)}{1-z} = \angle\lim_{z\to 1}\frac{1+\omega(z)}{4}\,\frac{(1-z)h'(z)}{\mu} = \frac{1}{2} \in (0,1).$$

Consider the function

$$h(z) = \exp\left[-\frac{1}{\mu}g(z)\right]. \tag{3.5.9}$$

It satisfies

$$-\frac{h'(z)}{h(z)}(1-z)^2 = 4\,\frac{1-\omega(z)}{1+\omega(z)}.$$

So, by a result of Lecko and Lyzzaik ([99]), the function h is univalent and starlike with respect to a boundary point. Since

$$\angle\lim_{z\to 1}\frac{(z-1)h'(z)}{h(z)} = 1,$$

we see that $h \in \mathrm{Star}_\lambda[1]$ with $\lambda = 1$. By Theorem 3.3 the smallest wedge that contains the image $h(\Delta)$ is defined by (3.4.1):

$$W^*(h) = \left\{w \in \mathbb{C}:\ -\frac{\pi}{2} + \theta^* < \arg w < \frac{\pi}{2} + \theta^*\right\},$$

where

$$\theta^* = \lim_{r\to 1^-}\arg h(r) = \lim_{r\to 1^-} -\frac{\operatorname{Im} g(r)}{\mu}.$$

Therefore, the smallest horizontal strip containing the image of function $g(z) = -\mu\log h(z)$, is defined by (3.5.7).

Suppose now that $g \in \Sigma[1]$ and its image is contained in a horizontal strip of finite width $\pi\mu$ $(0<\mu<\infty)$. In this case the function h defined by (3.5.9) is univalent and starlike with respect to a boundary point. Moreover, since the minimal strip, which contains the image $g(\Delta)$, is of width $\pi\mu$, the minimal wedge containing $h(\Delta)$ is of angle π. Once again by Theorem 3.3, $h \in \mathrm{Star}_\lambda[1]$ with $\lambda = 1$; hence it satisfies the Visser–Ostrowski condition

$$\angle \lim_{z\to 1} \frac{(z-1)h'(z)}{h(z)} = 1.$$

Since

$$\frac{(z-1)h'(z)}{h(z)} = \frac{(1-z)g'(z)}{\mu},$$

and $0 < \mu < \infty$, we have that $g \in \Sigma_\mu[1]$.

(ii) As above, define a starlike function h by (3.5.9). It is clear that $\lim_{r\to 1^-} h(r\eta) = \infty$. Since h is univalent, we conclude that $\angle \lim_{z\to\eta} h(z) = \infty$. Furthermore,

$$\angle \lim_{z\to\eta} \frac{(z-\eta)h'(z)}{h(z)} = \angle \lim_{z\to\eta} \frac{-(z-\eta)g'(z)}{\mu} = -\frac{\nu}{\mu}.$$

So, $h \in \mathrm{Star}_{1,-\nu/\mu}[1,\eta]$. As mentioned above, in this case $-1 \le -\frac{\nu}{\mu} \le 0$; that is, $\nu \in [0,\mu]$. By Theorem 3.4 the wedge

$$W_* = \left\{ w : \ |\arg w - \theta_*| < \frac{\nu\pi}{2\mu} \right\} \quad \text{with} \quad \theta_* = \lim_{r\to 1^-} \arg h(r\eta)$$

is contained in $h(\Delta)$, and there is no wedge $W \neq W_*$ with $W_* \subset W \subset h(\Delta)$. Since $g(z) = -\mu \log h(z)$, we have that the limit

$$a_* = \lim_{x\to -1^+} \operatorname{Im} g(x) = -\mu\theta_*$$

exists. In addition, the strip S_* defined by (3.5.8) is contained in $g(\Delta)$, and there is no strip $S \neq S_*$ with $S_* \subset S \subset g(\Delta)$.

Conversely, suppose that $g \in \Sigma_\mu[1]$, and its image $g(\Delta)$ contains an open strip. Then the function h defined by (3.5.9) belongs to $\mathrm{Star}_\mu[1]$ and its image $h(\Delta)$ contains an open wedge. By Corollary 3.4 we conclude that there is a boundary point η and a positive number ν such that $h \in \mathrm{Star}_{\mu,-\nu}[1,\eta]$. In this case $g(z) = -\mu \log h(z)$ belongs to $\Sigma_{\mu,\nu}[1,\eta]$. This completes our proof. □

Chapter 4

Spirallike Functions with Respect to a Boundary Point

In this chapter, we introduce the class of spirallike functions with respect to a boundary point. It turns out that, in fact, each such function is a complex power of a starlike function with respect to a boundary point. So, what we studied in the previous chapter forms the base for our present considerations. We start by looking at the images of spirallike functions with respect to a boundary point, i.e., spirallike domains, the boundary of which contains the origin.

4.1 Spirallike domains with respect to a boundary point

Definition 4.1 (cf. Definition 1.3). *A simply connected domain $\Omega \subset \mathbb{C}$, $0 \in \partial\Omega$, is called a* **spirallike domain with respect to a boundary point** *if there is a number $\mu \in \mathbb{C}$, $\operatorname{Re}\mu > 0$, such that for any point $w \in \Omega$ the spiral curve $\{e^{-t\mu}w,\ t \geq 0\}$ is contained in Ω.*

If, in particular, μ is real, then Ω is a starlike domain with respect to a boundary point.

Since we intend to study functions that map the unit disk Δ onto spirallike domains, the requirement for Ω to be simply connected is natural in view of the Riemann Mapping Theorem. For a simply connected domain Ω with $0 \in \partial\Omega$ it is possible to define on Ω a one-valued branch of the function $\arg w$. If, in addition, $w_0 = 1 \in \Omega$ then we can choose this branch in such a way that $\arg w_0 = 0$. In this manner, for any number $\lambda \in \mathbb{C}$ the function

$$w^{\lambda} = \exp\left[\lambda(\ln|w| + i\arg w)\right]$$

is well defined on Ω and attains the value 1 at the point $w_0 = 1$. We will denote the set of all spirallike (respectively, starlike) domains with respect to a boundary point that contain the point $w_0 = 1$ by $\mathcal{SP}$ (respectively, by $\mathcal{ST}$). It is clear that $\mathcal{ST} \subset \mathcal{SP}$.

To continue our discussion, we find a proper method to measure the "angular size" of spirallike domains. This is done as follows:

Let a domain Ω be spirallike ($\Omega \in \mathcal{SP}$), $w \in \Omega$ and $t \geq 0$. Denote the connected component of the set $\left\{\psi \in \mathbb{R} : e^{-\mu(t-i\psi)}w \in \Omega\right\}$ which contains the point $\psi = 0$ by $\Phi_\mu(w,t) = (a_\mu(w,t), b_\mu(w,t))$. In other words,

$$a_\mu(w,t) = \inf\left\{\phi \leq 0 : \ e^{-\mu(t-i\psi)}w \in \Omega \text{ for all } \psi \in (\phi,0)\right\}, \tag{4.1.1}$$

$$b_\mu(w,t) = \sup\left\{\phi \geq 0 : \ e^{-\mu(t-i\psi)}w \in \Omega \text{ for all } \psi \in (0,\phi)\right\}. \tag{4.1.2}$$

Proposition 4.1. *Let $\Omega \in \mathcal{SP}$ and μ be a complex number with* $\operatorname{Re}\mu > 0$ *such that the curve $\{e^{-t\mu}w,\ t \geq 0\} \subset \Omega$ for all $w \in \Omega$. Then the limit*

$$\alpha(w) := \lim_{t\to+\infty} \left(b_\mu(w,t) - a_\mu(w,t)\right) \tag{4.1.3}$$

exists finitely. Moreover, this limit does not depend on a point $w \in \Omega$, i.e., $\alpha(w) \equiv \alpha =$ constant. *In the particular case where $\mu \in \mathbb{R}$ and $\Omega \in \mathcal{ST}$ this limit is equal to the size θ of the minimal angle in which Ω lies divided by μ, i.e., $\alpha = \theta/\mu$.*

Proof. Definition 4.1 implies that if $e^{-\mu t_0}e^{i\mu\phi}w \in \Omega$ then for all $t \geq t_0$ the point $e^{-\mu t}e^{i\mu\phi}w$ is contained in Ω. Consequently, $a_\mu(w,t)$ is decreasing and $b_\mu(w,t)$ is increasing (with respect to t). So, the limit in (4.1.3) exists. To prove that α is finite, it is enough to show that the functions $a_\mu(w,t)$ and $b_\mu(w,t)$ are bounded. Fix $t \geq 0$. We show that

$$b_\mu(w,t) \leq \frac{2\pi \operatorname{Re}\mu}{|\mu|^2} \tag{4.1.4}$$

and

$$a_\mu(w,t) \geq -\frac{2\pi \operatorname{Re}\mu}{|\mu|^2} \tag{4.1.5}$$

in Ω.

This is clear if $\operatorname{Im}\mu = 0$. If $\phi = -\frac{2\pi}{\mu} \in \Phi_\mu(w,t)$, then Ω contains the circle $\left\{e^{-\mu(t-i\psi)}w,\ \psi \in [\phi,0]\right\}$ centered at the origin and Ω is not simply connected. Thus, without loss of generality, assume that $\operatorname{Im}\mu > 0$.

The spirallikeness of Ω implies that the curve Γ_1, defined by

$$\Gamma_1(t_1) = e^{-\mu(t+t_1)}w, \quad t_1 \in \left[0, \frac{2\pi \operatorname{Im}\mu}{|\mu|^2}\right],$$

lies in Ω. If inequality (4.1.4) is not satisfied, then $\phi = \dfrac{2\pi \operatorname{Re}\mu}{|\mu|^2} \in \Phi_\mu(w,t)$ and, therefore, the curve Γ_2, defined by

$$\Gamma_2(\psi) = e^{-\mu(t-i\psi)}w, \quad \psi \in [0,\phi],$$

also lies in Ω. If this were true, the curve $\Gamma_2\Gamma_1^{-1}$ lying in Ω, would wind once about the origin. This contradicts the simply connectedness of Ω, and condition (4.1.4) is proved.

As the function $a_\mu(w,t)$ is decreasing we can suppose that $t > \dfrac{2\pi \operatorname{Im}\mu}{|\mu|^2}$. Once again the spirallikeness of Ω implies that the curve Γ_3, defined by

$$\Gamma_3(t_1) = e^{-\mu(t+t_1)}w, \quad t_1 \in \left[-\frac{2\pi \operatorname{Im}\mu}{|\mu|^2}, 0\right],$$

lies in Ω. If inequality (4.1.5) is not satisfied, then $\phi = -\dfrac{2\pi \operatorname{Re}\mu}{|\mu|^2} \in \Phi_\mu(w,t)$ and, therefore, the curve Γ_4, defined by

$$\Gamma_4(\psi) = e^{-\mu(t-i\psi)}w, \quad \psi \in [\phi, 0],$$

also lies in Ω. If this were the case, the curve $\Gamma_4\Gamma_3^{-1}$ lying in Ω, would wind once about the origin. As above, this contradicts the simply connectedness of Ω, and condition (4.1.5) is also proved.

Now we show that $\alpha(w)$ does not depend on $w \in \Omega$.

Let K be any compact, connected subset of Ω. For each point $w_0 \in K$, there exists $\epsilon > 0$ such that the neighborhood

$$U(w_0,\epsilon) = \left\{e^{-\mu(t-i\psi)}w_0, \ -\epsilon < t < \epsilon, \ -\epsilon < \psi < \epsilon\right\}$$

is contained in Ω.

Let $w_1 \in U(w_0,\epsilon)$. Then

$$w_1 = e^{-\mu t'}\hat{w}, \text{ where } \hat{w} = e^{i\mu\psi'}w_0 \quad \text{with } |t'| < \epsilon, \ |\psi'| < \epsilon.$$

By formulae (4.1.1) and (4.1.2) we have

$$b_\mu(\hat{w},t) - a_\mu(\hat{w},t) = b_\mu(w_0,t) - a_\mu(w_0,t),$$

and thus $\alpha(\hat{w}) = \alpha(w_0)$.

Furthermore, it is clear that

$$b_\mu(w_1,t) - a_\mu(w_1,t) = b_\mu(\hat{w},t+t') - a_\mu(\hat{w},t+t').$$

Hence, the limits as $t \to \infty$ in both sides of the two latter equations coincide, that is, $\alpha(w_1) = \alpha(w_0)$, so it is a constant function on $U(w_0,\epsilon)$. Finding a finite covering system of neighborhoods $U_1, U_2, \dots U_n$ of K, we can conclude that $\alpha(w) \equiv$ constant on $U_1 \cup U_2 \cup \dots \cup U_n \supset K$, so α does not depend on $w \in \Omega$.

In the case when the domain Ω is starlike (i.e., $\mu \in \mathbb{R}$), the quantity $b_1(w,t) - a_1(w,t)$ equals exactly the size of the circle arch of the radius $e^{-t\mu}$ (which lies in Ω) divided by μ. The proposition is proved. □

Definition 4.2. *Let μ be a complex number with* $\operatorname{Re}\mu > 0$. *Also let* Ω *be a simply connected domain such that* $0 \in \partial\Omega$. Ω *will be called* μ**-spirallike (with respect to a boundary point)** *if for any point* $w \in \Omega$ *the following two conditions hold:*

(a) $\{e^{-t\mu}w,\ t \geq 0\} \subset \Omega$;

(b) *the limit* α *in (4.1.3) exists and is equal to* π:

$$\alpha = \lim_{t\to+\infty} (b_\mu(w,t) - a_\mu(w,t)) = \pi.$$

The set of all μ*-spirallike domains* $\Omega \in \mathcal{SP}$ *will be denoted by* μ-$\mathcal{SP}$.

It is clear that

$$\mathcal{ST} = \bigcup_{\mu\in(0,2]} \mu\text{-}\mathcal{SP}.$$

We investigate some properties of μ-spirallike domains.

Lemma 4.1. (i) *If* $\operatorname{Re}\mu > 0$ *and* Ω *is of the class* $\mathcal{SP}$, *then* $\Omega \in \mu$-$\mathcal{SP}$ *if and only if*

$$\widehat{\Omega} = \Omega^{\frac{1}{\mu}} := \left\{z^{\frac{1}{\mu}} :\ z \in \Omega\right\} \in 1\text{-}\mathcal{SP}.$$

Moreover, $\widehat{\Omega}$ *is starlike.*

(ii) *If there exists* $\Omega \in \mu$-$\mathcal{SP} \bigcap \nu$-$\mathcal{SP}$, *where* $\mu,\ \nu \in \mathbb{C}$ *with* $\operatorname{Re}\mu > 0,\ \operatorname{Re}\nu > 0$, *then* $\mu = \nu$.

Proof. In addition to formulae (4.1.1) and (4.1.2) let us write

$$a_1(\hat{w},t) = \inf\left\{\phi : e^{-(t-i\phi)}\hat{w} \in \widehat{\Omega}\right\},$$
$$b_1(\hat{w},t) = \sup\left\{\phi : e^{-(t-i\phi)}\hat{w} \in \widehat{\Omega}\right\},$$

where $\hat{w} = w^{\frac{1}{\mu}} \in \widehat{\Omega}$.

Since the inclusions $e^{-\mu(t-i\phi)}w \in \Omega$ and $e^{-(t-i\phi)}\hat{w} \in \widehat{\Omega}$ are one and the same, we have

$$b_\mu(w,t) - a_\mu(w,t) = b_1(\hat{w},t) - a_1(\hat{w},t).$$

Thus the limits as t goes to $+\infty$ in both sides of this equality are either equal to π or differ from π. Assertion (i) is proved.

In turn, (i) implies that the domains

$$\Omega_1 = \Omega^{\frac{1}{\mu}} := \left\{z^{\frac{1}{\mu}}, z \in \Omega\right\} \text{ and } \Omega_2 = \Omega^{\frac{1}{\nu}} := \left\{z^{\frac{1}{\nu}}, z \in \Omega\right\}$$

belong to the class 1-$\mathcal{SP}$. This means that for any point $w \in \Omega$ we have $w_1 = w^{\frac{1}{\mu}} \in \Omega_1$ and $w_2 = w^{\frac{1}{\nu}} \in \Omega_2$. So, we see: any point $w_2 \in \Omega_2$ if and only if the point $w_1 = w_2^{\frac{\nu}{\mu}}$ lies in Ω_1. In other words, $\Omega_1 = \Omega_2^{\frac{\nu}{\mu}}$.

Suppose now that $\arg \mu \neq \arg \nu$. Assertion (i) proved above implies that the domain Ω_1 is of the class $\frac{\mu}{\nu}$-$\mathcal{SP}$, i.e., by Definition 4.2, it contains the following spiral, which goes around the origin:

$$\left\{ e^{-t\frac{\mu}{\nu}} w, t \geq 0 \right\} \subset \Omega_1, \quad \text{when } w \in \Omega_1.$$

This contradicts the inclusion $\Omega_1 \in 1$-$\mathcal{SP}$ (see Proposition 4.1). So, $\arg \mu = \arg \nu$.

Suppose now that $|\mu| \neq |\nu|$; for example, $\mu = R\nu,\ R > 1$. Again we have $\Omega_1 = \Omega_2^{\frac{\nu}{\mu}} = \Omega_2^{\frac{1}{R}}$. Since the domain Ω_2 is contained in some wedge of angle π, then the domain Ω_1 is contained in the wedge of angle $\frac{\pi}{R} < \pi$ and this contradicts the inclusion $\Omega_1 \in 1$-$\mathcal{SP}$. Thus we have $\mu = \nu$. □

The proved Lemma 4.1 states that each spirallike domain (with respect to a boundary point) is μ-spirallike with a unique number μ, $\operatorname{Re} \mu > 0$. Now we show that μ cannot be arbitrary in the right half-plane.

Proposition 4.2. (i) *If $\Omega_1 \in 1$-$\mathcal{SP}$ and $|\mu - 1| \leq 1$, then $\Omega = \Omega_1^{\mu} \in \mu$-$\mathcal{SP}$.*

(ii) *In case for some $\mu \in \mathbb{C}$ there exists a Ω that belongs to μ-$\mathcal{SP}$, then $|\mu - 1| \leq 1$.*

Proof. Without loss of generality we assume that a domain $\Omega_1 \in 1$-$\mathcal{SP}$ lies in $\Pi_+ := \{z \in \mathbb{C} : \operatorname{Re} z > 0\}$. First, we will show that for any 1-spirallike domain $\Omega_1 \subset \Pi_+$, the domain $\Omega = \Omega_1^{\mu}$ is simply connected if $|\mu - 1| \leq 1,\ \mu \neq 0$; that is, $\operatorname{Re} \dfrac{1}{\mu} \geq \dfrac{1}{2}$.

Since the domain Ω_1 is simply connected and $0 \in \partial\Omega_1$, then Ω_1^{μ} is simply connected if and only if the mapping $z \mapsto z^{\mu}$ is one-to-one on Ω_1. This means that for any $w \in \Omega_1$ the equation

$$w^{\mu} = z^{\mu} \tag{4.1.6}$$

has no solution $z \in \Omega_1 \setminus \{w\}$.

Suppose that $z = re^{i\psi}$, $|\psi| < \frac{\pi}{2}$, is the solution of the above equation. Substituting $w = \rho e^{i\phi}$, $|\phi| < \frac{\pi}{2}$, we rewrite (4.1.6) in the following form:

$$\mu\Big(\ln \rho + i\phi\Big) = \mu\Big(\ln r + i\psi\Big) + 2\pi k i,\ k \in \mathbb{Z} \setminus \{0\},$$

or,

$$\frac{1}{\mu} = \frac{\ln \rho + i\phi - \ln r - i\psi}{2\pi k i} = \frac{\phi - \psi}{2\pi k} + i\frac{-\ln \rho + \ln r}{2\pi k}.$$

This equality implies that $\operatorname{Re} \dfrac{1}{\mu} < \dfrac{\pi}{2\pi|k|} = \dfrac{1}{2|k|} \leq \dfrac{1}{2}$. The latter inequality contradicts our supposition that $\operatorname{Re} \dfrac{1}{\mu} \geq \dfrac{1}{2}$. Thus the domain $\Omega = \Omega_1^{\mu}$ is simply connected. It is easy to see by Definition 4.1 that $\Omega \in \mathcal{SP}$. By Lemma 4.1, $\Omega \in \mu$-$\mathcal{SP}$. Assertion (i) is proved.

To prove assertion (ii), we suppose that $\Omega \in \mu\text{-}\mathcal{SP}$, where the number μ satisfies $\operatorname{Re}\dfrac{1}{\mu} < \dfrac{1}{2}$, and so $\operatorname{Re}\dfrac{1}{\mu} = \dfrac{1-\epsilon}{2}$ for some $\epsilon \in (0,1)$.

Given a point $w \in \Omega$ and t large enough, it follows by Definition 4.2 that

$$\pi\left(1-\frac{\epsilon}{2}\right) \le b_\mu(w,t) - a_\mu(w,t) \le \pi.$$

In other words, there exist values $\widetilde{\phi_1}$ and $\widetilde{\phi_2}$ such that

$$e^{-\mu(t-i\widetilde{\phi_j})}w \in \Omega, \quad j=1,2,$$

and

$$\pi(1-\epsilon) \le \widetilde{\phi_2} - \widetilde{\phi_1}.$$

Thus, for t big enough and for all $\phi \in \left[\widetilde{\phi_1}, \widetilde{\phi_2}\right]$,

$$e^{-\mu(t-i\phi)}w \in \Omega. \tag{4.1.7}$$

In particular, the points $e^{-\mu(t-i\phi_1)}w$ and $e^{-\mu(t-i\phi_2)}w$, where $\phi_1 = \widetilde{\phi_1}$ and $\phi_2 = \widetilde{\phi_1} + \pi(1-\epsilon)$, belong to Ω.

It follows by Definition 4.2 that

$$e^{-\mu(t_j-i\phi_j)}w \in \Omega, \quad j=1,2,$$

for all $t_1, t_2 \ge t$. Hence we can choose the numbers t_1 and t_2 such that

$$\frac{1}{\mu} = \frac{1-\epsilon}{2} + i\frac{t_2-t_1}{2}.$$

This implies that

$$e^{-\mu(t_1-i\phi_1)}w = e^{-\mu(t_2-i\phi_2)}w. \tag{4.1.8}$$

It follows from Lemma 4.1 that the domain $\Omega_1 = \Omega^{\frac{1}{\mu}} \in 1\text{-}\mathcal{SP}$. Thus the domain $\Omega_2 = \Omega_1^2 = \Omega^{\frac{2}{\mu}} \in 2\text{-}\mathcal{SP}$. Therefore, a one-valued branch of the function $\arg w$ is well-defined in the domain Ω_2. Further, equation (4.1.8) implies that

$$e^{-2(t_1-i\phi_1)}w_2 = e^{-2(t_2-i\phi_1)}e^{2\pi i(1-\epsilon)}w_2 \in \Omega_2, \tag{4.1.9}$$

where $w_2 = w^{\frac{2}{\mu}} \in \Omega_2$. Equality (4.1.9) means that for the same point of the simply connected domain Ω_1, its argument has two different values. This is a contradiction that proves assertion (ii). □

4.2 A characterization of spirallike functions with respect to a boundary point

Now we adapt Definition 1.4 to fit our purposes.

Definition 4.3. *A univalent function* $h : \Delta \mapsto \mathbb{C}$ *on the unit disk* Δ *is said to be of class* Spiral[1] *(respectively,* $\mathrm{Spiral}_\mu[1]$*) if*

(a) $h(0) = 1$ *and* $h(1) := \lim_{r\to 1^-} h(r) = 0$*;*

(b) $h(\Delta) \in \mathcal{SP}$ *(respectively,* $h(\Delta) \in \mu\text{-}\mathcal{SP}$*).*

Remark 4.1. It is clear that each function $h \in \mathrm{Spiral}_\mu(\Delta)$ is μ-spirallike in the sense of Definition 1.4. Moreover, Lemma 4.1 implies that if $h \in \mathrm{Spiral}_\mu[1]$ is ν-spirallike, then $\arg\nu = \arg\mu$. This fact is no longer true for spirallike functions with respect to an interior point.

Observe also that if μ is real then $\mathrm{Spiral}_\mu[1] = \mathrm{Star}_\mu[1]$, i.e., a function h normalized by $h(0) = 1,\ h(1) = 0$ belongs to $\mathrm{Spiral}_\mu[1]$ with a real positive number μ if and only if h is univalent, its image $h(\Delta)$ is starlike with respect to the origin, and the smallest wedge containing $h(\Delta)$ is of angle $\pi\mu$ (see Proposition 4.1).

Now we formulate the main result of this section.

Theorem 4.1. *Let* $h : \Delta \mapsto \mathbb{C}$ *be a holomorphic function and* $h(0) = 1$*. Let* $\mu \in \mathbb{C},\ |\mu - 1| \le 1,\ \mu \ne 0$*. The following assertions are equivalent.*

(I) $h \in \mathrm{Spiral}_\mu[1]$.

(II) $h_1(z) = h(z)^{1/\mu} \in \mathrm{Spiral}_1[1] = \mathrm{Star}_1[1]$*, i.e.,* h_1 *is starlike with respect to the boundary point* $h(1) = 0$ *function, and the smallest wedge that contains its image is of angle* π*.*

(III) *The function* h *satisfies the following condition*

$$\mathrm{Re}\left(\frac{2}{\mu}\cdot\frac{zh'(z)}{h(z)} + \frac{1+z}{1-z}\right) > 0, \quad z \in \Delta, \tag{4.2.1}$$

and it is possible to replace the number μ *in this inequality with a number* ν *only if* $\nu = R\mu,\ R > 1$*.*

(IV) *The function* $g(z) := \dfrac{zh(z)}{(1-z)^\mu}$ *is* ϕ*-spirallike of order* $\cos\phi - \dfrac{r}{2}$*, where* $\mu = r\,e^{i\phi}$*, i.e.,* g *is a univalent function satisfying the condition*

$$\mathrm{Re}\left(e^{-i\phi}\frac{zg'(z)}{g(z)}\right) > \cos\phi - \frac{r}{2}, \quad z \in \Delta, \tag{4.2.2}$$

and it is possible to replace the number μ *in this inequality with a number* ν *only if* $\nu = R\mu,\ R > 1$*.*

(V) *The function* h *satisfies the following three conditions:*

(a) h *is univalent in* Δ;

(b) $\operatorname{Re}\left(\mu \frac{h(z)}{h'(z)}\bar{z}\right) \geq \operatorname{Re}\left(\mu \frac{h(0)}{h'(0)}\bar{z}\right)(1-|z|^2)$;

(c) $\angle \lim_{z\to 1} \frac{h'(z)(z-1)}{h(z)} = \mu\,.$

Moreover, if h *is a univalent function on* Δ, *satisfying one of the assertions* (II)–(V) *with some complex number* μ, $\operatorname{Re}\mu > 0$, *then* μ *lies in the disk* $|\mu - 1| \leq 1$ *and* $h \in \mathrm{Spiral}_\mu[1]$.

Remark 4.2. Recall that if $f(1) := \angle \lim_{z\to 1} f(z)$ exists, then one can define $Q_f(1,z) := \frac{f'(z)(z-1)}{f(z)-f(1)}$, which is called the Visser–Ostrowski quotient (see (1.2.6)). Thus, it follows by the above assertion V(c) that $f \in \mathrm{Spiral}[1]$ is starlike whenever $\angle \lim_{z\to 1} Q_f(1,z)$ is a real number.

Proof. The proof of the theorem is done in several steps.

Step 1 (I)$\Longleftrightarrow$(II). By Lemma 4.1, it is immediate that if $h \in \mathrm{Spiral}_\mu[1]$, then $h_1(z) = h(z)^{1/\mu} \in \mathrm{Spiral}_1[1]$, and if $h_1(z) \in \mathrm{Spiral}_1[1]$, then $h(z) = h_1(z)^\mu \in \mathrm{Spiral}_\mu[1]$.

Step 2 (III)$\Longleftrightarrow$(IV). This equivalence is verified by substituting $g(z) = \frac{zh(z)}{(1-z)^\mu}$ in (4.2.2) and $h(z) = (1-z)^\mu \frac{g(z)}{z}$ in (4.2.1). Indeed, it is easy to see that

$$\left(\frac{2}{\mu}\cdot\frac{zh'(z)}{h(z)} + \frac{1+z}{1-z}\right) = \frac{2}{|\mu|}\left(e^{-i\phi}\cdot\frac{zg'(z)}{g(z)} + \frac{|\mu|}{2} - e^{-i\phi}\right),$$

and this equality proves our claim.

Step 3 (II)$\Longleftrightarrow$(III). Let us assume that (II) holds, i.e., $h_1 = h^{1/\mu} \in \mathrm{Spiral}_1[1]$. Then by Corollary 3.3, h_1 belongs to Robertson's class $G = G(1)$, so h_1 satisfies the inequality

$$\operatorname{Re}\left(\frac{2zh_1'(z)}{h_1(z)} + \frac{1+z}{1-z}\right) > 0, \quad z \in \Delta, \tag{4.2.3}$$

which coincides with (4.2.1). If inequality (4.2.1) holds for some $\nu \in \mathbb{C}$, then the function h_1 satisfies the inequality

$$\operatorname{Re}\left[\frac{2\mu}{\nu}\cdot\frac{zh_1'(z)}{h_1(z)} + \frac{1+z}{1-z}\right] > 0.$$

Therefore, the function $h_2(z) = h_1(z)^{\mu/\nu}$ belongs to the class G. Thus Corollary 3.3 implies that $h_2 \in \mathrm{Spiral}_l[1]$ for some positive number $l \leq 1$. Consequently, $h_1 =$

$h_2(z)^{\nu/\mu} \in \mathrm{Spiral}_{\frac{l\nu}{\mu}}[1]$. Hence, by Lemma 4.1, $\frac{l\nu}{\mu} = 1$ or $\nu = \frac{\mu}{l}$. As $l \le 1$, assertion (III) holds.

Assume now that assertion (III) holds. By substituting $h(z) = h_1(z)^\mu$ we get $h_1 \in G$. Using Corollary 3.3 we obtain: $h_1 \in \mathrm{Spiral}_l[1]$ with some $l \le 1$. Suppose that $l < 1$. Again, by Corollary 3.3, $h_1 \in G(l)$; that is,

$$\mathrm{Re}\left(\frac{2zh_1'(z)}{lh_1(z)} + \frac{1+z}{1-z}\right) > 0, \quad z \in \Delta.$$

Returning to the function $h(z) = h_1(z)^\mu$ we see that

$$\mathrm{Re}\left(\frac{2zh'(z)}{l\mu h(z)} + \frac{1+z}{1-z}\right) > 0,$$

which contradicts our assumption. Thus $l = 1$, i.e., $h_1 \in \mathrm{Spiral}_1[1]$, and we are done.

To proceed we note that by Definitions 4.2 and 4.3 the inclusion $h \in \mathrm{Spiral}_\mu[1]$ implies that, for any $z \in \Delta$ and $t \ge 0$.

$$e^{-t\mu}h(z) \in h(\Delta).$$

This means that the family $\mathcal{S} = \{F_t\}_{t\ge 0}$, defined by

$$F_t(z) := h^{-1}\left(e^{-t\mu}h(z)\right),$$

is a semigroup of holomorphic self-mappings of Δ. Differentiating $\mathcal{S}$ at $t = 0^+$ we find its generator:

$$f(z) = -\left.\frac{\partial F_t(z)}{\partial t}\right|_{t=0^+} = \frac{\mu h(z)}{h'(z)}. \tag{4.2.4}$$

Step 4 (I)$\Longrightarrow$(V). Let h be a μ-spirallike function. Then condition (a) of assertion (V) follows at once. Hence, as mentioned above, the function

$$f(z) = \mu\frac{h(z)}{h'(z)}$$

belongs to $\mathcal{G}$, and by Theorem 2.3

$$\mathrm{Re}\big[f(z)\bar{z}\big] \ge \mathrm{Re}\big[f(0)\bar{z}\big](1-|z|^2), \quad z \in \Delta.$$

Thus we get inequality V(b).

It remains to check condition V(c).

As shown above, (I) is equivalent to (III) (Steps 1 and 3). Then for any ν of the form $\nu = R\mu,\ R > 1$, the following inequality holds:

$$\mathrm{Re}\left(\frac{2z\mu}{\nu f(z)} + \frac{1+z}{1-z}\right) > 0, \quad z \in \Delta.$$

Note also that this inequality no longer holds for other values of ν. By the Riesz–Herglotz formula there exists a probability measure σ on the unit circle such that

$$\frac{2z\mu}{\nu f(z)} + \frac{1+z}{1-z} = \oint\limits_{|\zeta|=1} \frac{1+z\bar{\zeta}}{1-z\bar{\zeta}}\, d\sigma(\zeta), \quad z \in \Delta,$$

or, equivalently,

$$\frac{\mu(z-1)}{\nu f(z)} = \oint\limits_{|\zeta|=1} \frac{1-\bar{\zeta}}{1-z\bar{\zeta}}\, d\sigma(\zeta), \quad z \in \Delta. \tag{4.2.5}$$

Note that the integral representation (4.2.5) is not valid in cases when ν is different from $R\mu$, $R > 1$.

Now set $\nu = \mu$. Decomposing σ with respect to the Dirac measure δ at the point $\zeta = 1 \in \partial\Delta$, one can write $\sigma = (1-a)\sigma_1 + a\delta$, where $0 \le a \le 1$, and σ_1 and δ are mutually singular probability measures. If $a \neq 0$, equation (4.2.5) implies that

$$\frac{\mu(z-1)}{\lambda f(z)} = \oint\limits_{|\zeta|=1} \frac{1-\bar{\zeta}}{1-z\bar{\zeta}}\, d\sigma_1(\zeta), \quad z \in \Delta, \quad \lambda = (1-a)\mu,$$

which is valid only if $1 - a \ge 1$. This shows that $a = 0$ and $\sigma = \sigma_1$ is singular with respect to δ.

Let $\{z_n\}$ be any sequence in Δ non-tangentially convergent to 1. This means that there is a positive number K such that for all $n = 1, 2, \ldots$,

$$\frac{|1-z_n|}{1-\operatorname{Re} z_n} < K.$$

We now consider the functions $f_n : \partial\Delta \mapsto \mathbb{C}$ defined by

$$f_n(\zeta) := \frac{1-\bar{\zeta}}{1-z_n\bar{\zeta}}, \quad \zeta \in \partial\Delta.$$

It is easy to see that each function f_n maps the unit circle $\partial\Delta$ onto the circle $|w - c_n| = |c_n|$, where

$$c_n = c_n(\zeta) = \frac{1-\bar{z}_n}{1-|z_n|^2}, \quad n = 1, 2, \ldots .$$

Hence,

$$|f_n(\zeta)| \le 2|c_n| \le \frac{2|1-z_n|}{1-\operatorname{Re} z_n} < 2K$$

for all $n =$ big enough. Setting $\nu = \mu$ in (4.2.5) and applying Lebesgue's Bounded Convergence Theorem we obtain

$$\lim_{n\to\infty} \frac{z_n - 1}{f(z_n)} = \oint\limits_{|\zeta|=1} \lim_{n\to\infty} \frac{1-\bar{\zeta}}{1-z_n\bar{\zeta}}\, d\sigma(\zeta) = 1.$$

Therefore,

$$\angle\lim_{z\to 1}\frac{h'(z)(z-1)}{h(z)}=\angle\lim_{z\to 1}\frac{\mu(z-1)}{f(z)}=\mu,$$

and condition V(c) follows.

Step 5 (V)$\Longrightarrow$(I). Note that by condition V(a) the image $h(\Delta)$ is a simply connected domain. By Theorem 2.3, condition V(b) implies that the function $f(z)=\dfrac{\mu h(z)}{h'(z)}$ belongs to $\mathcal{G}$. Solving the Cauchy problem

$$\begin{cases}\dfrac{\partial F_t(z)}{\partial t}+\mu\,\dfrac{h(F_t(z))}{h'(F_t(z))}=0,\\[2mm] F_0(z)=z,\quad z\in\Delta,\end{cases}\tag{4.2.6}$$

we find the generated semigroup $\mathcal{S}=\{F_t\}_{t\geq 0}$, namely,

$$F_t(z):=h^{-1}\left(e^{-t\mu}h(z)\right)\in \mathrm{Hol}(\Delta).$$

Thus for all $z\in\Delta$ the curve $\{e^{-t\mu}h(z),\ t\geq 0\}$ is contained in $h(\Delta)$, i.e., $h\in\mathrm{Spiral}_\mu[1]$.

Assume that for some $\nu\in\mathbb{C}$ with $\mathrm{Re}\,\nu>0$ the function h belongs to $\mathrm{Spiral}_\nu[1]$. We have already seen in Step 4 that in this case

$$\angle\lim_{z\to 1}\frac{h'(z)(z-1)}{h(z)}=\nu.$$

Comparing this equality with (V c) we get $\nu=\mu$. This completes the proof of the theorem. □

4.3 Subordination criteria for the class $\mathrm{Spiral}_\mu[1]$

In this section we use the well-known notion of subordination.

Definition 4.4. *A function* $h_1\in\mathrm{Hol}(\Delta,\mathbb{C})$ *is said to be* **subordinate to** $h_2\in\mathrm{Hol}(\Delta,\mathbb{C})$ *(*$h_1\prec h_2$*) if there exists a holomorphic function* ω *with* $|\omega(z)|\leq|z|$, $z\in\Delta$, *such that* $h_1=h_2\circ\omega$.

The following description of spirallike functions of the class Spiral[0] is due to Ruscheweyh (see [124, Corollaries 1 and 2]).

Lemma 4.2. *Let* $g\in\mathrm{Hol}(\Delta,\mathbb{C})$ *with* $g(0)=g'(0)-1=0$. *Let* $\alpha\in(-\pi/2,\pi/2)$ *and* $0\leq\beta<\cos\alpha$. *Then*

$$\mathrm{Re}\left[\exp(i\alpha)\,\frac{z\,g'(z)}{g(z)}\right]>\beta,\quad z\in\Delta,\tag{4.3.1}$$

if and only if one of the following two conditions holds:

(a) *for all $u, v \in \overline{\Delta}$ we have*

$$\frac{ug(vz)}{vg(uz)} \prec \left(\frac{1-uz}{1-vz}\right)^{2(\cos\alpha-\beta)\exp(-i\alpha)}; \tag{4.3.2}$$

or

(b) *for all $t \in (0, 2\cos\alpha)$ the function g satisfies the inequality*

$$\left|g(z(1-\exp(i\alpha)t))\right| \le F(t,\alpha,\beta)|g(z)| \quad \textit{for all } z \in \Delta, \tag{4.3.3}$$

where

$$F(t,\alpha,\beta) = |1-\exp(i\alpha)t|\left(1-\frac{t}{2\cos\alpha}\right)^{2\cos\alpha(\beta-\cos\alpha)}. \tag{4.3.4}$$

Moreover, this bound is sharp.

By using this result and Theorem 4.1 one can characterize the class $\mathrm{Spiral}_\mu[1]$ in terms of subordination. Indeed, to do this we just have to substitute $g(z) = \dfrac{zh(z)}{(1-z)^\mu}$ in (4.3.2) and (4.3.3), where $h \in \mathrm{Spiral}_\mu[1]$. By Theorem 4.1 we already know that g satisfies the inequality

$$\mathrm{Re}\left(e^{-i\phi}\frac{zg'(z)}{g(z)}\right) > \cos\phi - \frac{|\mu|}{2}, \quad z \in \Delta,$$

where $\phi = \arg\mu$ if and only if $h \in \mathrm{Spiral}_\mu[1]$. Setting

$$\alpha := -\phi = -\arg\mu \quad \text{and} \quad \beta := \cos\phi - \frac{|\mu|}{2},$$

we get: $2(\cos\alpha-\beta) = |\mu|$. Thus one can rewrite conditions (a) and (b) of Lemma 4.2 in the form:

$$\frac{u\left(\dfrac{vzh(vz)}{(1-vz)^\mu}\right)}{v\left(\dfrac{uzh(uz)}{(1-uz)^\mu}\right)} = \frac{h(vz)(1-uz)^\mu}{h(uz)(1-vz)^\mu} \prec \left(\frac{1-uz}{1-vz}\right)^{|\mu|\exp(i\arg\mu)}, \tag{4.3.2$'$}$$

and

$$\left|\frac{z(1-\exp(-i\phi)t)h(z(1-\exp(-i\phi)t))}{(1-z(1-\exp(-i\phi)t))^\mu}\right| \le F(t,-\phi,\beta)\left|\frac{zh(z)}{(1-z)^\mu}\right|$$
$$= |1-\exp(-i\phi)t|\left(1-\frac{t}{2\cos\phi}\right)^{-|\mu|\cos\phi}\cdot\left|\frac{zh(z)}{(1-z)^\mu}\right|. \tag{4.3.3$'$}$$

So, we have proved the following characterization of the class $\mathrm{Spiral}_\mu[1]$.

Theorem 4.2. *Let $h : \Delta \mapsto \mathbb{C}$ be a holomorphic function and $h(0) = 1$. Let $\mu \in \mathbb{C}$, $|\mu - 1| \le 1$, $\mu \ne 0$ and $\phi = \arg \mu \in (-\pi/2, \pi/2)$. Then $h \in \mathrm{Spiral}_\mu[1]$ if and only if one of the following conditions holds:*

(a) *for all $u, v \in \overline{\Delta}$*

$$\left(\frac{1-uz}{1-vz}\right)^\mu \frac{h(vz)}{h(uz)} \prec \left(\frac{1-uz}{1-vz}\right)^\mu ;$$

or

(b) *for all $t \in (0, 2\cos\phi)$*

$$\left|\frac{h(z(1-e^{-i\phi}t))}{h(z)}\right| \le \left|\left(\frac{1-z(1-e^{-i\phi}t)}{1-z}\right)^\mu\right| \cdot \left(1 - \frac{t}{2\cos\phi}\right)^{-\operatorname{Re}\mu} .$$

Furthermore, setting $u = 0$, $v = 1$ in Theorem 4.2, we obtain the following corollary.

Corollary 4.1. *If $h \in \mathrm{Spiral}_\mu[1]$, then*

$$\left(\frac{1}{1-z}\right)^\mu h(z) \prec \left(\frac{1}{1-z}\right)^\mu .$$

In particular, if $h \in \mathrm{Star}_\mu[1]$ with $\mu \le 1$ (hence, $h \in G$), then $\operatorname{Re} \dfrac{h(z)}{(1-z)^\mu} > \dfrac{1}{2}$.

For the class of starlike functions, in particular, we also have the following consequence of Theorem 4.2.

Corollary 4.2. *Let $h : \Delta \mapsto \mathbb{C}$ be a holomorphic function and $h(0) = 1$. Let $\mu \in (0, 2)$. Then $h \in \mathrm{Star}_\mu[1]$ if and only if, for all $t \in (-1, 1)$,*

$$\left|\frac{h(zt)}{h(z)}\right| \le \left(\left|\frac{1-zt}{1-z}\right| \cdot \frac{2}{1+t}\right)^\mu .$$

4.4 Distortion Theorems

4.4.1 'Spiral angle' distortion theorems

First we observe that by Theorem 4.1 each spirallike function $h \in \mathrm{Spiral}[1] = \bigcup_{\mu \in \Omega} \mathrm{Spiral}_\mu[1]$, $\Omega = \{\lambda : |\lambda - 1| \le 1,\ \lambda \ne 0\}$, is a complex power of a starlike function with respect to a boundary point. Therefore, one can apply Theorems 3.3, 3.4 and 3.5 (see also Corollary 3.4) to give geometric characteristics for images of functions of the class $\mathrm{Spiral}_\mu[1]$. More precisely, let $\lambda \in \Omega = \{\mu \in \mathbb{C} : |\mu - 1| \le 1,\ \mu \ne 0\}$ and $\theta \in \left(-\frac{\pi}{2}, \frac{\pi}{2}\right)$ be given. Define the function $h_{\lambda,\theta} \in \mathrm{Hol}(\Delta, \mathbb{C})$ by

$$h_{\lambda,\theta}(z) = \left(\frac{1-z}{1+e^{-2i\theta}z}\right)^\lambda .$$

Here and later on, we choose a single-valued branch of the analytic function w^λ such that $1^\lambda = 1$. We will see that $h_{\lambda,\theta} \in \mathrm{Spiral}_\lambda[1]$.

Definition 4.5. *The set $W_{\lambda,\theta} = h_{\lambda,\theta}(\Delta)$ is called a* **canonical λ-spiral wedge** *with midline $l_{\theta,\lambda} = \{w \in \mathbb{C} : \ w = e^{(i\theta+t)\lambda}, \ t \in \mathbb{R}\}$ and vertex at the origin.*

To explain this definition, let us observe that $h = h_{\lambda,\theta}$ is a solution of the differential equation

$$\lambda h(z) = h'(z) f(z),$$

normalized by the conditions $h(0) = 1, \ h(1) = 0$, where f is given by

$$f(z) = \frac{(z-1)\left(1 + e^{-2i\theta} z\right)}{1 + e^{-2i\theta}}.$$

Since $f \in \mathcal{G}^+[1]$ with $f'(1) = 1$ and $\lambda \in \Omega$, it follows by Theorem 4.1 that h is a λ-spirallike function with respect to the boundary point $h(1) = 0$. Moreover, f is a generator of a one-parameter group of hyperbolic automorphisms of Δ having two boundary fixed points $z = 1$ and $z = -e^{-2i\theta}$. Hence, for each $w \in W_{\lambda,\theta}$, the spiral curve $\left\{e^{-t\lambda} w, \ t \in \mathbb{R}\right\}$ belongs to $W_{\lambda,\theta}$.

Using notions of Section 4.1, one can see that a canonical λ-spiral wedge belongs to λ-$\mathcal{SP}$.

Finally, we see that for real $\lambda \in (0, 2]$, the set $W_{\lambda,\theta}$ is a straight wedge (sector) of angle $\pi\lambda$, whose bisector is $l_{\theta,\lambda} = \{w \in \mathbb{C} : \ \arg w = \theta\lambda\}$.

An immediate consequence of Theorems 3.3 and 4.1 is the following assertion.

Proposition 4.3. *If $h \in \mathrm{Spiral}_\mu[1]$, then the canonical spiral wedge W_{μ,θ^*}, where*

$$\theta^* := \lim_{r \to 1^-} \arg h^{1/\mu}(r),$$

is the smallest one that contains the image $h(\Delta)$.

In other words, this proposition says that each spirallike function of class $\mathrm{Spiral}_\mu[1]$ is subordinate to h_{μ,θ^*}.

Analogously to Definition 3.4, we introduce the following subclasses of $\mathrm{Spiral}[\tau]$, $\tau \in \overline{\Delta}$.

Definition 4.6. *We say that a function $h \in \mathrm{Spiral}[\tau]$ belongs to the subclass $\mathrm{Spiral}[\tau, \eta]$, where $\eta \in \partial\Delta, \ \eta \neq \tau$, if the angular limit*

$$Q_h(\eta) := \angle \lim_{z \to \eta} \frac{(z - \eta) h'(z)}{h(z)} \tag{4.4.1}$$

exists finitely and is different from zero. In the case where $\tau = 1$, we say that $h \in \mathrm{Spiral}_{\mu,\nu}[1, \eta]$ if $h \in \mathrm{Spiral}_\mu[1]$ and the angular limit (4.4.1) at the boundary point η equals ν.

To proceed we recall that by Definition 1.4, a spirallike function h (with respect to an interior or a boundary point) is μ-spirallike if for each point $w \in h(\Delta)$ its image $h(\Delta)$ contains the spiral curve $\{e^{-t\mu}w : t \geq 0\}$.

Theorem 4.3. *Let $h \in \text{Spiral}[\tau]$, $\tau \in \overline{\Delta}$, be a μ-spirallike function on Δ. Then the image $h(\Delta)$ contains a canonical λ-spiral wedge with*

$$\arg \lambda = \arg \mu \tag{4.4.2}$$

if and only if $h \in \text{Spiral}[\tau, \eta]$ for some $\eta \in \partial\Delta$. Moreover, if $Q_h(\eta) = \nu$, then for some $\theta \in [0, 2\pi)$ the canonical spiral wedge $W_{-\nu,\theta} \subset h(\Delta)$; and it is maximal in the sense that there is no spiral wedge $W_{\lambda,\theta} \subset h(\Delta)$ with λ satisfying (4.4.2) that contains $W_{-\nu,\theta}$ properly.

Proof. First, given $h \in \text{Spiral}[\tau, \eta]$, we construct $h_1 \in \text{Spiral}[1, -1]$, which is spirallike with respect to a boundary point whose image eventually coincides at ∞ with $h(\Delta)$. If $\tau \in \partial\Delta$, we just set $h_1 = h \circ \Phi$, where $\Phi \in \text{Aut}(\Delta)$ is an automorphism of Δ such that $\Phi(1) = \tau$ and $\Phi(-1) = \eta$.

If $\tau \in \Delta$, we take any two points $z_1 = e^{i\theta_1}$ and $z_2 = e^{i\theta_2}$ such that $w_1 = h(z_1)$ and $w_2 = h(z_2)$ exist finitely and $\theta_1 \in (\arg \eta - \epsilon, \arg \eta)$, $\theta_2 \in (\arg \eta, \arg \eta + \epsilon)$, so the arc (θ_1, θ_2) on the unit circle contains the point η.

Since h is spirallike with respect to an interior point, it satisfies the equation

$$\beta h(z) = h'(z) f(z), \quad z \in \Delta,$$

where $f \in \mathcal{G}^+[\tau]$ and $\beta = f'(\tau)$, so $\arg \mu = \arg \beta$. This means that for each $w \in h(\Delta)$ the spiral curve $\{e^{-t\beta}w,\ t \geq 0\}$ belongs to $h(\Delta)$. In turn, the curves $\ell_1 = \{z = h^{-1}(e^{-t\beta}w_1),\ t \geq 0\}$ and $\ell_2 = \{z = h^{-1}(e^{-t\beta}w_2),\ t \geq 0\}$ lie in $\overline{\Delta}$ with ends in z_1 and τ, and z_2 and τ, respectively.

Since $z_1 \neq z_2$ and the interior points of ℓ_1 and ℓ_2 are semigroup trajectories in Δ, these curves do not intersect except at their common end point $z = \tau$. Consequently, the domain D bounded by ℓ_1, ℓ_2 and the arc (θ_1, θ_2) is simply connected, and there is a conformal mapping Φ of Δ such that $\Phi(\Delta) = D$ and $\Phi(-1) = \eta$, $\Phi(1) = \tau$. Now define $h_1 = h \circ \Phi$. It follows by our construction that $h_1(\Delta) \subset h(\Delta)$ and h_1 is spirallike with respect to a boundary point $h_1(1) = 0$. In addition, since Φ is conformal at the point $z = -1$, it satisfies the Visser–Ostrowski condition, and we have

$$\begin{aligned}
\angle \lim_{z \to -1} \frac{(z+1)h_1'(z)}{h_1(z)} &= \angle \lim_{z \to -1} \frac{(z+1)h'(\Phi(z))\Phi'(z)}{h(\Phi(z))} \\
&= \angle \lim_{z \to -1} \frac{(z+1)\Phi'(z)}{\Phi(z) - \eta} \cdot \angle \lim_{z \to -1} \frac{(\Phi(z) - \eta)h'(\Phi(z))}{h(\Phi(z))} \\
&= \angle \lim_{z \to -1} \frac{(\Phi(z) - \eta)h'(\Phi(z))}{h(\Phi(z))}.
\end{aligned} \tag{4.4.3}$$

Note also that Φ is a self-mapping of Δ mapping the point $z = -1$ to η and having a finite derivative at this point.

It follows by the Julia–Carathéodory Theorem, (see, for example, [40, 127] and [130]) that if z converges to -1 non-tangentially, then $\Phi(z)$ converges non-tangentially to $\eta = \Phi(-1)$. Then (4.4.3) implies that

$$Q_{h_1}(-1) = \angle \lim_{z\to -1} \frac{(z+1)h_1'(z)}{h_1(z)} = \nu \tag{4.4.4}$$

exists finitely if and only if $h \in \text{Spiral}[\tau, \eta]$ and

$$Q_{h_1}(-1) = Q_h(\eta). \tag{4.4.5}$$

We claim that this last relation implies that $h_1(\Delta)$ contains a canonical $(-\nu)$-spiral wedge $W_{-\nu,\theta}$ for some $\theta \in \left(-\frac{\pi}{2}, \frac{\pi}{2}\right)$.

To this end, observe that h_1 satisfies the equation

$$\beta h_1(z) = h_1'(z) \cdot f_1(z),$$

where $f_1(z) = \dfrac{f(\Phi(z))}{\Phi'(z)}$ is a generator of a semigroup of Δ with $f_1(1) = 0$ and $f_1'(1) = \beta_1$ for some $\beta_1 > 0$ such that

$$|\beta - \beta_1| \le \beta.$$

Therefore, h_1 is a complex power of the function $h_2 \in \text{Hol}(\Delta, \mathbb{C})$ defined by the equation

$$\beta_1 h_2(z) = h_2'(z) f_1(z), \quad h_2(1) = 0, \tag{4.4.6}$$

i.e.,

$$h_1(z) = h_2^{\mu}(z), \tag{4.4.7}$$

where $\mu = \frac{\beta}{\beta_1} \neq 0$, $|\mu - 1| \le 1$, hence $\arg \mu = \arg \beta$.

On the other hand, if we normalize h_2 by $h_2(0) = h_1^{1/\mu}(0)$, equation (4.4.7) has a unique solution h_2, which is a starlike function with respect to a boundary point $(h_2(1) = 0)$. Obviously,

$$Q_{h_2}(-1) = \frac{1}{\mu} Q_{h_1}(-1) \left(= \frac{1}{\mu} Q_h(\eta) \right). \tag{4.4.8}$$

Note that $\nu_1 := Q_{h_2}(-1)$ is a negative real number, while $\nu := Q_{h_1}(-1) = \nu_1 \mu$ is complex.

Now it follows by Theorem 3.4 that the starlike set $h_2(\Delta)$ contains a straight wedge (sector) of a nonzero angle $\pi|\nu_1|$. Namely, the maximal (straight) wedge $W_* \subset h_2(\Delta)$ is of the form

$$W_* = W_{-\nu_1,\theta_1} = \left\{ w \in \mathbb{C} : \ w = \left(\frac{1-z}{1+e^{-2i\theta_1} z} \right)^{-\nu_1}, \ z \in \Delta \right\},$$

with

$$\begin{aligned}-\nu_1\theta_1 &= \lim_{r\to 1^-} \arg h_2(-r) = \lim_{r\to 1^-} \arg h_1^{\nu_1/\nu}(-r) \\ &= \nu_1 \cdot \lim_{r\to 1^-} \arg h_1^{1/\nu}(-r) = \nu_1\theta_*,\end{aligned}$$

where

$$\theta_* = \lim_{r\to 1^-} \arg h_1^{1/\nu}(-r).$$

Therefore, $W_*^\mu = W_{-\nu_1\mu,\theta_1} = W_{-\nu,-\theta_*}$ is contained in $h_1(\Delta)$; hence in $h(\Delta)$. Finally, it follows by (4.4.8) that $\lambda := -\nu = |\nu_1|\mu$. This implies (4.4.2).

Conversely, let h be a μ-spirallike function on Δ such that $h(\Delta)$ contains a canonical λ-spiral wedge $W_{\lambda,\theta}$ for some λ satisfying (4.4.2) and $\theta \in \left(-\frac{\pi}{2}, \frac{\pi}{2}\right)$. Then for each $w_0 \in W_{\lambda,\theta}$, the curve

$$\ell := \left\{w \in \mathbb{C} : w = e^{-t\lambda}w_0,\ t \in \mathbb{R}\right\}$$

belongs to $h(\Delta)$. Hence the curve $h^{-1}(l) \subset \Delta$ joins the point $\tau \in \overline{\Delta}$ to a point $\eta \in \partial\Delta$. Again, as in the first step of the proof, one can find a conformal mapping $\Phi \in \mathrm{Hol}(\Delta)$ with $\Phi(1) = \tau$, $\Phi(-1) = \eta$ such that $h_1 = h \circ \Phi$ is a μ-spirallike function with respect to a boundary point, $h_1(1) = 0$, and

$$W_{\lambda,\theta} \subset h_1(\Delta) \subset h(\Delta). \tag{4.4.9}$$

Again the function $h_2 = h_1^{1/\mu}$ is starlike with respect to a boundary point, and $h_2(\Delta)$ contains the (straight) wedge $W_{\frac{\lambda}{\mu},\theta}$ because of (4.4.9).

Setting $\dfrac{\lambda}{\mu} = \kappa$, we see by (4.4.2) that κ is positive real. Hence, $h_2(\Delta)$ contains a straight canonical wedge $W_{\kappa,\theta}$ with $0 < \kappa \le |\nu_1|$, where $\nu_1 = Q_{h_2}(-1)$ exists finitely, and $W_{|\nu_1|,\theta}$ is the maximal wedge contained in $h_2(\Delta)$. But, as before, we have

$$\nu = Q_h(\eta) = \mu Q_{h_2}(-1) = \mu\nu_1.$$

The latter relations show that ν is finite and λ must satisfy the conditions $\arg\lambda = \arg\mu = \arg(-\nu)$ and $0 < |\lambda| \le |\nu|$. So the wedge $W_{-\nu,\theta}$ is a maximal canonical spiral wedge contained in $h(\Delta)$ satisfying condition (4.4.2). The theorem is proved. □

4.4.2 Growth estimates for semigroup generators

In this section we obtain growth estimates for semigroup generators from the classes considered in Section 2.4.

Let $f(z) \in \mathcal{G}^+[1]$ with $f'(1) = \beta_+ > 0$. It follows by (2.4.21) and Julia's Lemma that for some $F \in \mathrm{Hol}(\Delta)$

$$\begin{aligned}|f(z)| &= c|1-z|^2 \left|\frac{1+F(z)}{1-F(z)}\right| \\ &= c|1-z|^2 \frac{|1-F^2(z)|}{|1-F(z)|^2} \geq c|1-z|^2 \frac{1-|F(z)|^2}{|1-F(z)|^2} \\ &\geq \frac{c|1-z|^2}{\alpha_+} \cdot \frac{1-|z|^2}{|1-z|^2} = \frac{\beta_+}{2}(1-|z|^2), \qquad z \in \Delta.\end{aligned} \tag{4.4.10}$$

Similarly, for $x \in (-1,1)$ we obtain

$$\begin{aligned}\operatorname{Re} f(x) &= -c(1-x)^2 \operatorname{Re} \frac{1+F(x)}{1-F(x)} \\ &= -c(1-x)^2 \frac{1-|F(x)|^2}{|1-F(x)|^2} \leq -\frac{\beta_+}{2}(1-x^2).\end{aligned} \tag{4.4.11}$$

On the other hand, if $f \in \mathcal{G}[1,-1]$, then

$$\begin{aligned}\left|\frac{1}{f(z)}\right| &= \frac{1}{c|1-z|^2} \left|\frac{1-F(z)}{1+F(z)}\right| \\ &\geq \frac{1}{c|1-z|^2} \frac{1}{\alpha_-} \frac{1-|z|^2}{|1+z|^2} = -\frac{2}{\beta_-} \frac{1-|z|^2}{|1-z^2|^2},\end{aligned}$$

or

$$|f(z)| \leq -\frac{\beta_-}{2} \frac{\left|1-z^2\right|^2}{1-\left|z\right|^2}. \tag{4.4.12}$$

Similarly, for $x \in (-1,1)$,

$$\begin{aligned}\operatorname{Re} \frac{1}{f(x)} &= -\frac{1}{c(1-x)^2} \operatorname{Re} \frac{1-F(x)}{1+F(x)} \\ &= -\frac{1}{c(1-x)^2} \frac{1-|F(x)|^2}{|1+F(x)|^2} \leq \frac{2}{|\beta_-|} \frac{(1-x^2)}{(1-x^2)^2} = \frac{2}{|\beta_-|} \frac{1}{1-x^2}.\end{aligned} \tag{4.4.13}$$

Combining (4.4.10) and (4.4.12) we get the following assertion.

Theorem 4.4 (A distortion theorem). *Let $f \in \mathcal{G}[1,-1]$ with $\beta_+ = f'(1) \geq 0$ and $\beta_- = f'(-1) < 0$. Then for $z \in \Delta$,*

$$\frac{\beta_+}{2}(1-|z|^2) \leq |f(z)| \leq \frac{|\beta_-|}{2} \frac{|1-z^2|^2}{1-|z|^2}. \tag{4.4.14}$$

In particular, if z is in the circular lens

$$\Upsilon_M = \left\{ z \in \Delta : \ \frac{|1-z^2|}{1-|z|^2} \leq M \right\}, \quad M \geq 1,$$

then

$$\frac{\beta_+}{2M}|1-z^2| \le |f(z)| \le \frac{M\,|\beta_-|}{2}\left|1-z^2\right| . \tag{4.4.15}$$

Similarly, by using (4.4.11)–(4.4.13) we obtain the following.

Theorem 4.5. *Let $f \in \mathcal{G}^+[1,-1] \cap \mathcal{G}_h[1]$ with $f'(1) = \beta_+ > 0$ and $f'(-1) = \beta_- < 0$. Then for all $x \in (-1,1)$,*

$$\frac{\beta_-}{2}(1-x^2) \le \operatorname{Re} f(x) \le -\frac{\beta_+}{2}(1-x^2) < 0,$$

and

$$-\frac{2}{\beta_+}\frac{1}{1-x^2} \le \operatorname{Re}\frac{1}{f(x)} \le \frac{2}{\beta_-}\frac{1}{1-x^2} < 0.$$

4.4.3 Growth estimates for spirallike functions

In turn, the above assertions enable us to get distortion theorems for unbounded starlike and spirallike functions with respect to a boundary point of the classes $\mathrm{Spiral}_{\mu,\nu}[1,\eta]$ defined in Definition 4.6.

Recall that each function $h \in \mathrm{Spiral}_{\mu,\nu}[1,\eta]$ satisfies generalized Visser–Ostrowski conditions

$$Q_h(1) := \angle\lim_{z\to 1}\frac{(z-1)h'(z)}{h(z)} = \mu$$

and

$$Q_h(\eta) := \angle\lim_{z\to \eta}\frac{(z-\eta)h'(z)}{h(z)} = \nu$$

with

$$\mu \in \Omega := \{\lambda : \ |\lambda - 1| \le 1, \qquad \lambda \ne 0\}$$

and $0 \le -\frac{\nu}{\mu} \le 1$ (see the proof of Theorem 4.3).

In addition, the function $f \in \mathrm{Hol}(\Delta, \mathbb{C})$, defined by

$$f(z) = \frac{\mu h(z)}{h'(z)}, \tag{4.4.16}$$

is of class $\mathcal{G}^+[1]$ with $f'(1) = 1$ (see also [130]) and

$$h\left(F_t(z)\right) = e^{-\mu t}h(z),$$

where $\{F_t\}_{t\ge 0}$ is the semigroup generated by f (see Section 2.5).

Now for $h \in \mathrm{Spiral}_\mu[1]$ we obtain, from (4.4.10) and (4.4.16),

$$\left|\frac{h'(z)}{h(z)}\right| = |\mu|\frac{1}{|f(z)|} \le \frac{2|\mu|}{1-|z|^2}. \tag{4.4.17}$$

Integrating (4.4.17) we get

$$|\log h(z)| \le \log \left(\frac{1+|z|}{1-|z|}\right)^{2|\mu|}. \tag{4.4.18}$$

Assume now that $h \in \mathrm{Spiral}_{\mu,\nu}[1,\eta]$, i.e., $\angle \lim\limits_{z\to\eta} h(z) = \infty$ and

$$\nu := \angle \lim_{z\to\eta} \frac{(z-\eta)h'(z)}{h(z)}$$

is finite and different from zero. Without loss of generality we suppose that $\eta = -1$. Then again from (4.4.16) we have:

$$\angle \lim_{z\to\eta=-1} \frac{f(z)}{z+1} = \mu \lim_{z\to -1} \frac{h(z)}{(z+1)h'(z)} = \frac{\mu}{\nu}. \tag{4.4.19}$$

Hence, $\angle \lim\limits_{z\to -1} f(z) = 0$ and

$$\angle \lim_{z\to -1} f'(z) =: \beta_- \tag{4.4.20}$$

is finite, i.e., $f \in \mathcal{G}[1,-1]$. Moreover, it follows by (4.4.19) and (4.4.20) that

$$\nu\beta_- = \mu \tag{4.4.21}$$

and

$$\operatorname{Re}\nu < 0. \tag{4.4.22}$$

Then we get, by (4.4.14),

$$\left|\frac{h'(z)}{h(z)}\right| \ge 2|\nu| \frac{1-|z|^2}{|1-z^2|^2} \tag{4.4.23}$$

for all $z \in \Delta$. Thus we have obtained the following distortion theorem for a function of the class $\mathrm{Spiral}_{\mu,\nu}[1,-1]$.

Theorem 4.6. *Let $h \in \mathrm{Spiral}_{\mu,\nu}[1,-1]$. Then for all $z \in \Delta$ the following estimate holds:*

$$\frac{(1-|z|^2)2\,|\nu|}{|1-z^2|^2} \le \left|\frac{h'(z)}{h(z)}\right| \le \frac{2\,|\mu|}{1-|z|^2}.$$

In particular, if z is in the circular lens

$$\Upsilon_M = \left\{ z \in \Delta : \frac{|1-z^2|}{1-|z|^2} \le M \right\}, \quad M \ge 1,$$

then

$$\frac{1}{M}\frac{2|\nu|}{|1-z^2|} \le \left|\frac{h'(z)}{h(z)}\right| \le \frac{2|\mu|M}{|1-z|^2}.$$

In fact, we have shown also that if $h \in \mathrm{Spiral}_{\mu,\nu}[1,-1]$, then f defined by (4.4.16) belongs to $\mathcal{G}[1,-1]$ with $f'(1) > 0$.

Conversely, let $f \in \mathcal{G}[1,-1]$ be given with $f'(1) = \beta_+ = 1$. As we already know for each $\mu \in \Omega := \{|\lambda - 1| \le 1,\ \lambda \ne 0\}$, equation (4.4.16) has a unique univalent solution h satisfying conditions $h(0) = 1$ and $h(1) := \lim\limits_{r\to 1^-} h(r) = 0$. We want to show that actually $h \in \mathrm{Spiral}_{\mu,\nu}[1,-1]$ for some $\nu \in \mathbb{C}$ with $\operatorname{Re}\nu < 0$.

Indeed, because of relations (4.4.19)–(4.4.21), such ν exists, i.e., h satisfies a generalized Visser–Ostrowsi condition. Thus it remains to show that $\angle \lim\limits_{z\to -1} h(z) = \infty$, or, by Lindelöf's Theorem 1.6 that

$$\lim_{x\to -1} h(x) = \infty.$$

To this end we return to formula (2.4.21):

$$f(z) = -c(1-z)^2\, \frac{1+F(z)}{1-F(z)},$$

and consider equation

$$h_1(z) = h_1'(z) f(z). \tag{4.4.24}$$

This equation has a unique univalent solution h_1 satisfying conditions $h_1(0) = 1$ and $h_1(1) = 0$. Also, it is easy to see (comparing (4.4.16) and (4.4.24) that, in fact,

$$h(z) = h_1(z)^{\mu}.$$

Therefore, it is sufficient to prove our assertion for function h_1.

By (2.4.21) and (4.4.24) we get, for $x \in (-1,1)$,

$$\begin{aligned} -\operatorname{Re}\frac{h_1'(x)}{h_1(x)} &= \frac{1}{c(1-x)^2}\operatorname{Re}\frac{1-F(x)}{1+F(x)} \\ &= \frac{1}{c(1-x)^2}\frac{1-|F(x)|^2}{|1+F(x)|^2} \ge \frac{2}{|\beta_-|}\frac{1}{1-x^2}. \end{aligned} \tag{4.4.25}$$

Now if $x \in (-1,0)$ we obtain

$$\begin{aligned} -\log|h_1(x)| &= -\operatorname{Re}\log h_1(x) = -\int_0^x \operatorname{Re}\frac{h_1'(x)}{h_1(x)}\,dx \\ &\le \frac{2}{|\beta_-|}\int_0^x \frac{dx}{1-x^2} = \frac{2}{|\beta_-|}\cdot\frac{1}{2}\log\frac{1+x}{1-x}, \end{aligned} \tag{4.4.26}$$

and we have $\log|h_1(x)| \to \infty$ as $x \to -1$.

Note that inequality (4.4.26) is equivalent to

$$|h_1(x)| \ge \left(\frac{1-x}{1+x}\right)^{\frac{1}{|\beta_-|}}.$$

If $x \in (0,1)$, then (4.4.25) implies

$$-\log|h_1(x)| \geq \frac{1}{|\beta_-|}\log\frac{1-x}{1+x}$$

or

$$|h_1(x)| \leq \left(\frac{1-x}{1+x}\right)^{\frac{1}{|\beta_-|}}.$$

Finally, we observe that if $\nu = -\mu$, then $\beta_- = -1 = -\beta_+$, hence, $f(z) = \frac{\mu h(z)}{h'(z)}$ must be the generator of a group of hyperbolic automorphisms with fixed points at $z = 1$ and $z = -1$, such that $z = 1$ is the Denjoy–Wolff point, i.e., $f(z) = \frac{\beta_+}{2}\left(z^2 - 1\right)$. Thus, we have proved the following assertion.

Theorem 4.7. *A univalent function h on Δ belongs to the class* $\mathrm{Spiral}_{\mu,\nu}[1,-1]$ *if and only if the function*

$$f(z) = \frac{\mu h(z)}{h'(z)}$$

belongs to the class $\mathcal{G}^+[1,-1]$ with $f'(1) > 0$.

Moreover, if μ is real ($\mu \in (0,2]$), then so is ν ($\nu \leq -\mu < 0$) and h is, in fact, a starlike function on Δ satisfying the estimates:

$$|h(x)| \leq \left(\frac{1-x}{1+x}\right)^{-\nu}, \quad x \in [0,1),$$

and

$$|h(x)| \geq \left(\frac{1-x}{1+x}\right)^{-\nu}, \quad x \in (-1,0].$$

The equality $\nu = -\mu$ is possible if and only if

$$h(z) = \left(\frac{1-z}{1+z}\right)^{\mu}.$$

4.4.4 Classes $G(\mu,\beta)$

Here we establish a number of distortion results, namely, estimates for h and h' for holomorphic functions h belonging to some subclasses of the class $\mathrm{Spiral}_\mu[1]$. More precisely, we consider the following classes of functions.

Definition 4.7. *We say that a univalent function $h \in \mathrm{Hol}(\Delta,\mathbb{C})$, $h(0) = 1$, belongs to the class $G(\mu,\beta)$ (cf., Definition 3.3 and Theorem 4.1), where $\mu \in \Omega := \{\lambda \in \mathbb{C} : |\lambda - 1| \leq 0, \ \lambda \neq 0\}$ and $\beta \in [0,1)$, if h satisfies the inequality*

$$\mathrm{Re}\left(\frac{2}{\mu}\cdot\frac{zh'(z)}{h(z)} + \frac{1+z}{1-z}\right) > \beta, \quad z \in \Delta. \tag{4.4.27}$$

To study properties of these classes we start with an integral representation that generalizes Lemma 3.2.2.

Theorem 4.8. *Let h be a univalent function. Then $h \in G(\mu, \beta)$ if and only if h admits the following representation:*

$$h(z) = (1-z)^{\mu} \exp \left[-\mu(1-\beta) \oint_{|\zeta|=1} \log(1 - z\bar{\zeta}) d\sigma(\zeta) \right], \qquad (4.4.28)$$

where σ is a probability measure on the unit circle $|\zeta| = 1$.

Proof. First, suppose that the function h is represented by (4.4.28). Differentiating (4.4.28) and making a simple calculation, we check that inequality (4.4.27) holds.

Suppose now that $h \in G(\mu, \beta)$. Consider the function

$$p(z) = \frac{1}{1-\beta} \left[\frac{2}{\mu} \frac{zh'(z)}{h(z)} + \frac{1+z}{1-z} - \beta \right],$$

which obviously belongs to the Carathéodory class and thus can be represented by the Riesz–Herglotz formula:

$$p(z) = \oint_{|\zeta|=1} \frac{1 + z\bar{\zeta}}{1 - z\bar{\zeta}} d\sigma(\zeta),$$

where σ is a probability measure. This implies that

$$\frac{h'(z)}{\mu h(z)} = \oint_{|\zeta|=1} \left[\frac{(1-\beta)\bar{\zeta}}{1 - z\bar{\zeta}} - \frac{1}{1-z} \right] d\sigma(\zeta).$$

Integrating both sides of this equation we obtain (4.4.28). □

Remark 4.3. It follows by Theorem 4.1, Theorem 4.8 and Definition 4.7 that any function h, spirallike with respect to a boundary point, has the integral representation

$$h(z) = (1-z)^{\mu} \exp \left[-\mu \oint_{|\zeta|=1} \log(1 - z\bar{\zeta})\, d\sigma(\zeta) \right]$$

for some $\mu \in \Omega$ and a probability measure σ. Define another measure $\widetilde{\sigma}$ such that $\sigma = \beta\lambda + (1-\beta)\widetilde{\sigma}$, where λ is the normalized Lebesgue measure on the unit circle $\partial\Delta$. Obviously, $\int\limits_{\partial\Delta} d\widetilde{\sigma} = 1$. Since the integral of the antiholomorphic function $\log(1 - z\bar{\zeta})$ with respect to the Lebesgue measure λ is zero, we have that h belongs to the class $G(\mu, \beta)$ if and only if the measure $\widetilde{\sigma} = \frac{1}{1-\beta}(\sigma - \beta\lambda)$ is positive.

The following assertion is an immediate consequence of Theorem 4.8.

Corollary 4.3. *Let $\mu_1, \mu_2 \in \Omega$ and $\beta_1, \beta_2 \in [0,1)$. Let $h \in \mathrm{Hol}(\Delta, \mathbb{C})$ be univalent. Then $h \in G(\mu_1, \beta_1)$ if and only if the function*

$$\widetilde{h}(z) = (1-z)^{\mu_2(\beta_2-\beta_1)}(h(z))^{\frac{\mu_2(1-\beta_2)}{\mu_1(1-\beta_1)}}$$

belongs to $G(\mu_2, \beta_2)$. In particular, $h \in G(\mu, \beta)$ if and only if $\widetilde{h}(z) = (h(z))^{\frac{1}{\mu}}$ belongs to $G(1, \beta)$.

Corollary 4.4. *Let $\mu_1, \mu_2 \in \Omega$, $\mu_1 = r\mu_2$ with $r \le 1$, and let $\beta_2 \le r\beta_1$. Then $G(\mu_1, \beta_1) \subset G(\mu_2, \beta_2)$.*

Proof. Let $h \in G(\mu_1, \beta_1)$. We have

$$\begin{aligned} h(z) &= (1-z)^{\mu_1} \exp\left[-\mu_1(1-\beta_1) \oint_{|\zeta|=1} \log(1-z\bar{\zeta})d\sigma(\zeta)\right] \\ &= (1-z)^{\mu_2} \exp\left[-\mu_2(1-r\beta_1) \oint_{|\zeta|=1} \log(1-z\bar{\zeta})d\widetilde{\sigma}(\zeta)\right], \end{aligned}$$

where $\widetilde{\sigma}$ is a probability measure on the unit circle defined by

$$\widetilde{\sigma} = \frac{r(1-\beta_1)}{1-r\beta_1}\sigma + \frac{1-r}{1-r\beta_1}\delta,$$

with δ the Dirac measure at the point $\zeta = 1$. By Theorem 1.1, h satisfies the inequality

$$\mathrm{Re}\left(\frac{2}{\mu_2} \cdot \frac{zh'(z)}{h(z)} + \frac{1+z}{1-z}\right) > r\beta_1 \ge \beta_2.$$

Therefore, $h \in G(\mu_2, \beta_2)$. □

Corollary 4.5. *The set*

$$\bigcup_{n=1}^{\infty} \left\{ \left(\frac{1-z}{\prod_{j=1}^{n}(1-z\bar{\zeta}_j)^{\lambda_j}} \right)^{\mu} : \lambda_j \ge 0,\ \sum_{j=1}^{n} \lambda_j = 1-\beta,\ |\zeta_j| = 1 \right\}$$

is dense in $G(\mu, \beta)$ in the topology of uniform convergence on compact subsets of Δ.

Proof. Replacing the integral in (4.4.28) by approximating sums, we have for any $h \in G(\mu, \beta)$:

$$\begin{aligned} h(z) &= \lim_{n\to\infty} (1-z)^{\mu} \exp\left[-\mu(1-\beta)\sum_{j=1}^{n} \log(1-z\bar{\zeta}_j)\sigma_j\right] \\ &= \lim_{n\to\infty} (1-z)^{\mu} \prod_{j=1}^{n} (1-z\bar{\zeta}_j)^{-\mu(1-\beta)\sigma_j}, \end{aligned}$$

with $\sum_{j=1}^{n} \sigma_j = 1$. Putting $\lambda_j = (1-\beta)\sigma_j$, we complete the proof. $\square$

Corollary 4.6. *Let $h \in G(\mu, \beta)$. Then* $\dfrac{h(z)}{(1-z)^{\mu}} \prec \dfrac{1}{(1-z)^{\mu(1-\beta)}}$.

Proof. Since the function $\log(1-z)$ is convex, for any probability measure σ there exists an analytic function $\omega : \Delta \mapsto \Delta$ with $\omega(0) = 0$ such that

$$\oint_{|\zeta|=1} \log(1 - z\bar{\zeta}) d\sigma(\zeta) = \log(1 - \omega(z)).$$

Using Theorem 4.8 we obtain the equality

$$\frac{h(z)}{(1-z)^{\mu}} = \exp\left[-\mu(1-\beta) \oint_{|\zeta|=1} \log(1 - z\bar{\zeta}) d\sigma(\zeta)\right] = \frac{1}{(1-\omega(z))^{\mu(1-\beta)}},$$

which proves our assertion. $\square$

It turns out that for classes $G(\mu, \beta)$ one can improve the result of Proposition 4.3.

Corollary 4.7. *Let $f \in G(\nu, \beta)$. Then $h \prec h_{\mu,\theta^*}(z) := \left(\dfrac{1-z}{1+e^{-2i\theta^*}z}\right)^{\mu}$, where $\frac{\mu}{\nu}$ is a real number satisfying $\beta \leq \dfrac{\mu}{\nu} \leq 1$ and $|\theta^*| < \dfrac{\pi}{2}\dfrac{\nu}{\mu}(1-\beta)$. Consequently, $|\theta^*| < \dfrac{\pi}{2} \cdot \min\left(1, \dfrac{1-\beta}{\beta}\right)$.*

Proof. Using representation (4.4.28) one can estimate the numbers μ and θ^* from Proposition 4.3 (see also Theorem 4.1). Specifically,

$$\frac{h'(r)(r-1)}{\nu h(r)} = 1 - (1-\beta)(1-r) \int_{\partial\Delta} \frac{\bar{\zeta}}{1 - r\bar{\zeta}} d\sigma(\zeta).$$

Thus $\dfrac{\mu}{\nu} = \lim\limits_{r\to 1^-} \dfrac{h'(r)(r-1)}{\nu h(r)} \geq \beta.$

Further,

$$|\theta^*| = \lim_{r\to 1^-} \frac{\nu}{\mu}(1-\beta) \left| \int_{\partial\Delta} \arg(1 - r\bar{\zeta}) d\sigma(\zeta) \right|.$$

Thus the assertion follows. $\square$

Remark 4.4. For the case when $h \in G(\nu, \beta)$ with real ν (consequently, $h \in \text{Spiral}_{\mu}[1]$ with real μ), Proposition 4.3 (see also Theorem 3.3) asserts that the image of a function h is contained in the wedge of angle $\mu\pi$ with the midline $\arg w = \mu\theta$. The last corollary implies that if $h \in G(\nu, \beta)$, then the angle cannot be less than $\pi\nu\beta$, and the argument of the midline cannot be greater than $\frac{\pi}{2}\nu(1-\beta)$.

Now we are ready to establish desired estimates for h and h' when $h \in G(\mu, \beta)$. In particular, given a point $z \in \Delta$, we find the set of values for some functionals on these classes. For Robertson's class G (the class of starlike functions with respect to a boundary point having image in a half-plane) similar results were established by Todorov [141].

Theorem 4.9 (A distortion theorem for the class $G(\mu, \beta)$). *For each fixed $z \in \Delta$ we have*

$$\text{(i)} \qquad \left\{ \frac{1-z}{h(z)^{1/\mu}} : \ h \in G(\mu, \beta) \right\} = \left\{ (1+\lambda z)^{1-\beta}, \ |\lambda| \le 1 \right\} \tag{4.4.29}$$

and

$$\text{(ii)} \qquad \left\{ \frac{h'(z)}{\mu h(z)} + \frac{1}{1-z} : h \in G(\mu, \beta) \right\} = \left\{ w : \left| w - \frac{(1-\beta)\bar{z}}{1-|z|^2} \right| \le \frac{1-\beta}{1-|z|^2} \right\}. \tag{4.4.30}$$

Furthermore, if $h \in G(\mu, \beta)$ and $z \in \Delta, \ z \neq 0$, one of the relations

$$\frac{1-z}{h(z)^{1/\mu}} \in \left\{ (1+\lambda z)^{1-\beta}, \ |\lambda| = 1 \right\}$$

and

$$\frac{h'(z)}{\mu h(z)} + \frac{1}{1-z} \in \left\{ w : \left| w - \frac{(1-\beta)\bar{z}}{1-|z|^2} \right| = \frac{1-\beta}{1-|z|^2} \right\}$$

holds only if

$$h(z) = \frac{(1-z)^{\mu}}{(1-z\bar{\xi})^{(1-\beta)\mu}}, \qquad \xi \in \partial\Delta. \tag{4.4.31}$$

Proof. (i) By Theorem 4.8,

$$\log \left[\frac{1-z}{h(z)^{1/\mu}} \right] = (1-\beta) \oint_{\partial\Delta} \log(1 - z\bar{\zeta}) d\sigma(\zeta). \tag{4.4.32}$$

By the Carathéodory Principle,

$$\left\{ \log \left(\frac{1-z}{h(z)^{1/\mu}} \right) : \ h \in G(\mu, \beta) \right\} = \operatorname{Conv} \left\{ (1-\beta) \log(1 - z\bar{\zeta}), \ \zeta \in \partial\Delta \right\},$$

where Conv denotes the convex hull.

Since for each $z \in \Delta$ the function $g(w) := (1-\beta)\log(1-zw)$ maps Δ onto a strictly convex domain, this formula and (4.4.32) imply that the value $\log \left[\frac{1-z}{h(z)^{1/\mu}} \right]$

belongs to the image $g(\Delta)$ except for the case when the measure σ is the Dirac δ-function at some boundary point ξ. Hence the assertion follows.

(ii) Once again using representation (4.4.28) we get

$$\frac{h'(z)}{\mu h(z)} + \frac{1}{1-z} = (1-\beta)\oint_{\partial\Delta} \frac{\bar{\zeta}}{1-z\bar{\zeta}}\, d\sigma(\zeta).$$

For each fixed $z \in \Delta$, the function $g : \partial\Delta \mapsto \mathbb{C}$ defined by $g(\zeta) := \frac{\bar{\zeta}}{1-z\bar{\zeta}}$ maps the unit circle onto the circle $\left\{ w : \left| w - \frac{\bar{z}}{1-|z|^2} \right| = \frac{1}{1-|z|^2} \right\}$. Therefore, (4.4.30) follows from the Carathéodory Principle. $\square$

For real μ similar facts can be found in [29, 135, 42].

The following two assertions are immediate consequences of Theorem 4.9 (i).

Corollary 4.8. $\bigcup\limits_{z\in\Delta} \left\{ \frac{1-z}{h(z)^{1/\mu}} : \ h \in G(\mu,\beta) \right\} = \left\{ (1+\lambda)^{1-\beta}, \ |\lambda| \le 1 \right\}.$

Corollary 4.9. *For each $z \in \Delta$ and $h \in G(\mu,\beta)$,*

$$(1-|z|)^{1-\beta} \le \left| \frac{1-z}{h(z)^{1/\mu}} \right| \le (1+|z|)^{1-\beta},$$

and

$$\left| \arg \left(\frac{1-z}{h(z)^{1/\mu}} \right) \right| \le (1-\beta) \arcsin|z|,$$

where for $z \ne 0$, equality is achieved only for the functions (4.4.31) at the points $z = \pm|z|\xi$ and $z = |z|\xi e^{\pm i \arccos|z|}$, respectively. In particular, if μ is real, then

$$\frac{|1-z|^{\mu}}{(1+|z|)^{\mu(1-\beta)}} \le |h(z)| \le \frac{|1-z|^{\mu}}{(1-|z|)^{\mu(1-\beta)}}.$$

In the case when $\mu \in (0,2]$ (i.e., $h \in G(\mu,\beta)$ is, in fact, starlike), one obtains the following estimate for h'.

Corollary 4.10. *For each $z \in \Delta$ and $h \in G(\mu,\beta)$, $\mu \in (0,2]$, $\beta \in [0,1)$,*

$$\frac{\mu|1-z|^{\mu}}{(1-|z|^2)(1+|z|)^{\mu(1-\beta)}} \left(\left| \frac{1-\bar{z}}{1-z} + \beta\bar{z} \right| - 1 + \beta \right) \le |h'(z)|$$
$$\le \frac{\mu|1-z|^{\mu}}{(1-|z|^2)(1-|z|)^{\mu(1-\beta)}} \left(\left| \frac{1-\bar{z}}{1-z} + \beta\bar{z} \right| + 1 - \beta \right).$$

In particular,

$$|h'(z)| \le \frac{2\mu|1-z|^{\mu}}{(1-|z|^2)(1-|z|)^{\mu(1-\beta)}}.$$

The proof follows from Theorem 4.9 (ii) and Corollary 4.9.

4.5 Covering theorems for starlike and spirallike functions

In this section we prove a covering theorem for classes $G(\mu, \beta)$ of spirallike functions with respect to a boundary point. Note that if for some class of functions there exists a domain D such that $D \subset h(\Delta)$ for each function h of the class, and a function $h_0 \in \mathrm{Hol}(\Delta, \mathbb{C})$ maps the open unit disk onto D conformally, then the corresponding covering result can be expressed as the subordination $h_0 \prec h$ (see Definition 4.4).

Theorem 4.10. *Let $h \in G(\mu, \beta)$. Denote $h_0(z) = (1-z)^{\mu\beta}$. Then $h_0 \prec h$. This subordination is sharp since $h_0 \in G(\mu, \beta)$.*

Proof. By Corollary 4.3, it is sufficient to prove this assertion for the case $\mu = 1$. First we suppose that $h \in G(1, \beta)$ has the form

$$h(z) = \frac{1-z}{\prod_{j=1}^{n}(1-z\overline{\zeta_j})^{\lambda_j}}, \quad \sum_{j=1}^{n} \lambda_j = 1-\beta, \ |\zeta_j| = 1.$$

Let $z_0 = e^{i\phi} \in \partial\Delta, \ z_0 \neq \zeta_1, \zeta_2, \dots, \zeta_n$. Consider the two univalent convex functions on Δ,

$$g_0(z) = \log h_0(z) = \beta \log(1-z)$$

and

$$g(z) = \log(1-z_0) - (1-\beta)\log(1-z_0 z).$$

It is easy to see that $g_0(z_0) = g(1)$, i.e., this is their common boundary point. Furthermore,

$$z g_0'(z)\big|_{z=z_0} = \frac{-\beta z_0}{1-z_0} \quad \text{and} \quad z g'(z)\big|_{z=1} = \frac{(1-\beta)z_0}{1-z_0},$$

i.e., the exterior normal vectors to the images $g_0(\Delta)$ and $g(\Delta)$ at this common point have reverse directions. Because these functions are convex, $g_0(\Delta) \bigcap g(\Delta) = \emptyset$. Thus,

$$\begin{aligned} \log h(z_0) &= \log(1-z_0) - \sum_{j=1}^{n} \lambda_j \log(1-z_0\overline{\zeta_j}) \\ &= \sum_{j=1}^{n} \frac{\lambda_j}{1-\beta}\left[\log(1-z_0) - (1-\beta)\log(1-z_0\overline{\zeta_j})\right] \end{aligned}$$

belongs to $\overline{g(\Delta)}$ and does not belong to $g_0(\Delta) = \log h_0(\Delta)$. Since the point $h(z_0)$ is an arbitrary finite boundary point of the image $h(\Delta)$, we have proved that the boundary $\partial h(\Delta)$ does not intersect $h_0(\Delta)$. This fact implies that h_0 is subordinate to any function $h \in G(1, \beta)$ that has the form described above.

In light of Corollary 4.5 and by the Carathéodory Kernel Convergence Theorem we deduce the assertion of the theorem. □

Example 4.1. Consider the function

$$h(z) = \frac{1-z}{(1-0.9z-0.4iz)^{0.2}(1-0.9z+0.4iz)^{0.2}}.$$

A simple calculation shows that $h \in G(\mu,\beta)$ with $\mu = 1$ and $\beta = 0.6$. Hence by Theorem 4.10, its image $h(\Delta)$ contains the image $h_0(\Delta)$, where $h_0(z) = (1-z)^{0.6}$ (see Fig. 4.1).

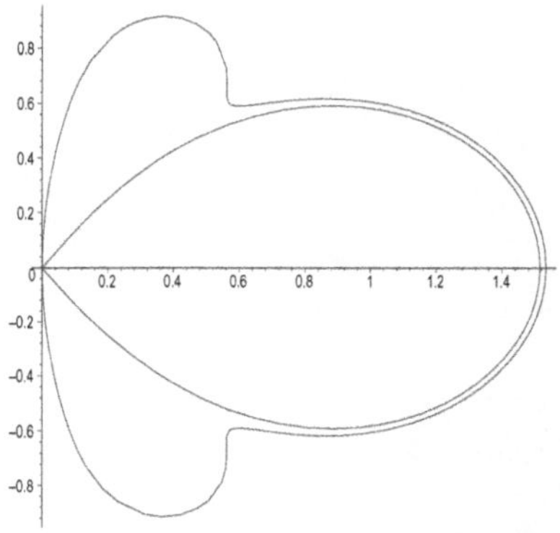

Figure 4.1: The image and the covered domain.

Corollary 4.11. *If $h \in G(\mu,\beta)$ with $\mu \in (0,2]$ and $\beta \in (0,1)$, then the image of h covers the disk $\{w : |w-1| < r(\mu\beta)\}$, where*

$$r(s) = \begin{cases} \sqrt{1+2^{2s}-2^{s+1}}, & \text{when } s \in (0,1] \\ 1, & \text{when } s \in (1,2] \end{cases}.$$

Proof. It suffices to find the distance between the point $z = 1$ and the curve $\{(1-z)^s,\ z \in \partial\Delta\}$, where $s = \mu\beta$. Indeed, setting $z = -e^{it}$, $t \in (-\pi,\pi)$, consider the function

$$a_s(t) = \left|(1+e^{it})^s - 1\right|^2 = \left(2\cos\frac{t}{2}\right)^{2s} + 1 - 2\left(2\cos\frac{t}{2}\right)^{s}\cos\frac{st}{2}.$$

This function is even in t, a_s increases on the segment $0 \le t \le \pi$ when $s < 1$, and a_s decreases on the same segment when $s > 1$. So,

$$\min_{-\pi<t<\pi} a_s(t) = \begin{cases} a_s(0), & \text{when } s \in (0,1] \\ a_s(\pi), & \text{when } s \in (1,2] \end{cases}.$$

Calculating these values we complete the proof. □

Remark 4.5. In [29] Chen and Owa proved a covering theorem for starlike functions with respect to a boundary point. Using our notation, their result can be quoted as follows: *If $h \in G(\mu, \beta)$ for some $\mu, \beta \in [0, 1]$ then its image covers the disk* $\left\{w : |w-1| < \frac{\mu\beta}{4}\right\}$. The radius of the covered disk, which was found in the last corollary, is considerably larger than the one due to Chen and Owa. In Fig. 4.2 we compare these two radii.

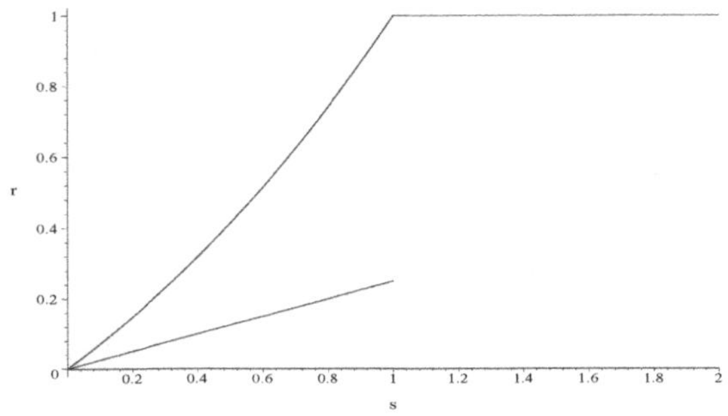

Figure 4.2: Radii of covered disks.

The following assertion connects the classes $G(\mu, \beta)$ with spirallike functions with respect to an interior point.

Proposition 4.4. *Let $\beta \in [0, 1)$, and let a complex number $\mu = r\,e^{i\phi}$ belong to $\Omega = \{\lambda : \ |\lambda - 1| \le 1, \ \lambda \neq 0\}$. Then a univalent function h, $h(0) = 1$, belongs to the class $G(\mu, \beta)$ if and only if the function g, defined by*

$$g(z) := \frac{zh(z)}{(1-z)^{\mu}}, \tag{4.5.1}$$

is ϕ-spirallike of order $\cos\phi - \dfrac{r(1-\beta)}{2}$*, i.e., g satisfies the condition*

$$\operatorname{Re}\left(e^{-i\phi}\frac{zg'(z)}{g(z)}\right) > \cos\phi - \frac{r(1-\beta)}{2}, \quad z \in \Delta. \tag{4.5.2}$$

Remark 4.6. Note that the relation $\mu = r\,e^{i\phi} \in \Omega$ implies that $\cos\phi - \frac{r(1-\beta)}{2} \ge 0$. By a result of Špaček (see [75]), the function g is univalent in Δ.

Proof. Let a function g be defined by (4.5.1). Then

$$\begin{aligned} e^{-i\phi}\frac{zg'(z)}{g(z)} &= \frac{r}{\mu}\left(1 + \frac{zh'(z)}{h(z)} + \frac{\mu z}{1-z}\right) \\ &= \frac{r}{2}\left(\frac{2zh'(z)}{\mu h(z)} + \frac{1+z}{1-z} - \beta\right) + e^{-i\phi} - \frac{r(1-\beta)}{2}. \end{aligned}$$

This equality implies the assertion. □

Repeating the proof of Theorem 4.2 and using Proposition 4.4 and Lemma 4.2, one can conclude the following.

Corollary 4.12. *Let $h : \Delta \mapsto \mathbb{C}$ be a holomorphic function and $h(0) = 1$. Let $\mu = re^{i\phi} \in \Omega$ and $\beta \in [0,1)$. Then $h \in G(\mu, \beta)$ if and only if one of the following conditions holds:*

(a) *for all $u, v \in \overline{\Delta}$,*

$$\left(\frac{1-uz}{1-vz}\right)^{\mu} \frac{h(vz)}{h(uz)} \prec \left(\frac{1-uz}{1-vz}\right)^{(1-\beta)\mu};$$

or

(b) *for all $t \in (0, 2\cos\phi)$,*

$$\left|\frac{h(z(1-e^{-i\phi}t))}{h(z)}\right| \leq \left|\left(\frac{1-z(1-e^{-i\phi}t)}{1-z}\right)^{\mu}\right| \cdot \left(1 - \frac{t}{2\cos\phi}\right)^{-\operatorname{Re}\mu(1-\beta)}.$$

Setting $u = 0$ and $v = 1$ in Corollary 4.12, we obtain an alternative proof of Corollary 4.6 above.

The next assertion follows from Theorem 4.10 and Proposition 4.4.

Corollary 4.13. *Let $\phi \in \left(-\frac{\pi}{2}, \frac{\pi}{2}\right)$, $\alpha < \cos\phi$, $\beta \in \left(0, \frac{\alpha}{\cos\phi}\right]$, and $\mu = \frac{2(\cos\phi - \alpha)}{1-\beta} e^{i\phi}$. For each univalent function g normalized by $g(0) = 0$, $g'(0) = 1$ and satisfying $\operatorname{Re}\left(e^{-i\phi}\frac{zg'(z)}{g(z)}\right) > \alpha$, the image of the function*

$$1 - (1-z)^{\frac{1}{\beta}} \left(\frac{g(z)}{z}\right)^{\frac{1}{\mu\beta}}$$

covers the open unit disk, and, consequently, the image of the function

$$\frac{(1-z)^{\frac{1}{\beta}} \left(\frac{g(z)}{z}\right)^{\frac{1}{\mu\beta}}}{2 - (1-z)^{\frac{1}{\beta}} \left(\frac{g(z)}{z}\right)^{\frac{1}{\mu\beta}}}$$

covers the right half-plane.

In particular, setting $\phi = 0$, $\alpha = \beta = \frac{1}{2}$, we get that for each function g starlike of order $\frac{1}{2}$, the image of the function $1 - (1-z)^2 \left(\frac{g(z)}{z}\right)$ contains the open unit disk, and the image of the function $\dfrac{(1-z)^2 g(z)}{2z - (1-z)^2 g(z)}$ contains the right half-plane.

Chapter 5

Kœnigs Type Starlike and Spirallike Functions

In this chapter we study linear-fractional models for discrete and continuous semigroups of holomorphic functions on the open unit disk Δ. This study is based on solutions of classical Schröder's and Abel's functional equations. It turns out that different problems of the semigroup theory (like the embedding problem, asymptotic behavior and rigidity) closely connected to geometrical properties of the solutions of Schröder's and Abel's equations. In order to emphasize more important conceptual issues, we refer the reader to related sources for the proofs of some technical results for discrete time semigroups.

5.1 Schröder's and Abel's equations

The circle of ideas related to functional equations has a very long history. In 1871, Schröder [125] considered the problem of fractional iteration and related it to the solution of various functional equations including the functional equation

$$h \circ F = \lambda h, \tag{5.1.1}$$

which bears his name. Here $F \in \text{Hol}\,(\Delta)$ and $\lambda \in \mathbb{C}$ are given. Note, in passing, that Schröder's equation can be considered the eigenvalue problem for the composition operator C_F defined by

$$C_F(h) = h \circ F.$$

In 1884, Kœnigs [93] constructed a solution of Schröder's equation near a fixed point $\tau \in \Delta$ of F when $F'(\tau) = \lambda$, $|\lambda| \neq 0, 1$. In the years since these classical works, much has been written; the interested reader might consult the bibliographies in [38], [95], [96] and [92].

The setting in that work was the analysis of a contractive map, analytic in a neighborhood of a fixed point. The solution $h \in \mathrm{Hol}(\Delta, \mathbb{C})$ of the basic equation was obtained as a concrete limit, now frequently called **Kœnigs' function** in this context.

A classical analysis problem is, given a function $F(z)$, find a function $F_t(z)$ with $F_1(z) = F(z)$, satisfying the semigroup property:

$$\begin{cases} F_{t+s}(z) = F_t\left(F_s(z)\right), \ t, s \geq 0, \\ F_0(z) = z. \end{cases} \tag{5.1.2}$$

This problem is called the **(semigroup) embedding problem.**

Definition 5.1. *A holomorphic self-mapping $F \in \mathrm{Hol}(\Delta)$ of the open unit disk Δ is said to be* **embeddable on a domain** $\Omega \subset \Delta$ *if there is a continuous one-parameter semigroup $\mathcal{S} = \{F_t\}_{t\geq 0} \subset \mathrm{Hol}(\Omega)$ such that $F\mid_\Omega = F_1$. In the case where $\Omega = \Delta$ we say that F is* **embeddable globally**[1].

In this situation $F^n = F_n$, where as above F^n denotes the n-fold iterate of F. So, the element F_t, $t > 0$, can be considered the **fractional iterate** of F.

It is easy to see that *each automorphism of Δ is always embeddable globally in a one-parameter group (of automorphisms) of Δ.*

Although for a given $F \in \mathrm{Hol}(\Delta)$ the discrete semigroup $\{F^n\}_{n=1}^{\infty}$ in general may not be embeddable in a continuous semigroup, at each $z \in \Delta$ fractional iterates F_t of F can be defined for all $t(= t(z))$ large enough (see, for example, [38, 40]).

Note that a necessary condition for $F \in \mathrm{Hol}(\Delta)$ to be embeddable on Ω is the univalence of F on Ω (see, for example, [17] or [130]). Therefore, if $F'(\xi) = 0$ at some point $\xi \in \Omega$, then F is not embeddable.

If F is analytic in a neighborhood of a fixed point τ, $F(\tau) = \tau$, and $0 < |F'(\tau)| < 1$, Kœnigs (1884) showed how the problem can be solved. The limit

$$\lim_{n\to\infty} \frac{F^n(z) - \tau}{(F'(\tau))^n} =: h\left(z\right) \tag{5.1.3}$$

can be shown to exist for z near τ. Then h satisfies Schröder's functional equation (5.1.1) with $\lambda = F'\left(\tau\right)$, whence $F(z) = h^{-1}[\lambda h(z)]$ and $F^n(z) = h^{-1}[\lambda^n h(z)]$. Moreover, the latter expression serves a local definition of $F_t(z)$ when t is not an integer.

If, for example, $\lambda = F'(\tau)$ is real, $\lambda \neq 0, 1$, and $h \in \mathrm{Hol}(\Delta, \mathbb{C})$ is a univalent starlike function on Δ, then the family $\{F_t(\cdot)\}_{t\geq 0}$ defined by

$$F_t(z) = h^{-1}[\lambda^t h(z)] \tag{5.1.4}$$

forms a one-parameter continuous semigroup of holomorphic mappings globally on Δ and $F_1(z) = F(z)$.

[1]Sometimes, one also says that F satisfies the Kœnigs embedding property; see Section 5.3.

Another approach when F has no interior fixed point is via **Abel's functional equation** (1881):

$$g(F(z)) = g(z) + c, \quad c \in \mathbb{C}. \tag{5.1.5}$$

If g can be found to satisfy (5.1.5) and is convex in the direction c (i.e., for each $z \in \Delta$ and $t > 0$ we have $g(z) + tc \in g(\Delta)$, cf. Section 3.5), then one may set $F_t(z) = g^{-1}[tc + g(z)]$ to embed F into a continuous semigroup. A summary of some of the classical work on the problem may be found in Cowen [38], Hadamard [78], Harris [79] and Siskakis [136].

As it turns out, both Schröder's and Abel's equations can be considered particular cases of a unified construction. Under very general conditions, a holomorphic mapping F of the disk Δ into itself can intertwine with a linear-fractional transformation Φ in the following sense.

Definition 5.2. *Suppose F is a holomorphic self-mapping of the unit disk Δ. If there exists a univalent mapping $\sigma : \Delta \mapsto \mathbb{C}$, and a linear-fractional mapping Φ such that $\Phi(\Delta) \subset \Delta$, $\Phi(\sigma(\Delta)) \subset \sigma(\Delta)$, and*

$$\sigma \circ F = \Phi \circ \sigma, \tag{5.1.6}$$

we call the pair (Φ, σ) a **linear-fractional (conjugation) model for** *F (shortly,* **LF-Model***). The mapping Φ related to F by (5.1.6) is said to be* **conjugate** *to F by σ.*

For Schröder's equation (5.1.1), one sets $\sigma = h$ and $\Phi(w) = \lambda w$, $w \in \mathbb{C}$, whilst for Abel's equation (5.1.5) $\sigma = g$ and $\Phi(w) = w + c$, $w \in \mathbb{C}$.

In turn, the induction method naturally involves the study of the whole discrete semigroup $\{F^n\}_{n=1}^{\infty}$ of iterates of F via the conjugation

$$\sigma \circ F^n = \Phi^n \circ \sigma. \tag{5.1.7}$$

If, in addition, the image $\sigma(\Delta)$ lies in a half-plane (or a disk) that is Φ-invariant, then F can be intertwining with a linear-fractional transformation Ψ of the unit disk, defined by the conjugation:

$$\varrho \circ F = \Psi \circ \varrho, \tag{5.1.8}$$

where ϱ is a self-mapping of Δ (cf. [21]).

Since many problems involving iteration or solution of functional equations can be explicitly solved for linear fractional transformations and the intertwining relates these answers to solutions of the problems for a given function F, we obtain the information we seek.

Regarding continuous-time semigroups, the following general observation is an important point in our further considerations.

Theorem 5.1. *Let $f \in \mathcal{G}$ generate a semigroup $\{F_t\}_{t\geq 0} \subset \mathrm{Hol}(\Delta)$. Let $\sigma : \Delta \mapsto \mathbb{C}$ be any univalent function on Δ with $\Omega = \sigma(\Delta)$, and let φ be a holomorphic function on Ω. Then σ satisfies the differential equation*

$$\varphi(\sigma(z)) = \sigma'(z)f(z), \quad z \in \Delta, \tag{5.1.9}$$

if and only if φ is the generator on Ω of the semigroup $\{\Phi_t\}_{t\geq 0} \subset \mathrm{Hol}(\Omega)$ defined by

$$\sigma(F_t(z)) = \Phi_t(\sigma(z)), \quad t \geq 0. \tag{5.1.10}$$

In other words, functional equation (5.1.10) *and differential equation* (5.1.9) *are equivalent.*

Proof. The sufficiency follows immediately, if we differentiate equation (5.1.10) at the point $t = 0^+$.

Conversely, let σ satisfy (5.1.9) for some $\Phi \in \mathrm{Hol}(\Omega, \mathbb{C})$. Define a one-parameter semigroup $\varphi_t : \Omega \mapsto \Omega$ by the formula

$$\Phi_t(w) = \sigma(F_t(\sigma^{-1}(w)), \quad t \geq 0. \tag{5.1.11}$$

Now we calculate

$$\begin{aligned}\frac{d\Phi_t(w)}{dt} &= \sigma'\left(F_t\left(\sigma^{-1}(w)\right)\right) \cdot \frac{dF_t\left(\sigma^{-1}(w)\right)}{dt}\\ &= -\sigma'\left(F_t\left(\sigma^{-1}(w)\right)\right) \cdot f\left(F_t\left(\sigma^{-1}(w)\right)\right), \quad t \geq 0.\end{aligned}$$

In turn, formulas (5.1.11) and (5.1.10) imply

$$\begin{aligned}\frac{d\Phi_t(w)}{dt} &= -\sigma'\left(\sigma^{-1}(\Phi_t(w))\right) \cdot f\left(\sigma^{-1}(\Phi_t(w))\right)\\ &= -\varphi(\sigma(\sigma^{-1}(\Phi_t(w)))) = -\varphi(\Phi_t(w)).\end{aligned}$$

So, $\varphi \in \mathrm{Hol}(\Omega, \mathbb{C})$ is the generator of the semigroup $\{\Phi_t\}_{t\geq 0}$ and we are done. □

Remark 5.1. Note also that solutions of equation (5.1.9) define a generalized class of the so-called φ-like functions in the sense of Brickman [22] (see Sections 1.1.5 and 2.5).

If, in particular, $\varphi(w) = \mu(w - w_0)$ for some $w_0 \in \mathbb{C}$ and $\mu \in \mathbb{C}$ with $\mathrm{Re}\,\mu > 0$, a solution σ of equation (5.1.9) (if it exists) defines a spirallike function with respect to w_0 (in this case w_0 must lie in $\overline{\sigma(\Delta)}$). In turn, (5.1.10) is an equation of Yaglom's type:

$$\sigma(F_t(z)) = e^{-t\mu}\sigma(z) + (1 - e^{-t\mu})w_0$$

(see, [79]), and it becomes Schröder's equation if $w_0 = 0$.

Similarly, if in (5.1.9) we set $\varphi(z) = c = const$, a solution of equation (5.1.10) (if it exists) defines a function convex in one direction (see Section 3.5), and solves Abel's functional equation (5.1.5).

To proceed we need the following classification.

Definition 5.3. *A holomorphic self-mapping F of Δ is said to be of*

- **dilation type** *if it has a fixed point in Δ;*
- **hyperbolic type** *if it has no fixed point in Δ and has the angular derivative $F'(\tau)$ at its Denjoy–Wolff point $\tau \in \partial\Delta$ strictly less than 1: $0 < F'(\tau) < 1$;*
- **parabolic type** *if it has no fixed point in Δ and has the angular derivative $F'(\tau) = 1$ at its Denjoy–Wolff point $\tau \in \partial\Delta$.*

All hyperbolic type mappings are **automorphic** in the sense that their orbits are separated in the hyperbolic metric of Δ (meaning that, for each $z \in \Delta$, the hyperbolic distance between successive points of the orbit $\{F^n(z)\}$ stays bounded away from zero).

Parabolic type mappings fall into two subclasses:

- **automorphic type:** those having an orbit that is separated in the hyperbolic metric of Δ;
- **non-automorphic type:** those for which no orbit is hyperbolically separated.

Remark 5.2. It can be shown that either all orbits are separated or none are separated; see details in Section 5.5.

Traditionally, the intertwining map σ in (5.1.6) is constructed by a limiting process from F, as illustrated by Kœnigs [93] in 1884 for the Denjoy–Wolff point inside the disk. The hyperbolic case was developed by Valiron [143]. In general, the boundary location of the Denjoy–Wolff point τ (in automorphic as well as in non-automorphic cases) has been investigated by Pommerenke [110] and Baker and Pommerenke [12]. A unified construction for parabolic type mappings is presented by Bourdon and Shapiro in [18] (for details see the recent paper [36] by Contreras, Díaz-Madrigal and Pommerenke). Additional information for all types of mappings can be found in Shapiro's book [127] and the survey of Bourdon and Shapiro [18]. Iteration of functions for which $F'(a) = 0$ was studied in [38] and [71] via Bottcher's functional equation $f(F(z)) = (f(z))^k$.

5.2 Remarks on stochastic branching processes

A model for iterations, which is related to the theory of branching processes in Probability Theory, goes back to the classical Galton–Watson processes through the probability generating functions. The probability generating function for a Galton–Watson process is the function $F(z) = \sum_{k=0}^{\infty} p_k z^k$, where $p_k \geq 0$ and $\sum_{k=0}^{\infty} p_k = 1$. (see, for example, [79]). The coefficient p_k is interpreted as the probability that an individual will have k offspring and the kth coefficient of F^n is interpreted as the probability that there will be k offspring in the nth generation.

In other words, if the number of particles in the nth generation is a random variable $Z(n)$, then one can show (see [10]) that its generating function

$$F^n(z) = \sum_{k=0}^{\infty} p_k(n) z^k \in \mathrm{Hol}(\Delta), \quad n \in \mathbb{N}, \tag{5.2.1}$$

with

$$p_k(n) = P\{Z(n) = k \mid Z(0) = 1\} \tag{5.2.2}$$

is the n-fold iterate of $F^1(z) = F(z)$. So, $\{F^n\}_{n=1}^{\infty}$ is a discrete-time semigroup of holomorphic self-mappings of Δ.

Questions concerning eventual population size are related to the asymptotic behavior of the iterates of F. We note that every probability generating function for the Galton–Watson process is an analytic function mapping Δ into itself.

If we replace the set of indexes $\mathbb{N} = \{0, 1, 2, \ldots\}$ by $\mathbb{R}^+ = \{t \in \mathbb{R} : t \geq 0\}$, a Markov branching process $\{Z(t),\ t \in \mathbb{R}^+\}$ can be considered a continuous-time analog of the Galton–Watson process and interpreted as the number of offspring produced at any (continuous) time $t \in \mathbb{R}^+$. We concentrate now on generating functions F_t, $t \geq 0$, of continuous branching processes defined by the same formulas (5.2.1)–(5.2.2) for all nonnegative t:

$$F_t(z) = \sum_{k=0}^{\infty} p_k(t) z^k \in \mathrm{Hol}(\Delta), \qquad t \geq 0, \tag{5.2.3}$$

where $p_k(t)$ are defined by

$$p_k(t) = P\{Z(t) = k \mid Z(0) = 1\}. \tag{5.2.4}$$

By definition $p_k(t) \geq 0$ and $\sum\limits_{k=0}^{\infty} p_k(t) \leq 1$. A branching process $Z(t)$ is called **regular** if the latter sum is equal to 1, and then the point $z = 1$ is a fixed point for all of functions F_t, $t \geq 0$.

In this case $\mathcal{S} := \{F_t\}_{t \geq 0}$ is a continuous-time semigroup of holomorphic self-mappings of Δ. The differential equation

$$\begin{cases} \dfrac{\partial F_t}{\partial t} + f(F_t) = 0, \quad t \geq 0, \\ F_0 = z, \quad z \in \Delta \end{cases} \tag{5.2.5}$$

in this situation is called the **Kolmogorov backward equation** (see [10]).

It is known that generators of such semigroups have the following form:

$$f(z) = a \left(z - \sum_{n=0}^{\infty} \tilde{p}_n z^n \right),$$

where the number $a > 0$ and the so-called infinitesimal probabilities $\tilde{p}_n$ satisfying $0 \leq \tilde{p}_n$ and $\sum\limits_{n=0}^{\infty} \tilde{p}_n = 1$ are the total data of the process. Thus,

$$f(z) = \sum_{n=0}^{\infty} p_n z^n, \tag{5.2.6}$$

$$\text{where } \ p_1 \geq 0, \ p_n \leq 0 \text{ for } n \neq 1 \text{ and } \sum_{n=0}^{\infty} p_n = 0.$$

Representation (5.2.6) implies immediately that $f(1) = 0$. The smallest root $q \in [0, 1]$ of the equation

$$f(x) = 0$$

has a clear probability meaning:

$$q = \lim_{t \to \infty} P\{Z(t) = 0\},$$

and is the probability that the given process becomes extinct. The number q is called the **extinction probability** of the branching process.

Since $F_t(0) = P\{Z(t) = 0\}$ we have that $q\ (= \lim\limits_{t\to\infty} F_t(0))$ is the Denjoy–Wolff point of the semigroup, so each generator of type (5.2.6) can be also represented by the Berkson–Porta formula (2.2.7) with $\tau = q$.

From the point of view of the theory of branching processes it is natural to assume that the first moment (the expectation)

$$E\left[Z(t)\right] = \sum_{n=1}^{\infty} n\tilde{p}_n$$

for a suitable Markov process exists, i.e., the series $\sum\limits_{n=1}^{\infty} n\tilde{p}_n$ converges. In our terms this means that the first derivative $f'(1) = \sum\limits_{n=1}^{\infty} np_n$ for a generator f having the form (5.2.6) exists. In this case $f \in \mathcal{G}[1]$.

We already know (see Theorem 2.11) that for $f \in \mathcal{G}[1]$ the number $\beta = f'(1)$ is real. It is now natural to differentiate among the following three cases: $\beta > 0$, $\beta = 0$ and $\beta < 0$.

If $\beta = f'(1) \geq 0$, then by Theorem 2.10, the point $\tau = 1$ is the Denjoy–Wolff point for the semigroup generated by f, and f can be represented by the Berkson–Porta formula:

$$f(z) = -(1 - z)^2 p(z),$$

where $\operatorname{Re} p(z) \geq 0$. If, in addition, f has the form of (5.2.6), then the continuous-time branching process related to the semigroup $\mathcal{S} := \{F_t\}_{t\geq 0}$ generated by f has the extinction probability

$$q = \lim_{t\to\infty} P\{Z(t) = 0\} = 1.$$

In the case when $\beta > 0$, the semigroup generated by f converges to $\tau = 1$ with exponential rate of convergence, namely

$$|F_t(z) - 1| \le C(z)\exp(-t\beta)$$

(see Section 7.2.1). In this situation all elements of the semigroup are of hyperbolic type and the corresponding branching process is called **subcritical**.

If $\beta = 0$, the rate of convergence is more slow than exponential. In this case the semigroup elements are of parabolic type; in the theory of branching processes this case is called **critical**.

If $\beta = f'(1) < 0$, then the point $\zeta = 1$ is not an attractive point of the semigroup generated by f. Hence, it follows by the Denjoy–Wolff Theorem and the Berkson–Porta representation that f must have a different null point $\tau \neq 1$, which is the Denjoy–Wolff point for the semigroup $\mathcal{S}$ generated by f. If, in addition, f is of the form (5.2.6), this point (the extinction probability q) must lie on the interval $[0, 1)$; this case is called **supercritical**. Note also that in the last situation the Berkson–Porta representation is not related to the boundary null point $\zeta = 1$ of f. However, formula (2.4.4) proven above, can be used for all values of β. In particular, we have the following consequence of Theorem 2.11.

Corollary 5.1. *Let*

$$f(z) = -(1-z)^2 p(z) + \frac{\beta}{2}(z^2 - 1)$$

be an infinitesimal generating function of a supercritical branching process with the first moment $\beta < 0$. Then the extinction probability q is the unique, on the interval $[0, 1)$, root of the equation

$$p(x) = \frac{\beta}{2}\frac{x+1}{x-1}.$$

For convenience we summarize the above facts in the following table.

	$\beta = f'(1)$	General case	Probabilistic case
1.	$\beta > 0$	hyperbolic, $\zeta = 1$ is the Denjoy–Wolff point	subcritical case
2.	$\beta = 0$	parabolic, $\zeta = 1$ is the Denjoy–Wolff point	critical case
3.	$\beta < 0$	there is an additional root of $f(z) = 0$ in $\overline{\Delta}$, $\zeta = 1$ is not the Denjoy–Wolff point	supercritical case, there is $0 \le q < 1$ such that $f(q) = 0$

The bibliographies of Harris [79] and Athreya and Ney [10] are guides to classical work on branching processes.

Solutions of Schröder's and Abel's equations (as general linearization models) play a crucial role in the study of branching processes and their asymptotics (see Section 7.2.1). Here we also mention Yaglom's equation

$$\hat{g}(F(z)) = m\hat{g}(z) + 1 - m,$$

which is a combination of Schröder's and Abel's equations.

5.3 Kœnigs' linearization model for dilation type semigroups. Embeddings

Kœnigs' approach to solution of Schröder's equation is an important complement of assertion (i) of the Denjoy–Wolff Theorem mentioned above and plays a crucial role in the study of asymptotic behavior of discrete complex dynamics. It shows that any holomorphic self-mapping of Δ that has a non-zero derivative $F'(\tau)$ of module less than 1 at a fixed point τ can be "linearized" in the sense that the original mapping is analytically conjugate in a neighborhood of τ to its derivative mapping $w \mapsto \lambda w$, where (necessarily) $\lambda = F'(\tau)$.

Here we give Kœnigs' solution of Schröder's equation for holomorphic mappings $F \in \mathrm{Hol}(\Delta)$ with an interior fixed point.

Theorem 5.2 (see [127, 130]). *Suppose F is a holomorphic self-mapping of Δ that fixes a point $\tau \in \Delta$, is not an (elliptic) automorphism of Δ, and has non-vanishing derivative λ at τ. Then there exists the limit*

$$\lim_{n\to\infty} \frac{F^n(z)-\tau}{\lambda^n} =: h(z), \tag{5.3.1}$$

which satisfies Schröder's equation (5.1.1). *Moreover, if F is univalent, then so is h.*

Remark 5.3. Note that in this case the iterates F^n have the following asymptotic behavior:

$$F^n(z) - \tau \sim \lambda^n.$$

If F has an interior fixed point $\tau \in \Delta$ with $\lambda = F'(\tau) \neq 0$, the global embedding property of F can be described in terms of its Kœnigs function.

Theorem 5.3. *Let $F \in \mathrm{Hol}(\Delta)$ be a univalent function on the open unit disk Δ with an interior fixed point $\tau \in \Delta$, and $\lambda = F'(\tau)$. Let $h \in \mathrm{Hol}(\Delta, \mathbb{C})$ be the Kœnigs function for F defined by* (5.3.1). *The following assertions hold:*

(i) *If for some domain $\Omega \subset \Delta$, $\Omega \ni \tau$, the restriction $h|_\Omega$ is a $(-\log\lambda)$-spirallike function, then F is embeddable on Ω.*

(ii) *There exists a domain $\Omega \subset \Delta$, $\tau \in \Omega$, which satisfies (i), whence $F(\Omega) \subset \Omega$ and F is embeddable on Ω.*

(iii) *If $\lambda = F'(\tau)$ is a real number, then Ω can be chosen as the pseudohyperbolic disk of non-euclidean radius* $\tanh\frac{\pi}{4}$*:*

$$\Omega = \left\{ z \in \Delta : \ |M_\tau(z)| < \tanh\frac{\pi}{4} \right\},$$

where $M_\tau(z) = \frac{\tau - z}{1 - z\bar{\tau}}$ is the Möbius transformation of the unit disk Δ.

Proof. (i) Let $F \in \mathrm{Hol}(\Delta)$ be a univalent function on the open unit disk Δ with an interior fixed point $\tau \in \Delta$, $\lambda = F'(\tau)$. Then by Theorem 5.2, its Kœnigs function h is also univalent. Suppose that h is a $(-\log\lambda)$-spirallike function on Ω. Then one can define a semigroup $\mathcal{S} = \{F_t\}_{t\geq 0} \subset \mathrm{Hol}\,(\Omega)$ by the formula

$$F_t(z) = h^{-1}\left(\lambda^t h(z)\right), \quad z \in \Omega,\ t \geq 0.$$

On the other hand, by Schröder's equation (5.1.1) we have $F_1(z) = h^{-1}\left(\lambda h(z)\right) = F(z)$ for all $z \in \Omega \subset \Delta$, so F is embeddable on Ω.

(ii) Since $h(\tau) = 0$ and $h'(\tau) = 1$, we have by the Koebe One-Quarter Theorem (see, for example, [75]) that the image $h(\Delta)$ contains the disk $\Delta_1 := \left\{w : |w| < \frac{1}{4}\right\}$. By the Schwarz–Pick Lemma $|\lambda| \leq 1$, and we also have that for each $w \in \Delta_1$ and for each $t \geq 0$ the point $\lambda^t w$ also belongs to Δ_1. Define $\Omega = h^{-1}(\Delta_1)$. Since $\Delta_1 = h(\Omega)$ is λ-spirallike we have that F is embeddable on Ω.

(iii) By Theorem 5.2, h is a univalent function. Define a univalent function h_1 by $h_1(z) = h\left(M_\tau(z)\right)$. Then $h_1(0) = 0$ and $h_1'(0) = h'(\tau) = 1$. Hence h_1 maps the disk $\Delta_2 = \left\{w : |w| < \tanh\frac{\pi}{4}\right\}$ onto a starlike domain (see, for example [74]). Since $h(\Omega) = h_1(\Delta_1)$ and $\lambda = F'(\tau)$ is real, we conclude by (i) that F is embeddable on Ω. □

Theorem 5.4. *Let $F \in \mathrm{Hol}(\Delta)$ be a univalent function on the open unit disk Δ with an interior fixed point $\tau \in \Delta$, and $\lambda = F'(\tau)$ with $0 < |\lambda| < 1$. Let $h \in \mathrm{Hol}(\Delta, \mathbb{C})$ be the Kœnigs function for F defined by (5.3.1). Then F is embeddable globally if and only if h is a univalent spirallike function on Δ satisfying the condition*

$$\mathrm{Re}\,\frac{(z-\tau)\left(1-z\overline{\tau}\right)h'(z)}{h(z)\log\lambda} < 0, \quad z \in \Delta. \tag{5.3.2}$$

Proof. If h satisfies (5.3.2), then h is $(-\log\lambda)$-spirallike. So, by Theorem 5.3, F can be globally embeddable.

Suppose now that F is embeddable globally in a one-parameter semigroup $\mathcal{S} = \{F_t\}_{t\geq 0}$. By the Berkson–Porta formula the infinitesimal generator f of $\mathcal{S}$ admits the representation

$$f(z) = (z-\tau)(1-z\overline{\tau})p(z),$$

where p is a holomorphic function in Δ with a positive real part. Hence, $f'(\tau) = (1-|\tau|^2)p(\tau)$. On the other hand, it follows from the Cauchy problem (2.1.3) that $\lambda = F'(\tau)\left(= F_1'(\tau)\right) = \exp\left(-f'(\tau)\right)$. So, $\log\lambda = -(1-|\tau|^2)p(\tau)$.

We now define the function

$$h_1(z) := (z-\tau)\exp\left\{\int_\tau^z \frac{1}{\zeta-\tau}\left(\frac{(1-|\tau|^2)p(\tau)}{(1-\zeta\overline{\tau})p(\zeta)} - 1\right)d\zeta\right\}.$$

This function is well-defined, because p does not vanish in Δ.

For a fixed $z \in \Delta$, the function $\xi(t) = h_1(F_t(z))$ is defined on $[0, \infty)$. By the chain rule,

$$\frac{d\xi(t)}{dt} = -h_1'(F_t(z))f(F_t(z)) = h_1(F_t(z)) \log \lambda = \xi(t) \log \lambda.$$

Thus $\xi(t)$ is a solution of the Cauchy problem:

$$\frac{d\xi(t)}{dt} = \xi(t) \log \lambda, \quad \xi(0) = h_1(z).$$

Solving this problem, we obtain $\xi(t) = h_1(F_t(z)) = e^{t \log \lambda} h_1 F(z)$. Setting $t = 1$ we find that h_1 is a solution of Schröder's equation. By the uniqueness theorem for Schröder's equation (see, for example [127]), $h(z) = ch_1(z)$. Therefore,

$$\frac{h'(z)}{h(z)} = \frac{h_1'(z)}{h_1(z)} = \frac{-\log \lambda}{(z-\tau)(1-z\bar{\tau})p(z)},$$

where $\operatorname{Re} p(z) > 0$; that is, h satisfies (5.3.2). □

Remark 5.4. Thus if f is the generator of a one-parameter semigroup $\mathcal{S} = \{F_t\}_{t \geq 0}$ such that $f(\tau) = 0$ for some $\tau \in \Delta$, then the limit (5.3.1) with $F = F_1$, $F^n = F_n$ exists if and only if $\operatorname{Re} \mu > 0$, where $\mu = f'(\tau)$ and $\lambda = \exp(-\mu)$. Moreover, in this case the following asymptotic formula holds:

$$h(z) = \lim_{t \to \infty} e^{t\mu} (F_t(z) - \tau)$$

(see, for example, [58]). This formula is a continuous analog of (5.3.1).

5.4 Valiron's type linearization models for hyperbolic type semigroups. Embeddings

Even in the discrete case the situation becomes more complicated if $F \in \operatorname{Hol}(\Delta)$ has no fixed point in Δ. If this is the case, the Denjoy–Wolff Theorem 1.11 asserts that there is a unique boundary point $\tau \in \partial\Delta$ such that

$$\lim_{n \to \infty} F^n(z) = \tau, \quad z \in \Delta.$$

Without loss of generality one can set $\tau = 1$. We have

$$F(1) \left(:= \angle \lim_{z \to 1} F(z) \right) = 1 \quad \text{and} \quad 0 < \alpha = F'(1) \left(:= \angle \lim_{z \to 1} F'(z) \right) \leq 1.$$

Recall that F is classified to be of hyperbolic type if $\alpha = F'(1) < 1$.

If F is a generating function of a Galton–Watson branching process, this case is also called subcritical.

In this situation one can consider the limit

$$\lim_{n\to\infty} \frac{1 - F^n(z)}{\alpha^n} = \varphi(z), \qquad z \in \Delta.$$

This function (if it exists) is an analog of the Kœnigs function (5.3.1) for the boundary case. The problem is that φ *might be identically zero and then cannot serve as a characterization of the asymptotic behavior of* F *as well as its embedding criteria.*

Valiron [143] showed that $\varphi(z)$ is a well-defined non-constant holomorphic function on Δ if and only if

$$\sum_{n=1}^{\infty} \left| \frac{1 - F^{n+1}(0)}{1 - F^n(0)} \right| < \infty. \tag{5.4.1}$$

Another somewhat more explicit necessary and sufficient condition was established by Pommerenke [113]:

$$\int_0^1 \left| \frac{\alpha}{1 - F(x)} - \frac{1}{1 - x} \right| dx < \infty. \tag{5.4.2}$$

In addition, he showed that φ must be conformal at the boundary point $\tau = 1$, i.e., the angular limit $\angle \lim_{z\to 1} \frac{\varphi(z)}{z-1}$ exists finitely and is not zero.

Earlier Wolff and Valiron suggested another normalization and proved the following limit result.

Theorem 5.5 (see, for example, [143]). *Let* $F \in \mathrm{Hol}(\Delta)$ *be such that* $F(1) = 1$ *and* $\alpha = F'(1) < 1$. *Then the limit*

$$\lim_{n\to\infty} \frac{1 - F^n(z)}{|1 - F^n(0)|} = \psi(z), \quad z \in \Delta,$$

always exists and satisfies Schröder's equation

$$\psi \circ F = \alpha\psi.$$

This function ψ is also sometimes called the Kœnigs function for F (see [49, 32]).

Remark 5.5. Once the existence of the above limit is proved, it is easy to see that the limit

$$\lim_{n\to\infty} \frac{1 - F^n(z)}{1 - F^n(0)} = h(z), \qquad z \in \Delta,$$

also exists with $h(z) = \kappa\psi(z)$ for some κ, $|\kappa| = 1$. For our purpose sometimes it is more convenient to consider this function because of the normalization $h(0) = 1$. We call it the **Kœnigs–Valiron function** for F (or simply the (K-V)–function). Moreover, each other solution to Schröder's equation of positive real part must be a positive multiple of ψ (see [20]).

We will see that in our settings a hyperbolic type mapping F has the (semigroup) embedding property if and only if its (K-V)–function is starlike with respect to a boundary point (see Section 3) and belongs to the class $\mathrm{Star}_1[1]$.

Theorem 5.6. *Let $F \in \mathrm{Hol}(\Delta)$ be embeddable. Suppose that F has its Denjoy–Wolff point at $\tau = 1$ with $\alpha = F'(1) < 1$. Then for each $\xi \in \Delta$ the limit*

$$h_\xi(z) = \lim_{n\to\infty} \frac{1 - F^n(z)}{1 - F^n(\xi)}$$

exists and the Kœnigs–Valiron function $h_0(z) = \frac{h_\xi(z)}{h_\xi(0)}$ is a function of the class $\mathrm{Star}_1[1]$.

Proof. First we note that for a given point $\xi \in \Delta$ the family of functions

$$h_{\xi,n}(z) = \frac{1 - F^n(z)}{1 - F^n(\xi)} \tag{5.4.3}$$

is a normal family. Indeed, since

$$-\pi < \arg(1 - F^n(z)) - \arg(1 - F^n(\xi)) < \pi,$$

we have that each $h_{\xi,n}$ cannot obtain values in the left-hand side of the real axis.

Therefore, by Montel's Theorem, $\{h_{\xi,n}\}_{n=0}^{\infty}$ is normal, i.e., there is a subsequence $\{h_{\xi,n_k}\}_{k=1}^{\infty}$ which converges locally uniformly on Δ either to a holomorphic function h_ξ or to infinity.

However, the last case is impossible because of the obvious equality

$$h_{\xi,n}(\xi) = 1. \tag{5.4.4}$$

Setting, in particular, $\xi = 0$ we have

$$h_0(0) = 1. \tag{5.4.5}$$

Now it follows by the Hurwitz Theorem that h_0 is either a univalent function on Δ, or a constant. To see that the last case is again impossible, we differentiate $h_{\xi,n}(z)$ to obtain

$$h'(z) = \lim_{k\to\infty} \frac{(F^{n_k})'(z)}{F^{n_k}(\xi) - 1} \tag{5.4.6}$$

and

$$\frac{h'(z)}{h(z)} = \lim_{k\to\infty} \frac{(F^{n_k})'(z)}{F^{n_k}(z) - 1}. \tag{5.4.7}$$

Further, since F is embeddable, we consider the net $\{h_t\}_{t\geq 0}$ defined by

$$h_t(z) = \frac{1 - F_t(z)}{1 - F_t(\xi)}. \tag{5.4.8}$$

Since $F_n = F^n$, we have by the previous considerations that there is a sequence $\{h_{t_k}\}_{k=1}^{\infty}$ that converges to h. We will show that, in fact, the net $\{h_t\}$ itself converges to h, and h is of class $\mathrm{Star}_1[1]$. To do this, consider now the expression

$$q_t(z) = \frac{(F_t)'(z)}{F_t(z) - 1}. \tag{5.4.9}$$

If f is the infinitesimal generator of the semigroup $\{F_t\}_{t\geq 0}$, then

$$\frac{\partial F_t(z)}{\partial t} = -f(F_t(z)). \tag{5.4.10}$$

On the other hand, it is known (see, for example, [115]) that $\{F_t\}_{t\geq 0}$ also satisfies the differential equation

$$\frac{\partial F_t(z)}{\partial t} = -\frac{\partial F_t(z)}{\partial z} f(z). \tag{5.4.11}$$

Comparing (5.4.10) and (5.4.11) we obtain the functional equation

$$f(F_t(z)) = \frac{\partial F_t(z)}{\partial z} \cdot f(z). \tag{5.4.12}$$

By using (5.4.9) and (5.4.12) we get now

$$q_t(z) = \frac{f(F_t(z))}{f(z)} \cdot \frac{1}{F_t(z) - 1}. \tag{5.4.13}$$

Since $\tau = 1$ is the Denjoy–Wolff point for the semigroup $\{F_t\}_{t\geq 0}$, it follows by a continuous version of the Julia–Wolff–Carathéodory Theorem (see Section 1.3) that the angular derivative

$$\beta := f'(1) = \angle \lim_{z\to 1} f'(z) = \angle \lim_{z\to 1} \frac{f(z)}{z-1} \tag{5.4.14}$$

exists and is a nonnegative real number. Moreover,

$$\left.\frac{\partial F_t(z)}{\partial z}\right|_{z=1} = \angle \lim_{z\to 1} \frac{\partial F_t(z)}{\partial z} = e^{-t\beta}. \tag{5.4.15}$$

Since $F_1 = F$ and $\alpha = F'(1) < 1$, we obtain from (5.4.15)

$$\beta = f'(1) = -\ln \alpha > 0. \tag{5.4.16}$$

In addition, it follows by (5.4.7), (5.4.9) and (5.4.13) that

$$\frac{h'(z)}{h(z)} = \lim_{k\to\infty} q_{t_k}(z) = \frac{1}{f(z)} \lim_{k\to\infty} \frac{f(F_{t_k}(z))}{F_{t_k}(z) - 1}. \tag{5.4.17}$$

Since $F_1 \in \mathrm{Hol}(\Delta)$ is of hyperbolic type, it follows by Lemma 2.66 in [40] that for each compact $K \Subset \Delta$ there is a non-tangential approach region containing all the iterates $F_n(K)$. By using compactness arguments we have that $\{F_t(z)\}_{t\geq 0}$ converges to $\tau = 1$ non-tangentially for each $z \in \Delta$. Thus the limit

$$\lim_{k\to\infty} \frac{f(F_{t_k}(z))}{F_{t_k}(z) - 1} = f'(1) = \beta > 0.$$

So, we have by (5.4.17) that the function $h(z)$ defined by (5.4.6) satisfies the differential equation

$$\beta h(z) = h'(z) f(z). \tag{5.4.18}$$

Finally, it follows by Theorem 2.14 that h is a starlike function with respect to a boundary point $h(1) = 0 \in \partial\Delta$. Moreover, (5.4.18) and (5.4.16) imply that the Visser–Ostrowski condition

$$Q_h(1) := \angle \lim_{z\to 1} \frac{(z-1)h'(z)}{h(z)} = 1$$

holds. In other words, $h_0 \left(= \dfrac{h}{h(0)} \right) \in \mathrm{Star}_1[1]$.

In addition, it follows by the uniqueness of the solution of (5.4.18) that the limit in (5.4.6) does not depend on the choice of a convergent sequence $\{h_{n_k}\}_{k=1}^{\infty}$.

Thus we arrive at

$$h_0(z) = \lim_{n\to\infty} \frac{1 - F^n(z)}{1 - F^n(0)} = \lim_{t\to\infty} \frac{1 - F_t(z)}{1 - F_t(0)}, \tag{5.4.19}$$

and we are done. □

The following is a converse assertion.

Theorem 5.7. *Let $F \in \mathrm{Hol}(\Delta)$ have the Denjoy–Wolff point at $\tau = 1$, and let its (K-V)-function $h\,(= h_0)$ be not a constant on Δ. Then*

(i) $h(F(0)) = \alpha = F'(1) = \angle \lim\limits_{z\to 1} F'(z) < 1$;

(ii) *if $h \in \mathrm{Star}_1[1]$, then F is embeddable.*

Proof. Since h is not constant, F must be of hyperbolic type (see [110, p. 440]). Hence $\alpha = F'(1) < 1$. We have, by (5.4.19),

$$\begin{aligned} h(F(z)) &= \lim_{n\to\infty} \frac{1 - F^{n+1}(z)}{1 - F^n(0)} \\ &= \lim_{n\to\infty} \frac{1 - F^{n+1}(z)}{1 - F^{n+1}(0)} \cdot \frac{1 - F^n(F(0))}{1 - F^n(0)} = h(F(0)) \cdot h(z). \end{aligned}$$

On the other hand, since $F^n(0) = z_n$ converges to $\tau = 1$ non-tangentially, we have by the Julia–Carathéodory Theorem that

$$h(F(0)) = \lim_{n\to\infty} \frac{F^n(F(0)) - 1}{F^n(0) - 1} = \lim_{n\to\infty} \frac{F(z_n) - 1}{z_n - 1} = F'(1) =: \alpha < 1.$$

Thus h satisfies Schröder's equation

$$h(F(z)) = \alpha h(z). \tag{5.4.20}$$

Now assume that $h \in \mathrm{Star}_1[1]$. Since $h(\Delta)$ is a starlike domain, one can define the semigroup $\mathcal{S} = \{F_t\}_{t\geq 0}$ by

$$F_t(z) = h^{-1}(e^{-\beta t} h(z)), \tag{5.4.21}$$

where $\beta = -\ln \alpha > 0$. Then it follows by (5.4.20) and (5.4.21) that

$$F_1(z) = h^{-1}(e^{-\beta} h(z)) = h^{-1}(\alpha h(z)) = F(z).$$

The theorem is proved. □

Remark 5.6. Actually, in the proof of Theorem 2 (ii) we used only the fact that the (K-V)–function h of F is a starlike function with respect to a boundary point. Using, in particular, Lemma 3.2, one can formulate the following characterization of embeddable hyperbolic type self-mappings of the unit disk.

Corollary 5.2. *Let $F \in \mathrm{Hol}(\Delta)$ have the Denjoy–Wolff point at $\tau = 1$. The following are equivalent:*

(i) $F'(1) = \angle \lim_{z\to 1} F(z) < 1$ *and F is embeddable;*

(ii) *there is $\lambda \in \left[0, \frac{1}{2}\right]$ such that the function $s \in \mathrm{Hol}(\Delta, \mathbb{C})$ defined by*

$$s(z) = \frac{z}{(1-z)^{2-2\lambda}} \cdot h(z),$$

where h is the (K-V)–function for F, is a normalized starlike function of order λ, i.e., $s \in S^*_\lambda$.

Remark 5.7. The extremal case $\lambda = 0$ described in Corollary 5.2 (ii) is of special interest because it is connected to the Koebe function, the classical extremal function for class S^*. Indeed, for $\lambda = 0$ we have that assertion (ii) is equivalent to the following one:

(ii′) *the function s defined by*

$$s(z) = k(z)h(z),$$

where $k(z) = \frac{z}{(1-z)^2}$ is the Koebe function, is starlike with respect to the origin.

Remark 5.8. We have seen already that for $F \in \mathrm{Hol}(\Delta)$ with $F(1) = 1$, $\alpha = F'(1) < 1$, the (K-V)–function $h(z)$ is a solution of Schröder's functional equation

$$h(F(z)) = \alpha h(z).$$

If F is embeddable into a semigroup $\{F_t\}_{t\geq 0}$ generated by $f \in \mathcal{G}^+[1]$, then h is also a unique solution of the equation

$$h(F_t(z)) = e^{-t\beta} h(z), \quad t \geq 0,$$

where $\beta = f'(1) = -\log \alpha$, normalized by the condition $h(0) = 1$. Also $h(1) := \angle \lim\limits_{z\to 1} h(z) = 0$.

The uniqueness can be seen by the uniqueness of the solution of (5.4.18) represented by

$$h(z) = \exp\left(\int_0^z \frac{\beta dz}{f(z)} \right). \tag{5.4.22}$$

Since (5.4.22) is equivalent (5.4.18), this representation of the (K-V)–function with some $f \in \mathcal{G}[1]$ with $\beta = f'(1) \, (= -\ln \alpha) > 0$ is also sufficient for F to be embeddable.

By using the representations of class $\mathcal{G}[1]$ given in Theorems 2.7 and 2.13, one can formulate the following criteria for a self-mapping F of hyperbolic type to be embeddable.

Corollary 5.3. *Let $F \in \mathrm{Hol}(\Delta)$ with $F(1) = 1$, $\alpha = F'(1) < 1$, and let $h \in \mathrm{Hol}(\Delta, \mathbb{C})$ be its (K-V)–function defined by*

$$h(z) = \lim_{n\to\infty} \frac{1 - F^n(z)}{1 - F^n(0)}.$$

The following assertions are equivalent:

(i) *F is embeddable;*

(ii) *there exists $p \in \mathrm{Hol}(\Delta, \mathbb{C})$ with $\operatorname{Re} p \geq 0$ and $\angle \lim\limits_{z\to 1} \frac{p(z)}{z-1} = p'(1) = (\ln \alpha)^{-1}$ such that h admits the representation*

$$h(z) = \exp\left(-\int_0^z \frac{\beta p(z) dz}{(1-z)^2} \right), \quad \beta = -\log \alpha;$$

(iii) *for any $c \in (0, -\frac{1}{2}\ln \alpha)$ there is $G \in \mathrm{Hol}(\Delta)$ with $G(1) = 1$ and $\gamma = G'(1) = \angle \lim\limits_{z\to 1} G(z) = \frac{2c}{\beta}$, $\beta = -\ln \alpha$, such that the (K-V)–function h admits the representation*

$$h(z) = \exp\left(-\frac{2}{\gamma} \int_0^z \frac{1 - G(z)}{(1-z)^2(1 + G(z))} dz \right).$$

Proof. The equivalence of assertions (i) and (ii) is just an interpretation of formula (5.4.22) in the spirit of the Berkson–Porta representation of class $\mathcal{G}[1]$:

$$f(z) = -(1-z)^2 q(z) \tag{5.4.23}$$

with $\operatorname{Re} q \geq 0$ everywhere.

Setting $p(z) = \frac{1}{q(z)}$, we get by (5.4.23) that

$$0 < \beta := f'(1) = \lim_{z\to 1} \frac{f(z)}{z-1} = -\lim_{z\to 1} \frac{(z-1)}{p(z)} = -\frac{1}{p'(1)} = -\ln \alpha.$$

In turn, (5.4.23) implies that (ii) is equivalent to (iii) when we present $q(z)$ in the form

$$q(z) = c\,\frac{1+G(z)}{1-G(z)},$$

where $G \in \operatorname{Hol}(\Delta)$ is a self-mapping of Δ.

Since $c \in (0, \frac{\beta}{2})$ we have

$$0 < \beta = \angle \lim_{z\to 1} \frac{f(z)}{z-1} = c \cdot \lim_{z\to 1} \frac{(1-z)}{1-G(z)} \cdot (1+G(z)).$$

Since $|G(z)| < 1,\ z \in \Delta$, we have that the limit

$$\gamma = \angle \lim_{z\to 1} \frac{1-G(z)}{1-z} \tag{5.4.24}$$

exists and $\gamma = \frac{2c}{\beta} < 1$.

The reverse considerations complete our proof. □

5.5 Pommerenke's and Baker–Pommerenke's linearization models for semigroups with a boundary sink point

5.5.1 Pommerenke's linearization model for automorphic type mappings

For a pair z, w in Δ by $\rho(z,w)$ we denote the hyperbolic distance between z and w, where $\rho(\cdot,\cdot)$ is defined by (2.2.3). Assume that $F \in \operatorname{Hol}(\Delta)$ has a boundary Denjoy–Wolff point. It can be shown that if for some point $z_0 \in \Delta$ the orbit $\{F^n(z_0)\}_{n=0}^{\infty}$ satisfies the condition $\lim_{n\to\infty} \rho\left(F^{n+1}(z_0), F^n(z_0)\right) = 0$, then the same condition holds for all $z \in \Delta$, i.e., $\lim_{n\to\infty} \rho\left(F^{n+1}(z), F^n(z)\right) = 0$.

Definition 5.4. *Given a parabolic mapping $F \in \mathrm{Hol}(\Delta)$, one says that F is of* **automorphic type** *(or of* **positive hyperbolic step***) if for some $z \in \Delta$ (hence for all $z \in \Delta$)*

$$\lim_{n\to\infty} \rho\left(F^{n+1}(z), F^n(z)\right) > 0.$$

Otherwise, if for some $z \in \Delta$ (hence for all $z \in \Delta$)

$$\lim_{n\to\infty} \rho\left(F^{n+1}(z), F^n(z)\right) = 0,$$

F is said to be of **non-automorphic type** *(or of* **zero hyperbolic step***).*

Let $F \in \mathrm{Hol}(\Delta)$ have the boundary Denjoy–Wolff point $\tau = 1$. We study a linearization model

$$h(F(z)) = \varphi(h(z)), \tag{5.5.1}$$

where φ is an affine mapping in $\Pi_+ = \{w \in \mathbb{C} : \ \mathrm{Re}\, w \geq 0\}$ of the form $\varphi(w) = aw + b$ and $h : \Delta \mapsto \Pi_+$ is a conjugation for F and φ.

Theorem 5.8 (Pommerenke). *Let F be a holomorphic self-mapping of Δ of automorphic type with $F(1) = 1, \ F'(1) = \alpha \leq 1$. Then*

(i) *the limit $h(z) = \lim_{n\to\infty} h_n(z), \ z \in \Delta$ exists, where*

$$h_n(z) = d(F^n(0)) \left[\frac{1 + F^n(z)}{1 - F^n(z)} - \frac{2i \,\mathrm{Im}\, F^n(0)}{|1 - F^n(0)|^2}\right]$$

$$\text{with} \quad d(z) = \frac{|1 - z|^2}{1 - |z|^2},$$

and $h : \Delta \mapsto \Pi_+$ with $h(0) = 1$;

(ii) *the function h satisfies the following (Abel–Schröder) functional equation*

$$h(F(z)) = \frac{1}{\alpha} h(z) + ib, \quad b \in \mathbb{R},$$

where $b = \lim_{n\to\infty} 2d(F^n(0)) \left(\frac{1}{1-F^{n+1}(0)} - \frac{1}{1-F^n(0)}\right) \neq 0$.

Remark 5.9. As a matter of fact, if F is of non-automorphic type, then the limit $h(z) = \lim_{n\to\infty} h_n(z)$ also exists, but $h(z) \equiv 1$ and b in (ii) is zero. So one concludes that F *is of automorphic type if and only if h is not constant.*

Remark 5.10. If F is of hyperbolic type, one can show that the function h defined in Theorem 5.8 is connected to the function

$$\tilde{h}(z) = \lim_{n\to\infty} \frac{1 - F^n(0)}{1 - F^n(z)}$$

(compare with the (K-V)-function defined by (5.4.19)) by the equality

$$h(z) = 1 + A(\tilde{h}(z) - 1),$$

where

$$A = \lim_{n\to\infty} \frac{2\left(1 - \overline{F^n(0)}\right)}{1 - |F^n(0)|^2}.$$

Hence, number b in Theorem 5.8 is defined by

$$b = \operatorname{Im}\left(\frac{1}{\alpha} - 1\right)(A - 1) = \left(\frac{1}{\alpha} - 1\right) \lim_{n\to\infty} \frac{2 \operatorname{Im} F^n(0)}{1 - |F^n(0)|^2}.$$

Remark 5.11. We observe that originally Pommerenke's limit scheme as well as Valiron's limit scheme were considered for automorphic type self-mappings of the right half-plane. From a certain point of view this makes the model more transparent. Indeed, using the Cayley transform $C(z) = \frac{1+z}{1-z}$, which maps Δ onto Π_+, one can translate the problem for a holomorphic self-mapping $\phi : \Pi_+ \mapsto \Pi_+$ defined by

$$\phi = C \circ F \circ C^{-1}. \tag{5.5.2}$$

For our aims it is also sometimes useful to pass to the associated model in Π_+. This mapping $\phi : \Pi_+ \mapsto \Pi_+$ defined by (5.5.2) is called the **iteration mapping associated with** $F : \Delta \mapsto \Delta$ (see [36]).

In this case equation (5.5.1) becomes

$$g(\phi(w)) = \varphi(g(w)), \tag{5.5.3}$$

where

$$g(w) = h(C^{-1}(w)), \tag{5.5.4}$$

$g \in \operatorname{Hol}(\Pi_+)$ is a conjugation for ϕ and φ.

It turns out that technically, problem (5.5.3) is simpler than (5.5.1).

Note that in our settings the point $w = \infty$ is the Denjoy–Wolff point for ϕ, that is, ϕ^n converges to ∞ locally uniformly on Π_+ and

$$\phi'(\infty) = \frac{1}{F'(1)} \geq 1, \tag{5.5.5}$$

where $\phi'(\infty)$ is understood as the angular derivative at ∞:

$$\phi'(\infty) = \angle \lim_{\omega\to\infty} \frac{\phi(\omega)}{\omega}. \tag{5.5.6}$$

Finally, if $\kappa = \phi'(\infty)\left(= \frac{1}{F'(1)}\right)$, then the boundary Schwarz–Wolff Lemma can be rewritten in the form

$$\operatorname{Re}\phi(\omega) \geq \kappa \operatorname{Re}\omega, \quad \omega \in \Pi_+. \tag{5.5.7}$$

We also have that

$$\kappa = \inf_{u>0} \frac{\operatorname{Re}\phi(u+iv)}{u}, \tag{5.5.8}$$

where $u + iv = \omega \in \Pi_+$.

Moreover, if for some $\omega_0 \in \Pi_+$ equality (5.5.7) holds, then ϕ must be an automorphism of Π_+.

Let $\phi : \Pi_+ \mapsto \Pi_+$ be such that $\phi(\infty) = \infty$ and $\phi'(\infty) := \angle \lim\limits_{w\to\infty} \frac{\phi(w)}{w} = \kappa \geq 1$, i.e., $\phi(w) = \kappa w + p(w)$ with $\operatorname{Re} p \geq 0$ and $\angle \lim\limits_{w\to\infty} \frac{p(w)}{w} = 0$.

If $\kappa > 1$, then ϕ (respectively, F in (5.5.2)) is of hyperbolic type. If $\kappa = 1$, then ϕ (respectively, F in (5.5.2)) is of parabolic type.

As in the disk, hyperbolic type mappings are of **automorphic type**. At the same time, the class of parabolic mappings falls into two subclasses of **automorphic type** and **non-automorphic type**, see[110].

Let us explain more precisely what this means. Denote $w_n = \phi^n(1) = u_n + v_n i$ and consider $q_n = \dfrac{w_{n+1} - w_n}{w_{n+1} + \overline{w_n}} = \dfrac{w_{n+1} - w_n}{(w_{n+1} - w_n) + 2u_n}$.

Since $|q_n|$ is the pseudohyperbolic distance in Π_+, it follows by the Schwarz–Pick Lemma that the sequence $\{|q_n|\}_{n=0}^{\infty}$ is non-increasing: $|q_n| \geq |q_{n+1}|$, hence $\lim\limits_{n\to\infty} |q_n|$ exists.

In this case *a mapping* $\phi : \Pi_+ \mapsto \Pi_+$ *is of automorphic type (respectively, non-automorphic type) if* $\lim\limits_{n\to\infty} q_n \neq 0$ *(respectively,* $\lim\limits_{n\to\infty} q_n = 0$*).*

The property of $\phi : \Pi_+ \mapsto \Pi_+$ to be of automorphic type means that the pseudohyperbolic distance of its orbit at $w = 1$ (hence, at any $w \in \Pi_+$) is separated from zero. In this case $|\arg w_n|$ must tend to $\pi/2$. However, it is not clear whether the sequence $u_n = \operatorname{Re} w_n$ is bounded. Note that by the Julia–Wolff–Carathéodory Theorem the sequence $\{u_n\}$ is increasing. Its boundedness means that w_n converges to infinity in a sense strongly tangentially. We will see in Section 7.3 that under some smoothness conditions, the strongly tangential convergence of a semigroup is equivalent to its property to be of automorphic type.

Now we consider the expression

$$L(w_0) := \lim_{n\to\infty} \operatorname{Re}\phi_n(w_0) \in (0, +\infty].$$

It is just a computation to show that $L(w_0) < \infty$ if and only if

$$\inf_n \frac{|1 - F^n(z_0)|^2}{1 - |F^n(z_0)|^2} > 0,$$

where $F = C^{-1} \circ \phi \circ C$ and $z_0 = C^{-1}(w_0)$. In other words, the forward orbit $\{F^n(z_0)\}_{n=0}^{\infty}$ converges to $\tau = 1$ **strongly tangentially** in the sense: it lies outside a circle internally tangent to $\partial\Delta$ at $\tau = 1$.

The following assertion shows that the property of $L(w_0)$ to be finite (or, equivalently, the property of strongly tangential convergence) does not depend on the initial point of the orbit.

Theorem 5.9 ([36]). *Let $\phi \in \mathrm{Hol}(\Pi_+)$ be of parabolic type with the Denjoy–Wolff point at ∞. Then for each pair $w_1, w_2 \in \Pi_+$ the following limit always exists and is positive:*

$$L\,(= L(w_1, w_2) := \lim_{n\to\infty} \frac{\operatorname{Re}\phi(w_1)}{\operatorname{Re}\phi(w_2)} > 0.$$

Consequently, if $L(w_0) < \infty$ for some $w_0 \in \Pi_+$, then $L(w) < \infty$ for each $w \in \Pi_+$.

In addition, if F is of non-automorphic type (of zero hyperbolic step), then $L = 1$.

5.5.2 Baker–Pommerenke's model for non-automorphic type self-mappings

Now let $F : \Delta \mapsto \Delta$ be of non-automorphic type, that is, $F(\tau) = \tau$ for some $\tau \in \partial\Delta$ and $\rho(F^{n+1}(z), F^n(z)) \to 0$ as $n \to \infty$ for some z (hence, for all z) in Δ. In this situation F must be of parabolic type, i.e., $F'(\tau) = 1$. Again for simplicity we set $\tau = 1$. As we already know, the limit function defined in Theorem 5.8 is just a constant; therefore, this model does not work. Moreover, sometimes there is no way to construct a linear-fractional transformation of the unit disk that leaves it invariant and is conjugate to F.

Baker and Pommerenke [12] constructed a linear-fractional model by using another limit function that is not necessarily a self-mapping of a disk or a half-plane.

Their original model was again given for Π_+. We, however, formulate it for Δ, according to our object of interest.

Theorem 5.10 (Baker–Pommerenke). *Let $F : \Delta \mapsto \Delta$ be a parabolic mapping of non-automorphic type with $F(1) = F'(1) = 1$. Then the limit*

$$g(z) = \lim_{n\to\infty} \frac{1 - F^{n+1}(0)}{1 - F^n(0)} \cdot \frac{F^n(z) - F^n(0)}{F^{n+1}(0) - F^n(0)} \tag{5.5.9}$$

exists, and g is a solution of Abel's equation

$$g(F(z)) = g(z) + 1.$$

Remark 5.12. In each one of the following two cases:

(a) the unrestricted limit $\lim\limits_{z\to 1} \dfrac{F(z) - 1}{z - 1}$ exists (so, it is equal to 1);

(b) the orbit $\{F^n(0)\}_{n=0}^{\infty}$ converges to 1 non-tangentially,

the limit in (5.5.9) coincides with

$$g(z) = \lim_{n\to\infty} \frac{F^n(z) - F^n(0)}{F^{n+1}(0) - F^n(0)} \tag{5.5.10}$$

(see also Sections 5.5.3 and 7.3.1 below). In particular, this limit scheme is applicable to the study of generating functions of regular branching processes. In this study not only the first derivative at 1, $F'(1) = 1$, has a probabilistic interpretation (the expectation); the second derivative $F''(1)$ plays role of the variance of the branching process. The following result indicates the strong relationship between the variance and the Kœnigs type function g defined by (5.5.9) or (5.5.10).

Theorem 5.11 (see [111]). *Let the generating function F of a branching process be not of the form $F(z) = \tilde{F}(z^m)$, $m \geq 2$. Assume that $F'(1) = 1$ and $F''(1) \in [0, +\infty]$. Then $\frac{g^{(k)}(0)}{k!}$ tends to $\frac{2}{F''(1)}$ as $k \to \infty$. If $F''(1)$ is finite, then*

$$\lim_{z\to 1} (1 - z)g(z) = \frac{2}{F''(1)}.$$

Explicit uniform estimates in terms of $p_0 = F(0) = P(Z(1) = 0)$, instead of the variance $F''(1)$, are given in [37].

Proposition 5.1. *Let g be defined by (5.5.10), where F is the generating function of a branching process. Then for all $x \in [0, 1)$,*

$$\frac{1}{1 - F^n(x)} \geq \frac{1}{1 - x} + np_0,$$

and for all $z \in \Delta$,

$$|g(z)| \leq \frac{1}{p_0(1 - |z|)}.$$

5.5.3 Higher order angular differentiability at boundary fixed points. A unified model

Definition 5.5. *Let $f \in \mathrm{Hol}(\Delta, \mathbb{C})$ and let $\tau \in \partial\Delta$. We say that f is of class $C_A^{p+\varepsilon}(\tau)$, where $p \in \mathbb{N} \cup \{0\}$ and $\varepsilon \in [0, 1)$, if it admits the expansion*

$$f(z) = \sum_{j=0}^{p} a_j (z - \tau)^j + \gamma(z), \quad z \in \Delta, \tag{5.5.11}$$

where $a_0, \dots, a_p \in \mathbb{C}$ and $\gamma \in \mathrm{Hol}(\Delta, \mathbb{C})$ with

$$\angle \lim_{z\to\tau} \frac{\gamma(z)}{(z - \tau)^{p+\varepsilon}} = 0.$$

We say that $f \in C^{p+\varepsilon}(\tau)$ if this limit is taken in the whole disk.

It is clear that the numbers $a_0, \ldots, a_p$ appearing in the above expression are necessarily unique. If $f \in C_A^{p+\varepsilon}(\tau)$, it is natural to define (angular) boundary derivatives $f^{(j)}(\tau) = j!a_j$, $j = 0, \ldots, p$. Moreover, it can be shown by induction that $f \in C_A^n(\tau)$ $(n > 1)$ if and only if the limit $\angle \lim_{z \to \tau} f^{(n)}(z) := f^{(n)}(\tau)$ exists finitely (see, for example, [18] and [51]).

Consider a mapping $F \in \text{Hol}(\Delta)$ with a boundary fixed point $\tau \in \partial\Delta$. According to the Julia–Wolff–Carathéodory Theorem (see Section 1.3), we easily see that $F \in C_A^1(\tau)$ if and only if F has a finite angular derivative at τ. Clearly, when this happens, the corresponding number a_1 is exactly $F'(\tau)$. In particular, parabolic and hyperbolic members of $\text{Hol}(\Delta)$ are of class $C_A^1(\tau)$ at their Denjoy–Wolff point τ (see Theorem 1.10).

It is easy to verify that F belongs to $C_A^2(\tau)$ if and only if $F \in C_A^1(\tau)$ and the following angular limit exists finitely:

$$L := \angle \lim_{z \to \tau} \frac{F(z) - \tau - F'(\tau)(z - \tau)}{(z - \tau)^2}.$$

Now the corresponding numbers a_1 and a_2 are just $a_1 = F'(\tau)$ and $a_2 = L = \frac{F''(\tau)}{2}$. Note that if F is of parabolic type with the Denjoy–Wolff point $\tau = 1 \in \partial\Delta$, then $F \in C_A^2(1)$ if and only if the following angular limit exists finitely:

$$\angle \lim_{z \to 1} \frac{F(z) - z}{(z - 1)^2}.$$

For such mappings Bourdon and Shapiro have suggested in [18] a unified model for both classes of automorphic and non-automorphic parabolic mappings. This method was studied and developed in detail by Contreras, Díaz-Madrigal and Pommerenke. Namely, they proved the following results (see [36]).

Theorem 5.12. *Let $F \in \text{Hol}(\Delta)$ be a parabolic self-mapping of automorphic type (of positive hyperbolic step) with the Denjoy–Wolff point $\tau = 1 \in \partial\Delta$. Likewise, let ϕ be the associated mapping with F in Π_+. Then the following are equivalent:*

(1) *F belongs to $C_A^2(1)$.*

(2) *For some (hence, any) $w_0 \in \Pi_+$, the normalized sequence*

$$g_n(w) := \phi_n(w) - \phi_n(w_0), \quad w \in \Pi_+,$$

converges in the compact open topology to some non-constant mapping $g \in \text{Hol}(\Pi_+, \mathbb{C})$ (of course, g depends on w_0).

If any of the above statements hold, then there exists a real number $\nu \neq 0$, such that $g \circ \phi = g + i\nu$.

Theorem 5.13. *Let $F \in \text{Hol}(\Delta)$ be of non-automorphic type (of zero hyperbolic step) with the Denjoy–Wolff point $\tau = 1 \in \partial\Delta$. Likewise, let ϕ be the map associated with F in Π_+. Then the following are equivalent:*

(1) *F belongs to* $C_A^2(1)$ *and* $\operatorname{Re} F''(1) > 0$.

(2) *For some (hence, any)* $w_0 \in \Pi_+$, *the normalized sequence*

$$g_n(w) := \phi_n(w) - \phi_n(w_0), \quad w \in \Pi_+,$$

converges in the compact open topology to a non-constant function $g \in \operatorname{Hol}(\Pi_+, \mathbb{C})$ *(of course, g depends on w_0) such that* $g \circ \phi = g + a$ *with* $\operatorname{Re} a > 0$.

Combining these two assertions with Theorem 4.4 (p. 52) in [18] one can conclude the following.

Theorem 5.14. *Let $F \in \operatorname{Hol}(\Delta)$ be a self-mapping of Δ of parabolic type with the Denjoy–Wolff point $\tau = 1 \in \partial\Delta$ and let F belong to $C^3(1)$ with $F''(1) \neq 0$. Then the sequence of functions*

$$h_n(z) = \frac{1 + F^n(z)}{1 - F^n(z)} - \frac{1 + F^n(z_0)}{1 - F^n(z_0)}$$

converges in the compact open topology to a non-constant function $h \in \operatorname{Hol}(\Delta, \mathbb{C})$ (of course, h depends on z_0), which satisfies Abel's equation

$$h \circ F = h + a$$

with $a = F''(1)$. Moreover,

(a) *if F is automorphic, then* $\operatorname{Re} a = 0$.

(b) *if F is non-automorphic, then* $\operatorname{Re} a > 0$.

5.6 Embedding property via Abel's equation

We know that if $\tau = 1$ is the Denjoy–Wolff point of $F \in \operatorname{Hol}(\Delta)$ (so, $F'(1) \le 1$), then in both hyperbolic ($F'(1) < 1$) and parabolic ($F'(1) = 1$) cases, Abel's functional equation

$$g(F(z)) = g(z) + 1 \tag{5.6.1}$$

can be solved. This fact may be used to study the semigroup embedding problem.

Theorem 5.15. *Let $F \in \operatorname{Hol}(\Delta)$, $F(1) = 1$, be not an automorphism. Then F is embeddable if and only if there exists a solution g of Abel's functional equation* (5.6.1)

$$g(F(z)) = g(z) + 1,$$

which is holomorphic in Δ function satisfying the condition

$$\operatorname{Re}\left[(1 - z)^2 g'(z)\right] > 0$$

for all $z \in \Delta$.

Proof. Let $F \in \mathrm{Hol}(\Delta)$ be embedded into a one-parameter semigroup $\{F_t\}_{t\geq 0} \subset \mathrm{Hol}(\Delta)$. Its infinitesimal generator can be represented in the form

$$f(z) = -(1-z)^2 p(z),$$

where p is a holomorphic function on Δ with a positive real part. In this case we define

$$g(z) = \int_0^z \frac{d\zeta}{(1-\zeta)^2 p(\zeta)}.$$

Note that

$$g'(z) = \frac{1}{(1-z)^2 p(z)},$$

and

$$\mathrm{Re}\left[(1-z)^2 g'(z)\right] = \mathrm{Re}\,\frac{1}{p(z)} > 0$$

for $z \in \Delta$. For fixed $z \in \Delta$ we consider $\zeta(t) = g(F_t(z)),\ t \geq 0$. Now we derive

$$\frac{d\zeta}{dt} = g'(F_t(z))\frac{\partial}{\partial t}F_t(z) = g'(F_t(z))(1-F_t(z))^2 p(F_t(z)) = 1.$$

From here and the initial condition $\zeta(0) = g(z)$, it follows that

$$g(F_t(z)) = g(z) + t.$$

This equality means that g is the required solution of Abel's equation (5.6.1).

Conversely, suppose that g is holomorphic in Δ function satisfying equation (5.6.1) and $\mathrm{Re}[(1-z)^2 g'(z)] > 0$ for $z \in \Delta$. Noting that for each c the function $g(z) + c$ satisfies the same conditions, we can assume that $g(0) = 0$. Therefore, $g \in \Sigma[1]$ by Lemma 3.8. This means that together with each point w_0, the image domain $g(\Delta)$ contains the ray $w(t) = w_0 + t,\ t \geq 0$. Consequently, the family $\{F_t\}_{t\geq 0}$, where

$$F_t(z) = g^{-1}(g(z) + t),$$

is well defined on Δ and is a one-parameter continuous semigroup on Δ. Since $F_1(z) = g^{-1}(g(z) + 1) = F(z)$, we are done. □

A detailed survey on Abel's equation for discrete time semigroups of parabolic type has recently presented in [37] (see also [108]). The existence, uniqueness and construction problems are discussed there.

Chapter 6

Rigidity of Holomorphic Mappings and Commuting Semigroups

The problem of finding conditions for a holomorphic mapping F to coincide identically with a given holomorphic mapping G when they behave similarly on some subset of their common domain of definition, has been studied by many mathematicians.

The following assertions are classical:

♦ *If F and G are holomorphic in Δ and $F = G$ on a subset of Δ that has a non-isolated point, then $F \equiv G$ on Δ (Vitali's uniqueness principle).*

♦ *If F and G are holomorphic in Δ and continuous on $\overline{\Delta}$, and $F = G$ on some arc γ of the boundary $\partial\Delta$, then $F \equiv G$ on Δ.*

From the point of view of complex dynamics it is natural to look for conditions on derivatives of F and G at specific fixed points that will yield the conclusion that $F \equiv G$.

If $G = I$ is the identity mapping on Δ, then the classical Schwarz–Pick lemma asserts:

♦ *Let $F \in \mathrm{Hol}(\Delta)$ and let $\tau \in \Delta$ be a fixed point of F. Then the equalities $F(\tau) = G(\tau)$ $(= \tau)$ and $F'(\tau) = G'(\tau)$ $(= 1)$ imply that $F \equiv G$ on Δ.*

The same conclusion holds for an arbitrary holomorphic mapping G on Δ if F commutes with G and satisfies the conditions $F(\tau) = G(\tau) = \tau \in \Delta$ and $F'(\tau) = G'(\tau) \neq 0$ (see, for instance, [21]).

Different "identity principles" have recently been studied by several mathematicians under suitable boundary conditions (see, for example, [25, 13, 21, 139, 140]). Generally speaking, the following cases have been considered:

(A) G is a holomorphic self-mapping of Δ of a specific form, e.g., the identity, a constant mapping (say, zero), an affine mapping, an automorphism of Δ or a linear fractional transformation.

(B) G is an arbitrary self-mapping of Δ and F commutes with G, i.e., $F \circ G = G \circ F$.

(C) In addition, one can consider certain families of holomorphic mappings and look for conditions which ensure that they either coincide or have similar structures.

6.1 The Burns–Krantz theorem

In this section, we give a brief review on rigidity of holomorphic functions related to the remarkable theorem of Burns and Krantz (see [25]).

In the next section, we will show how to prove these results by using the generation theory for one-parameter semigroups.

Let $F \in \mathrm{Hol}(\Delta)$ and let $\tau \in \Delta$ be an interior fixed point of F. In this case, if F coincides with the identity mapping I at τ up to the first derivative, i.e., $F(\tau) = \tau$ and $F'(\tau) = 1$, then $F \equiv I$ on Δ by the classical Schwarz–Pick lemma. Moreover, if $|F'(\tau)| = 1$ and $F'(\tau) \neq 1$, then F is an elliptic automorphism of Δ.

In the boundary case ($\tau \in \partial\Delta$) a uniqueness theorem in the spirit of the uniqueness part of the classical Schwarz lemma was established by Burns and Krantz [25]. We formulate their result as follows:

Theorem 6.1. *Let $F \in \mathrm{Hol}(\Delta)$. If the unrestricted limit*

$$\lim_{z\to 1} \frac{F(z) - z}{(z-1)^3} = 0,$$

then $F \equiv I$ on Δ.

This result was improved by Tauraso and Vlacci [140] (see also Bracci, Tauraso and Vlacci [21]). They have shown that it is sufficient to require the vanishing of such a limit at a boundary point τ only in the radial sense, i.e., $\lim\limits_{r\to 1^-} \dfrac{F(r\tau) - r\tau}{(1-r)^3} = 0$. Furthermore, it has been shown by Kriete and MacCluer [94] that the function F in this limit can be replaced with its real part. Namely, for $\tau = 1$, the equality $\liminf\limits_{r\to 1^-} \dfrac{\mathrm{Re}\, F(r) - r}{(1-r)^3} = 0$ implies that $F \equiv I$ on Δ. More recently, Baracco, Zaitsev and Zampieri [13] have proved that it is sufficient to consider such a limit only on a sequence in Δ. We quote their result as follows:

Theorem 6.2. *Let $F \in \mathrm{Hol}(\Delta)$. If for some sequence $\{z_n\}_{n=1}^{\infty} \subset \Delta$, converging non-tangentially to $\tau = 1$,*

$$\lim_{n\to\infty} \frac{F(z_n) - z_n}{(1-z_n)^3} = 0,$$

then $F \equiv I$ *on* Δ.

A continuous version of the Burns–Krantz theorem is presented in [134]:

♦ *Let* $F \in \mathrm{Hol}(\Delta)$ *and let*

$$\liminf_{r \to 1^-} \frac{\operatorname{Re} F(r) - r}{(1-r)^3} = \mu.$$

Then $\mu \geq 0$ *and*

$$|F(z) - z|^2 \leq \mu \cdot K(z), \quad z \in \Delta,$$

for some continuous function $K : \Delta \mapsto \mathbb{R}^+$.

In view of these results, the following problem arises:

♢ find conditions on the first derivatives of a holomorphic self-mapping F of the unit disk at its boundary Denjoy–Wolff point τ (say, $\tau = 1$) which ensure that F is an affine self-mapping of Δ. From the point of view of the Burns–Krantz theorem, it seems plausible to conjecture that such conditions (if they do exist) might be the following: $F \in C_A^3(1)$, $F'(1) = \alpha$, $F''(1) = F'''(1) = 0$, and then $F(z) = \alpha z + (1 - \alpha)$. However, the following counterexample shows that this conjecture is false.

Example 6.1. Let $F(z) = \frac{1}{2}(z+1) + \frac{1}{20}(z-1)^4$. Then $F \in \mathrm{Hol}(\Delta) \cap C_A^3(1)$ and $\tau = 1$ is its Denjoy–Wolff point. Moreover, $F''(1) = F'''(1) = 0$. However, F is not affine.

At the same time, if a function $F \in \mathrm{Hol}(\Delta)$ is such that

$$\frac{|1 - F(z)|^2}{1 - |F(z)|^2} \leq \frac{\alpha |1 - z|^2}{\alpha(1 - |z|^2) + (1 - \alpha)|1 - z|^2}, \quad z \in \Delta, \tag{6.1.1}$$

where $\alpha = F'(1)$, then these conditions do imply that F is affine. Note that, for a parabolic self-mapping F (i.e., if $\alpha = 1$), this inequality holds automatically.

Theorem 6.3 ([134]). *Let* $F \in \mathrm{Hol}(\Delta) \cap C_A^3(\tau)$ *be a holomorphic self-mapping of* Δ *with the Denjoy-Wolff point* $\tau = 1$ *and let* $F''(1) = F'''(1) = 0$. *Then* F *is an affine mapping of the form* $F(z) = \alpha z + 1 - \alpha$ $(\alpha = F'(1))$ *if and only it satisfies condition* (6.1.1), *i.e.,*

$$F(D(1,k)) \subseteq D\left(1, \frac{k\alpha}{1 + k(1-\alpha)}\right), \quad k > 0. \tag{6.1.2}$$

It follows from the Julia–Wolff–Carathéodory theorem that the geometrical condition (6.1.2) is equivalent to the fact that F maps the whole disk Δ into the horocycle $D\left(1, \frac{\alpha}{1-\alpha}\right)$.

Sufficient conditions regarding the behavior of a holomorphic self-mapping of Δ in a neighborhood of its boundary Denjoy–Wolff point which ensure that it is an automorphism of Δ can be found in [36] and [140]. To formulate them, we denote by S_F **the Schwarzian derivative** of F, i.e., $S_F(z) := \frac{F'''(z)}{F'(z)} - \frac{3}{2}\left(\frac{F''(z)}{F'(z)}\right)^2$. Tauraso and Vlacci mentioned in [140] that if $F \in \mathrm{Hol}(\Delta) \cap C_A^3(1)$, then the equalities $F'(1) = 1$, $\operatorname{Re} F''(1) = 0$ and $\operatorname{Re} S_F(1) = 0$ imply that F is nothing but the parabolic automorphism given by

$$F(z) = \lambda \frac{z+a}{1+\bar{a}z},$$

where $\lambda = \frac{2-ib}{2+ib}$, $b = \operatorname{Im} F''(1)$ and $a = \frac{ib}{2-ib}$. A more general assertion, valid for all types of automorphisms of Δ, is pointed out in [36] by Contreras, Díaz-Madrigal and Pommerenke.

Theorem 6.4. *Let $F \in \mathrm{Hol}(\Delta) \cap C_A^3(\tau)$ be a holomorphic self-mapping of Δ with the Denjoy–Wolff point $\tau \in \partial\Delta$ and let $\operatorname{Re}[\tau F''(\tau)] = \alpha(\alpha - 1)$. Then the Schwarzian derivative $S_F(\tau) = 0$ if and only if $F = \Phi$ is the automorphism*

$$\Phi(z) := \lambda \frac{z+a}{1+\bar{a}z},$$

where $a = \dfrac{\overline{F''(\tau)}}{\tau F''(\tau) - 2\alpha^2}$ *and* $\lambda = \dfrac{\tau F''(\tau) - 2\alpha^2}{\overline{\tau F''(\tau) - 2\alpha^2}}$.

In particular,

(1) *a nontrivial (i.e., $F \neq I$) holomorphic map $F \in \mathrm{Hol}(\Delta)$ is a parabolic automorphism if and only if there exists $\tau \in \partial\Delta$ such that $F \in C_A^3(\tau)$ and*

$$F(\tau) = \tau, \quad F'(\tau) = 1, \quad \operatorname{Re}(\tau F''(\tau)) = 0, \quad \text{and} \quad S_F(\tau) = 0;$$

(2) *a holomorphic map $F \in \mathrm{Hol}(\Delta)$ is a hyperbolic automorphism if and only if there exist $\tau \in \partial\Delta$ and $\alpha \in (0,1)$ such that $F \in C_A^3(\tau)$ and*

$$F(\tau) = \tau, \quad F'(\tau) = \alpha, \quad \operatorname{Re}(\tau F''(\tau)) = \alpha(\alpha - 1), \quad \text{and} \quad S_F(\tau) = 0.$$

Remark 6.1. As a matter of fact, since Φ is the general form of automorphisms with their Denjoy–Wolff point at τ, Theorem 6.4 can be reduced to Theorem 6.2 when we apply to F the automorphism Φ^{-1}.

Since each automorphism of Δ is a linear-fractional transformation (LFT), the following more general problem also arises:

◇ find conditions concerning the behavior of $F \in \mathrm{Hol}(\Delta)$ in a neighborhood of a boundary point $\tau \in \partial\Delta$ which ensure that F is an LFT, i.e., that it is of the form $F(z) = \frac{az+b}{cz+d}$, with complex coefficients a, b, c, d.

Linear-fractional transformations are also of intrinsic interest because of their deep connections with and applications to Markov branching processes, the theory of composition operators, and operator theory in Pontryagin and Krein spaces.

An answer to the above problem was independently given in [140] and [134].

Theorem 6.5. *Let* $F \in \mathrm{Hol}(\Delta) \cap C_A^3(1)$ *with* $F(1) = 1$, *and let* $a := \frac{1}{\alpha^2}(F''(1) + \alpha(1-\alpha))$, *where* $\alpha = F'(1)$. *Then*

(i) $\mathrm{Re}\, a \geq 0$;

(ii) $F(\Delta) \subset D\left(1, \frac{1}{\mathrm{Re}\, a}\right)$ *if and only if*

$$F(D(1,k)) \subseteq D\left(1, \frac{\alpha k}{1 + \alpha k \,\mathrm{Re}\, a}\right) \quad \textit{for all} \quad k > 0; \tag{6.1.3}$$

(iii) *if the inclusions in* (ii) *hold, then* $\mathrm{Re}\, S_F(1) \leq 0$ *and* $\mathrm{Im}\, S_F(1) = 0$. *Moreover,* $S_F(1) = 0$ *if and only if* F *is an LFT.*

Thus F is an automorphism of Δ if and only if $S_F(1) = 0$ and $\mathrm{Re}\, a = 0$, i.e.,

$$\mathrm{Re}\, F''(1) = \alpha(\alpha - 1). \tag{6.1.4}$$

In this case, condition (ii) just means that $F(\Delta) \subset \Delta$, i.e., it holds automatically.

Remark 6.2. Note also that the latter assertion gives conditions on a holomorphic self-mapping F of Δ at **a boundary regular fixed point** $\tau \in \partial\Delta$, **which is not necessarily the Denjoy–Wolff point** of F, to be a linear-fractional transformation. At the same time, if $\tau = 1$ is the Denjoy–Wolff point of F, then this theorem can be essentially improved. Namely, one can show (see [134]) that **it is enough to require that condition (6.1.3) holds for at least one $k > 0$ to ensure that equality $S_F(1) = 0$ implies that F must be an LFT.** Moreover, if F is a hyperbolic type mapping ($\alpha < 1$), then one can even require a weaker inclusion than (6.1.3) to obtain the same conclusion.

Remark 6.3. Observe that if equality (6.1.4) does not hold and $\alpha > 1$, then in general F can be either a hyperbolic automorphism of Δ or a dilation type mapping.

Indeed, if $\tau = 1$ is a repelling fixed point ($\alpha > 1$), then there is a point $\zeta \in \overline{\Delta}$ which is the Denjoy–Wolff point of F. If $\zeta \in \partial\Delta$, then $F'(\zeta) = \frac{1}{\alpha} < 1$ and F is an LFT having two boundary fixed points. Hence, F must be a hyperbolic automorphism of Δ. In this case, $\mathrm{Re}\, a = 0$. If $\zeta \in \Delta$, then $\mathrm{Re}\, a \neq 0$. It can be shown that, in fact, $\mathrm{Re}\, a > 0$ and each horocycle $D(1,k)$ with $k \geq \frac{\alpha - 1}{\alpha \,\mathrm{Re}\, a}$ is F-invariant. Moreover, $\frac{|1-\zeta|^2}{1-|\zeta|^2} = \frac{\alpha-1}{\alpha \,\mathrm{Re}\, a} =: k_0$, that is, the Denjoy–Wolff point ζ lies on the boundary of the F-invariant horocycle $D(1, k_0)$. Hence the mapping F is a hyperbolic automorphism of the disk $D(1, k_0)$.

Remark 6.4. Actually, one can show that the condition $F \in C_A^3(1)$ in Theorems 6.3, 6.4 and 6.5 can be replaced by the existence of the finite radial derivatives at τ up to the third order.

More general rigidity results in the spirit of Theorem 6.5 which ensure that $F \in \mathrm{Hol}(\Delta)$ is a rational self-mapping of Δ are presented in [140]. Other generalizations of the Burns–Krantz theorem can be found in Arslan [9] and Chelst [28].

To proceed, we note that many rigidity principles for holomorphic functions (even not necessarily self-mappings of Δ) can be obtained by manipulating the Julia–Wolff–Carathéodory theorem. For example, the following fact is a simple consequence of this theorem (see [103]):

♦ *If $F \in \mathrm{Hol}(\Delta, \overline{\Delta})$, then the conditions $\lim_{r\to 1^-} F(r\tau) = \tau$ and $\lim_{r\to 1^-} F'(r\tau) = 0$ at some $\tau \in \partial\Delta$ imply that $F \equiv \tau$.*

Applying this fact to the function $p \in \mathrm{Hol}(\Delta, \overline{\Pi_+})$ defined by $p(z) := \dfrac{1-F(z)}{1+F(z)}$, where $\Pi_+ := \{z : \operatorname{Re} z > 0\}$ is the right half-plane and $F \in \mathrm{Hol}(\Delta, \overline{\Delta})$, we get the following assertion (cf. [103]):

♦ *Let $f \in \mathrm{Hol}(\Delta, \overline{\Pi_+})$. Then the equality*

$$\angle \lim_{z\to\tau} \frac{f(z)}{z-\tau} = 0$$

for some $\tau \in \partial\Delta$ implies that $f \equiv 0$.

As a matter of fact, these two assertions follow immediately from Hopf's lemma (see [44, p. 116]). The second assertion can also be stated in the following form:

♦ *Let $f, g \in \mathrm{Hol}(\Delta, \mathbb{C})$ be such that*

$$\operatorname{Re} f(z) \geq \operatorname{Re} g(z), \quad z \in \Delta,$$

and

$$\angle \lim_{z\to\tau} \frac{f(z)-g(z)}{z-\tau} = 0$$

for some $\tau \in \partial\Delta$; then $f \equiv g$ in Δ.

A generalization of these facts can be given as follows.

Denote by

$$A_\alpha := \left\{ |\arg w| < \frac{\alpha\pi}{2}, \quad \alpha \in (0,2] \right\}$$

the sector of opening $\alpha\pi$ with vertex at $z = 0$.

Theorem 6.6 (see [103]). *If $f \in \mathrm{Hol}(\Delta, \overline{A_\alpha})$, then the equality*

$$\angle \lim_{z\to\tau} \frac{f(z)}{(z-\tau)^\alpha} = 0$$

for some $\tau \in \partial\Delta$ implies that $f \equiv 0$.

Now we consider the class of functions $F \in \mathrm{Hol}(\Delta, \mathbb{C})$ that satisfy the so-called boundary flow-invariance condition

$$\limsup_{z \to \partial\Delta} \mathrm{Re}(F(z)\overline{z}) \leq 1. \tag{6.1.5}$$

In particular, each holomorphic self-mapping F of Δ obviously satisfies condition (6.1.5).

Theorem 6.7 (cf. [53]). *Let $F \in \mathrm{Hol}(\Delta, \mathbb{C})$ satisfy condition* (6.1.5). *If for some $\tau \in \partial\Delta$,*

$$\angle \lim_{z \to \tau} \frac{F(z) - z}{(z - \tau)^3} = 0,$$

then $F \equiv I$ on Δ.

This assertion is in a sense a generalization of the Burns–Krantz theorem. If, in addition, F can be continuously extended to the boundary $\partial\Delta$, then by setting $f = I - F$, one can reformulate it as follows:

Theorem 6.8. *Let $f \in \mathrm{Hol}(\Delta, \mathbb{C}) \cap C(\overline{\Delta})$ be such that*

$$\mathrm{Re}(f(z)\overline{z}) \geq 0, \tag{6.1.6}$$

and assume that for some $\tau \in \partial\Delta$,

$$\angle \lim_{z \to \tau} \frac{f(z)}{(z - \tau)^3} = 0.$$

Then $f \equiv 0$.

Note that inequality (6.1.6) is a criterion for a function $f \in \mathrm{Hol}(\Delta, \mathbb{C}) \cap C(\overline{\Delta})$ to be a semigroup generator. Therefore, it is natural to look for rigidity properties for generators which are not necessarily continuous on $\partial\Delta$. We will tackle this question in the next section.

An extension of Theorem 6.8 to a wider class of functions $f \in \mathrm{Hol}(\Delta, \mathbb{C})$ which are continuous on $\overline{\Delta}$ and satisfy for some $\alpha \in (0, 2]$ the boundary condition

$$\mathrm{Re}(f(z)\overline{z}) \geq |f(z)| \cos \frac{\alpha\pi}{2}, \quad z \in \partial\Delta, \tag{6.1.7}$$

was given in [53]. This inequality coincides with (6.1.6) when $\alpha = 1$.

Theorem 6.9. *Let $f \in \mathrm{Hol}(\Delta, \mathbb{C})$ be continuous on $\overline{\Delta}$ and satisfy condition* (6.1.7). *Then the equality*

$$\lim_{\substack{z \to \tau \\ z \in \overline{\Delta}}} \frac{f(z)}{(z - \tau)^{2+\alpha}} = 0 \quad \textit{for some} \quad \tau \in \partial\Delta \tag{6.1.8}$$

implies that $f \equiv 0$.

Observe that if $F \in \mathrm{Hol}(\Delta)$, then the mapping f defined by $f(z) = z - F(z)$ belongs to the class $\mathcal{G}$. Therefore any rigidity result for semigroup generators provides immediately a rigidity theorem for holomorphic self-mappings. Another useful relation between classes $\mathrm{Hol}(\Delta)$ and $\mathcal{G}$ is the following consequence of the Berkson–Porta formula: $f(z) = (z - \tau)(1 - z\bar{\tau})\dfrac{1 + F(z)}{1 - F(z)}$. In the next section, we prove some results for generators which imply, in turn, the Burns–Krantz Theorem 6.1 as well as its extensions given in Theorems 6.3–6.8.

6.2 Rigidity of semigroup generators

The following consequence of the Berkson–Porta formula establishes a one-to-one correspondence between the classes $\mathcal{G}[0]$ and $\mathcal{G}^+[1]$.

Lemma 6.1. *Let f and g be two holomorphic functions on Δ connected by the formula*

$$k(z) \cdot f(z) + g(z) = 0, \tag{6.2.1}$$

where $k(z) = \dfrac{z}{(1-z)^2}$ is the Koebe function. Then $f \in \mathcal{G}^+[1]$ if and only if $g \in \mathcal{G}[0]$.

This formula implies, in its turn, infinitesimal versions of the classical Schwarz lemma and the Burns–Krantz rigidity theorem.

Proposition 6.1. *Let $g \in \mathcal{G}[0]$. Then $g(z) \equiv 0$ if and only if $g'(0) = 0$.*

Proof. If $g'(0) = 0$, then we get by (6.2.1) that

$$\lim_{z\to 0} \frac{f(z)}{(1-z)^2} = -\lim_{z\to 0} \frac{g(z)}{z} = 0,$$

i.e., $f(0) = 0$, which is impossible, when $g \neq 0$, since f has then no null point in Δ. □

Corollary 6.1. *Let $f \in \mathcal{G}$ be such that $f'(0) = 0$. Then f generates a group $\mathcal{S} = \{F_t\}_{t\in\mathbb{R}}$ of automorphisms of Δ. Moreover, if $f(0) \neq 0$, then $\mathcal{S}$ consists of hyperbolic automorphisms of Δ.*

Proof. Since $f \in \mathcal{G}$, it can be represented as

$$f(z) = f(0) - \overline{f(0)}z^2 + zq(z),$$

where $\operatorname{Re} q(z) \geq 0$ (see Theorem 2.6). At the same time, the function $g \in \mathrm{Hol}(\Delta, \mathbb{C})$, defined by (6.2.1), also belongs to the class $\mathcal{G}$ and satisfies the conditions $g(0) = 0$ and $g'(0) = 0$. Hence $g(z) = 0$ identically, and we are done. □

Proposition 6.2. *Let $f \in \mathcal{G}_p[1]$ be such that $f''(1) = 0$. If $f \in C^3_A(1)$, then $f'''(1)$ is a nonnegative real number. Moreover, $f'''(1) = 0$ if and only if $f(z) \equiv 0$.*

Proof. Let $f \in \mathcal{G}_p[1]$ be such that $f''(1) = \angle\lim_{z\to 1} \dfrac{f(z)}{(z-1)^2} = 0$. Then again by (6.2.1), we obtain that $\angle\lim_{z\to 1} g(z) = 0$. Hence $z = 1$ is a boundary null point of the generator g. Since g has an interior null point, it cannot belong to $\mathcal{G}^+[1]$. In other words, the limit

$$\angle\lim_{z\to 1} \frac{g(z)}{z-1} = -\angle\lim_{z\to 1} \frac{zf(z)}{(z-1)^3} = -\frac{f'''(1)}{3}$$

must be a negative number whenever g is not identically zero. □

The last part of this section follows the results obtained in [134].

Theorem 6.10. *Let f be the generator of a semigroup $\mathcal{S} = \{F_t\}_{t\geq 0}$ having the Denjoy–Wolff point $\tau = 1$. Let $f \in C_A^3(1)$ and let $g(z) = f'(1)(z-1) + \frac{1}{2}f''(1)(z-1)^2$ be its quadratic part. Then*

(i) *g is the generator of a semigroup of linear-fractional transformations on Δ;*

(ii) *if $\operatorname{Re} f'''(1) = 0$, then $f = g$ if and only if $h := f - g$ belongs to the class $\mathcal{G}$.*

Proof. Since $f \in \mathcal{G}^+[1]$, it follows from the Berkson–Porta representation that

$$f(z) = -(1-z)^2 p(z),$$

where $\operatorname{Re} p(z) \geq 0$.

Since $f \in C_A^2(1)$, we can also write that

$$p(z) = \frac{f(z)}{-(1-z)^2} = \frac{f'(1)(z-1) + \frac{1}{2}f''(1)(z-1)^2}{-(1-z)^2} - \frac{h(z)}{(z-1)^2},$$

where $\angle\lim_{z\to 1} \dfrac{h(z)}{(z-1)^2} = 0$, or

$$\begin{aligned} p(z) &= f'(1)\frac{1}{1-z} - \frac{1}{2}f''(1) - \frac{h(z)}{(z-1)^2} = \\ &= \frac{f'(1)}{2}\frac{1+z}{1-z} + \frac{1}{2}\left[f'(1) - f''(1)\right] - \frac{h(z)}{(z-1)^2}. \end{aligned}$$

Now it follows from Theorem 1.14 that the function

$$p_1(z) := p(z) - \frac{1}{2}f'(1)\frac{1+z}{1-z}$$

is of nonnegative real part. Therefore,

$$\frac{1}{2}\left[f'(1) - \operatorname{Re} f''(1)\right] = \angle\lim_{z\to 1} \operatorname{Re} p_1(z) \geq 0.$$

This implies that the function

$$q(z) := \frac{1}{2}f'(z)\frac{1+z}{1-z} + \frac{1}{2}\left[f'(1) - f''(1)\right]$$

is also of nonnegative real part.

On the other hand,

$$\begin{aligned}-(1-z)^2 q(z) &= \frac{1}{2}f'(1)(z^2-1) - \frac{1}{2}\left[f'(1) - f''(1)\right](z-1)^2 = \\ &= f'(1)(z-1) + \frac{1}{2}f''(1)(z-1)^2 = g(z).\end{aligned}$$

Now assertion (i) follows again from the Berkson–Porta representation and assertion (ii) is seen to be a direct consequence of Proposition 6.2. □

Now we formulate our main rigidity result.

Theorem 6.11. *Let* $f \in \mathrm{Hol}(\Delta, \mathbb{C}) \cap C_A^3(1)$ *with* $\beta := f'(1)$ *satisfy representation* $f(z) = -(1-z)^2 p(z)$ *with*

$$\inf_{z\in\Delta}\left[\mathrm{Re}\, p(z) - \frac{\beta}{2}\frac{1-|z|^2}{|1-z|^2}\right] =: m \geq 0. \tag{6.2.2}$$

Then f *generates a semigroup* $\mathcal{S} = \{F_t\}_{t\geq 0}$ *of linear-fractional transformations if and only if the following two conditions hold:*

(i) $f'(1) - \mathrm{Re}\, f''(1) \leq 2m$;

(ii) $f'''(1) = 0$.

Moreover, in this case, $m = 0$ *if and only if* f *is a generator of a group of automorphisms of* Δ.

Remark 6.5. We will see below that, in fact, under our setting,

$$f'(1) - \mathrm{Re}\, f''(1) = 2m \geq 0.$$

So, we have the following consequences of Theorem 6.11:

Corollary 6.2. *Let* $f \in \mathcal{G}(\Delta) \cap C_A^3(1)$ *with* $f(1) = 0$. *Then* f *is a generator of a group* $\mathcal{S}$ *of automorphisms of* Δ *if and only if*

$$f'(1) - \mathrm{Re}\, f''(1) = f'''(1) = 0. \tag{6.2.3}$$

In addition, if (6.2.3) *holds, then the group* $\mathcal{S}$ *consists of hyperbolic automorphisms if and only if* $f'(1) \neq 0$. *Otherwise* ($f'(1) = 0$), $\mathcal{S}$ *consists either of parabolic automorphisms* ($\mathrm{Im}\, f''(1) \neq 0$) *or identity mappings* ($\mathrm{Im}\, f''(1) = 0$, *hence,* $f''(1) = 0$).

Corollary 6.3. *Let F be a self-mapping of Δ with the Denjoy–Wolff point $\tau = 1$, and let $f \in \mathcal{G}$ have the form*

$$f(z) = -(1-z)^2 \frac{1+F(z)}{1-F(z)} . \tag{6.2.4}$$

Then F is an automorphism of Δ if and only if f generates a group of hyperbolic automorphisms of Δ. Moreover, F is parabolic if and only if $f'(1) = 2$.

Corollary 6.4. *Let $f \in \mathcal{G} \cap C_A^3(1)$ be of the form $f(z) = -(1-z)^2 p(z)$. Then f is a generator of a semigroup $\mathcal{S}$ of affine self-mappings of Δ if and only if the following two conditions hold:*

(i) $\operatorname{Re} p(z) \geq \frac{1}{2} f'(1)$;

(ii) $f''(1) = f'''(1) = 0$.

In this case $\beta := f'(1) \geq 0$ and $f(z) = \beta(z-1)$.

The sufficient part of Theorem 6.11 can be improved as follows:

Theorem 6.12. *Let $f \in \operatorname{Hol}(\Delta, \mathbb{C}) \cap C_A^3(1)$ with $\beta := f'(1)$ have the form $f(z) = -(1-z)^2 p(z)$, where function $p \in \mathcal{P}$ satisfies (6.2.2):*

$$\inf_{z \in \Delta} \left[\operatorname{Re} p(z) - \frac{\beta}{2} \frac{1-|z|^2}{|1-z|^2} \right] =: m \geq 0.$$

Let $g(z) = f'(1)(z-1) + \frac{1}{2} f''(1)(z-1)^2$ be its Taylor's polynomial of degree 2 at $z = 1$. The following assertions hold:

(i) $f'(1) - \operatorname{Re} f''(1) \geq 0$;

(ii) *if $f'(1) - \operatorname{Re} f''(1) \leq 2m$, then $f'''(1)$ is a nonnegative real number and*

$$|f(z) - g(z)|^2 \leq \frac{1}{6} f'''(1) \frac{\operatorname{Re}\left[(f(z) - g(z))(1-\bar{z})^2 \right]}{1-|z|^2}, \quad z \in \Delta. \tag{6.2.5}$$

In particular, $f(z) = g(z)$, i.e., g generates a semigroup of LFT's which are self-mappings of Δ if and only if $f'''(1) = 0$.

Let $F \in \operatorname{Hol}(\Delta)$ be a holomorphic self-mapping of Δ with the fixed point $\tau = 1$, i.e., $F(1) = 1$ and $0 < \alpha = F'(1) < \infty$. Then the function f defined by (6.2.4) belongs to $\mathcal{G}$. By using direct calculations, one proves the following assertion.

Lemma 6.2. *The function f defined by (6.2.4) belongs to the class $C_A^3(1)$ if and only if F belongs to this class. Moreover, f generates a hyperbolic type semigroup* *with*

$$f'(1) = \frac{2}{\alpha} > 0, \quad \alpha = F'(1), \tag{6.2.6}$$

$$f''(1) = 2\frac{(F'(1))]^2 - F''(1)}{(F'(1))^2} \tag{6.2.7}$$

$$= 2\left[\frac{1}{\alpha} - \frac{1}{\alpha^2}\left[F''(1) - \alpha(\alpha-1)\right]\right],$$

$$f'''(1) = -\frac{2}{\alpha}S_F(1). \tag{6.2.8}$$

We are now at the point to prove our main results.

Proof of Theorem 6.12. Let $f \in \mathrm{Hol}(\Delta, \mathbb{C}) \cap C_A^3(1)$ i.e., f admits the representation

$$f(z) = f'(1)(z-1) + \frac{1}{2}f''(1)(z-1)^2 + +\frac{1}{3!}f'''(1)(z-1)^3 + \gamma_f(z),$$

where

$$\angle \lim_{z\to 1} \frac{\gamma_f(z)}{(z-1)^3} = 0. \tag{6.2.9}$$

Then $p(z) = -f(z)(1-z)^{-2}$ is of the form

$$p(z) = f'(1)\frac{1}{1-z} - \frac{1}{2}f''(1) - \frac{1}{3!}f'''(1)(z-1) + \gamma_p(z)$$

where, by (6.2.9),

$$\angle \lim_{z\to 1} \frac{\gamma_p(z)}{(z-1)} = -\angle \lim_{z\to 1} \frac{\gamma_f(z)}{(z-1)^3} = 0.$$

We can also write

$$p(z) = \frac{1}{2}f'(1)\frac{1+z}{1-z} + \frac{1}{2}\left(f'(1) - f''(1)\right) + q(z),$$

where

$$q(z) = -\frac{1}{3!}f'''(1)(z-1) + \gamma_p(z).$$

Noting that the function $p_1(z) := p(z) - \frac{\beta}{2}\frac{1+z}{1-z}$ is of positive real part and setting

$$b = \frac{1}{2}\left(f'(1) - f''(1)\right), \tag{6.2.10}$$

we get that

$$\operatorname{Re} b = \angle \lim_{z\to 1} \operatorname{Re}\left[p_1(z) - q(z)\right] = \angle \lim_{z\to 1} \operatorname{Re} p_1(z) \geq 0, \tag{6.2.11}$$

which proves assertion (i) of the theorem.

Assume now that $\operatorname{Re} b \le m$. Then $\operatorname{Re} q(z) \ge 0$ and

$$q'(1) = -\frac{1}{3!} f'''(1).$$

Applying now Theorem 1.14 to the function $\frac{1}{q(z)}$ and going back to the difference $f(z) - g(z) = -(1-z)^2 q(z)$, we obtain inequality (6.2.5).

To complete the proof, we note that $g(z)\left(= f'(1)(z-1) + \frac{1}{2} f''(1)(z-1)^2\right)$ can be written in the form

$$g(z) = -b(1-z)^2 + \frac{1}{2}\beta(z^2 - 1).$$

It follows from the Berkson–Porta formula that the first term of this sum is a generator of a parabolic type semigroup of Δ, while the second term is a generator of a group of hyperbolic automorphisms. Since $\mathcal{G}$ is a real cone, we get that g must belong to $\mathcal{G}$. Finally, we have by (6.2.5) that $f \equiv g$ if and only if $f'''(1) = 0$. □

Remark 6.6. Note that, actually, we have from (6.2.11) that $\operatorname{Re} b = \inf_{z \in \Delta} p_1(z) = m \ge 0$. This proves the assertion in Remark 6.5.

Since the necessary part of Theorem 6.11 is obvious, Theorem 6.11 is now a direct consequence of Theorem 6.12.

By using Lemma 6.2 and Theorems 6.11 and 6.12, one obtains Theorems 6.3–6.5.

6.3 Commuting semigroups of holomorphic mappings

6.3.1 Identity principles for commuting semigroups

In this section, we apply the above results to establish conditions on semigroup generators that provide that the generated semigroups commute.

Definition 6.1. *We say that two semigroups $\mathcal{S}_1 = \{F_t\}_{t\ge 0}$ and $\mathcal{S}_2 = \{G_t\}_{t\ge 0}$ of holomorphic self-mappings on Δ are commuting if for each pair $s, t \ge 0$, the elements $F_t \in \mathcal{S}_1$ and $G_s \in \mathcal{S}_2$ commute:*

$$F_t \circ G_s = G_s \circ F_t.$$

Theorem 6.13. *Let f and g be generators of one-parameter commuting semigroups $\mathcal{S}_f = \{F_t\}_{t\ge 0}$ and $\mathcal{S}_g = \{G_t\}_{t\ge 0}$, respectively, and $f(\tau) = 0$ at some point $\tau \in \overline{\Delta}$.*

(i) *Let $\tau \in \Delta$. If $f'(\tau) = g'(\tau)$, then $f \equiv g$.*

(ii) *Let $\tau \in \partial\Delta$. Suppose f and g admit the following representations:*

$$f(z) = f'(\tau)(z-\tau) + \ldots + \frac{f^{(m)}(\tau)}{m!}(z-\tau)^m + \gamma_1(z) \tag{6.3.1}$$

and

$$g(z) = g(\tau) + g'(\tau)(z-\tau) + \ldots + \frac{g^{(m)}(\tau)}{m!}(z-\tau)^m + \gamma_2(z), \qquad (6.3.2)$$

where $\dfrac{\gamma_1(z)}{(z-\tau)^m}$ *and* $\dfrac{\gamma_2(z)}{(z-\tau)^m}$ *tend to* 0 *as* $z \to \tau$ *along some curve lying in* Δ *and ending at* τ. *If* $f^{(m)}(\tau) = g^{(m)}(\tau) \neq 0$, *then* $f \equiv g$.

Remark 6.7. If $\tau \in \partial\Delta$ is the Denjoy–Wolff point of a semigroup generated by a mapping $h \in \mathcal{G}(\Delta)$, then h admits the expansion

$$h(z) = h'(\tau)(z-\tau) + o(z-\tau)$$

when $z \to \tau$ in each non-tangential approach region at τ and $h'(\tau) = \angle \lim\limits_{z\to\tau} h'(z)$. Moreover, in this case, $h'(\tau)$ is a nonnegative real number which is zero if and only if h generates a semigroup of parabolic type (see Theorem 2.10).

Therefore, if f (or g) in Theorem 6.13 generates a semigroup of hyperbolic type with the Denjoy–Wolff point $\tau \in \partial\Delta$, then the condition $f'(\tau) = g'(\tau)$ is enough to provide that $f \equiv g$.

Remark 6.8. As a matter of fact, if f and g have expansion (6.3.1) and (6.3.2) when $z \to \tau$ in each non-tangential approach region at $\tau \in \partial\Delta$ up to the third order $m = 3$, such that $f'(\tau) = g'(\tau)$, $f''(\tau) = g''(\tau)$ and $f'''(\tau) = g'''(\tau)$, then $f \equiv g$.

If, in particular, $f^{(i)}(\tau) = g^{(i)}(\tau) = 0$, $i = 1, 2, 3$, then both f and g equal zero identically by Proposition 6.2.

Theorem 6.13 is a consequence of the following more general assertion.

Define two linear semigroups $\{A_t\}_{t\geq 0}$ and $\{B_t\}_{t\geq 0}$ of composition operators on $\mathrm{Hol}(\Delta, \mathbb{C})$ by

$$A_t(h) = h \circ F_t \quad \text{and} \quad B_t(h) = h \circ G_t, \quad t \geq 0. \qquad (6.3.3)$$

The operators Γ_f and Γ_g defined by

$$\Gamma_f(h) = h'f \quad \text{and} \quad \Gamma_g(h) = h'g \qquad (6.3.4)$$

are their generators, respectively.

Theorem 6.14 ([53]). *Let* f *and* $g \in \mathrm{Hol}(\Delta, \mathbb{C})$ *be generators of one-parameter semigroups* $\mathcal{S}_f = \{F_t\}_{t\geq 0}$ *and* $\mathcal{S}_g = \{G_t\}_{t\geq 0}$, *respectively. Let* A_t *and* B_t *be defined by* (6.3.3). *Then the following are equivalent:*

(i) $F_t \circ G_s = G_s \circ F_t$, $s, t \geq 0$, *i.e., the semigroups* $\mathcal{S}_f$ *and* $\mathcal{S}_g$ *are commuting;*

(ii) $A_t \circ B_s = B_s \circ A_t$, $s, t \geq 0$, *i.e., the linear semigroups* $\{A_t\}_{t\geq 0}$ *and* $\{B_t\}_{t\geq 0}$ *are commuting;*

(iii) $\Gamma_f \circ \Gamma_g = \Gamma_g \circ \Gamma_f$, *i.e., the linear semigroup generators* Γ_f *and* Γ_g *are commuting;*

(iv) *the Lie commutator*

$$[f, g] = f'g - g'f = 0;$$

(v) $f = \alpha g$ *for some* $\alpha \in \mathbb{C}$.

Proof. Suppose that $f \not\equiv 0$. First we prove the equivalence of assertions (i) and (v).

Let (i) hold. If $f(\tau) = 0$, $\tau \in \Delta$, then τ is the unique common fixed point for the semigroup $\mathcal{S}_f$ generated by f, i.e., $F_t(\tau) = \tau$ for all $t \geq 0$.

If F_t and G_s are commuting for all $s, t \geq 0$, then we have

$$G_s(\tau) = G_s\left(F_t(\tau)\right) = F_t\left(G_s(\tau)\right).$$

Hence, it follows by the uniqueness of the fixed point τ that $G_s(\tau) = \tau$ for all $s \geq 0$, and so $g(\tau) = 0$.

Consider the function $h \in \mathrm{Hol}(\Delta, \mathbb{C})$ defined by the differential equation

$$\mu h(z) = h'(z) f(z). \tag{6.3.5}$$

It is known that if $\mu = f'(\tau)$, then equation (6.3.5) has a unique solution $h \in \mathrm{Hol}(\Delta, \mathbb{C})$ normalized by the condition $h'(\tau) = 1$. In addition, this function h solves Schroeder's functional equation

$$h\left(F_t(z)\right) = e^{-\mu t} h(z) \tag{6.3.6}$$

(see Theorem 5.1). Now, for any $s, t \geq 0$, we get, from (6.3.6),

$$h\left(G_s\left(F_t(z)\right)\right) = h\left(F_t\left(G_s(z)\right)\right) = e^{-\mu t} h\left(G_s(z)\right).$$

Let $h_s = h \circ G_s$. Then we have

$$h_s\left(F_t(z)\right) = e^{-\mu t} h_s(z). \tag{6.3.7}$$

Differentiating (6.3.7) at $t = 0^+$, we get

$$\mu h_s(z) = h_s'(z) f(z). \tag{6.3.8}$$

Comparing (6.3.5) and (6.3.8) implies $h_s(z) = \lambda(s) h(z)$ for some $\lambda(s) \in \mathbb{C}$, or

$$h\left(G_s(z)\right) = \lambda(s) h(z). \tag{6.3.9}$$

Since the left-hand side of the latter equality is differentiable in $s \geq 0$, the scalar function $\lambda(s)$ is differentiable too. Differentiating (6.3.9) at $s = 0^+$, we get

$$\lambda'(0) h(z) = -h'(z) g(z). \tag{6.3.10}$$

Note that $h(\tau) = 0$ while $h(z) \neq 0$ for all $z \in \Delta$, $z \neq \tau$. In addition, by Theorem 5.2, h is univalent. Hence, $h'(z) \neq 0$ for all $z \in \Delta$.

Finally, we obtain from (6.3.5) and (6.3.10) that

$$f(z) = \alpha g(z), \quad \text{where} \quad \alpha = -\frac{\mu}{\lambda'(0)}.$$

Now, let us suppose that f has no null point in Δ. Then the function $p : \Delta \mapsto \mathbb{C}$ given by

$$p(z) = -\int_0^z \frac{d\zeta}{f(\zeta)} \tag{6.3.11}$$

is a well-defined holomorphic function on Δ with $p(0) = 0$.

Recall that the semigroup $\{F_t\}_{t\geq 0}$ generated by f can be defined by the Cauchy problem

$$\begin{cases} \dfrac{dF_t(z)}{dt} + f\left(F_t(z)\right) = 0, \quad t \geq 0, \\ F_0(z) = z, \quad z \in \Delta. \end{cases} \tag{6.3.12}$$

Substituting here $f(z) = -\dfrac{1}{p'(z)}$, we obtain

$$p'\left(F_t(z)\right) dF_t(z) = dt.$$

Integrating the latter equality on the interval $[0, t]$, we get that p is a solution of Abel's functional equation

$$p\left(F_t(z)\right) = p(z) + t. \tag{6.3.13}$$

Now, for any fixed $s \geq 0$, we have

$$p\left(G_s\left(F_t(z)\right)\right) = p\left(F_t\left(G_s(z)\right)\right) = p\left(G_s(z)\right) + t.$$

Once again, setting $p_s = p \circ G_s$, we have

$$p_s\left(F_t(z)\right) = p_s(z) + t. \tag{6.3.14}$$

Differentiating (6.3.14) at $t = 0^+$, we get

$$p_s'(z) = -\frac{1}{f(z)}, \tag{6.3.15}$$

and by (6.3.11), $p_s(z) = p(z) + \kappa(s)$, $\kappa(s) \in \mathbb{C}$, or

$$p\left(G_s(z)\right) = p(z) + \kappa(s). \tag{6.3.16}$$

Differentiating (6.3.16) at $s = 0^+$, we obtain the equality

$$p'(z) = -\frac{\kappa'(0)}{g(z)}. \tag{6.3.17}$$

Comparing (6.3.15) and (6.3.17) gives

$$f = \alpha g \quad \text{with} \quad \alpha = \frac{1}{\kappa'(0)}. \tag{6.3.18}$$

Now we prove that (v)⇒(i). Let $f = \alpha g$ for some $\alpha \in \mathbb{C}$.

First we assume that g has an interior null-point $\tau \in \Delta$. In this case, there is a univalent solution of the differential equation

$$\mu h(z) = h'(z) g(z) \tag{6.3.19}$$

with some $\mu \in \mathbb{C}, \quad \operatorname{Re}\mu \geq 0$.

Since $f = \alpha g$, we have that h is also a solution of the equation

$$\nu h(z) = h'(z) f(z), \quad \nu = \alpha\mu. \tag{6.3.20}$$

In turn, equations (6.3.19) and (6.3.20) are equivalent to Schroeder's functional equations

$$h\left(G_s(z)\right) = e^{-\mu s} h(z), \quad s \geq 0 \tag{6.3.21}$$

and

$$h\left(F_t(z)\right) = e^{-\nu t} h(z), \quad t \geq 0, \quad \nu = \alpha\mu, \tag{6.3.22}$$

respectively, where $\{F_t\}_{t\geq 0}$ is the semigroup generated by f.

Consequently,

$$\begin{aligned} F_t\left(G_s(z)\right) &= h^{-1}\left(e^{-\nu t} h\left(G_s(z)\right)\right) = h^{-1}\left(e^{-\nu t} \cdot e^{-\mu s} h(z)\right) \\ &= h^{-1}\left(e^{-\mu s} h\left(F_t(z)\right)\right) = G_s\left(F_t(z)\right) \end{aligned}$$

for all $s, t \geq 0$ and we are done.

Now let us assume that g has a boundary null-point $\tau \in \partial\Delta$ with $g'(\tau) \geq 0$. In this case, for each $c \in \mathbb{C}, \ c \neq 0$, Abel's equations

$$p\left(G_s(z)\right) = p(z) + cs$$

and

$$p\left(F_t(z)\right) = p(z) + c\alpha t$$

have the same solution

$$p(z) = -c \int_0^z \frac{d\zeta}{g(\zeta)} = -c\alpha \int_0^z \frac{d\zeta}{f(\zeta)},$$

which is univalent on Δ.

Once again we calculate

$$\begin{aligned} F_t\left(G_s(z)\right) &= p^{-1}\left(p\left(G_s(z)\right) + c\alpha t\right) = p^{-1}\left(p(z) + c\alpha t + cs\right) \\ &= p^{-1}\left(p\left(F_t(z)\right) + cs\right) = G_s\left(F_t(z)\right). \end{aligned}$$

The implication (v)⇒(i) is proved.

The equivalence of (i) and (ii) is obvious.

To verify the equivalence of (iii) and (iv), we just calculate:

$$\Gamma_f\left(\Gamma_g(h)\right) = h''gf + h'g'f,$$
$$\Gamma_g\left(\Gamma_f(h)\right) = h''fg + h'f'g.$$

Hence, $\Gamma_f \circ \Gamma_g = \Gamma_g \circ \Gamma_f$ if and only if $f'g - g'f = 0$.

Now, it is clear, that (v) implies (iv).

Finally we prove the implication (iv) ⇒ (v). Obviously, (iv) implies that if f has no null points in Δ, then g also has no null points in Δ and, hence, (v) follows. If $f(\tau) = 0$ for some $\tau \in \Delta$, then also $g(\tau) = 0$. Using the Berkson–Porta formula, one can write $f(z) = (z-\tau)p(z)$ and $g(z) = (z-\tau)q(z)$, where p and q do not vanish in Δ. Now it follows that

$$[f, g] = (z-\tau)^2[p, q].$$

Hence, again we have $p = aq$, and hence $f = ag$ for some $a \in \mathbb{C}$, $a \neq 0$. □

Proof of Theorem 6.13. First we note that, by Theorem 6.14,

$$f = \alpha g, \quad \alpha \in \mathbb{C}. \tag{6.3.23}$$

(i) Let $f'(\tau) = g'(\tau) = 0$. By Theorem 2.7, f admits the representation

$$f(z) = (z-\tau)(1-\bar{\tau}z)p(z), \quad z \in \Delta,$$

where $p \in \mathrm{Hol}(\Delta, \mathbb{C})$, $\mathrm{Re}\, p(z) \geq 0$.

Since $f'(\tau) = (1-|\tau|^2)p(\tau) = 0$, we have $p(\tau) = 0$. It follows from the Maximum Principle that $p \equiv 0$. Hence, $f \equiv 0$ and by (6.3.23) also $g \equiv 0$.

Assume now $f'(\tau) = g'(\tau) \neq 0$. Then it follows from (6.3.23) that $\alpha = 1$ and so $f \equiv g$.

(ii) In general, by (6.3.23) we have $f^{(k)}(\tau) = \alpha g^{(k)}(\tau)$, $0 < k \leq m$. Hence, the condition $f^{(k)}(\tau) = g^{(k)}(\tau) \neq 0$ for some $0 < k \leq m$ implies that $\alpha = 1$ and, consequently, $f \equiv g$. □

Let $\mathcal{S}_f = \{F_t\}_{t\geq 0}$ be the semigroup generated by $f \in \mathcal{G}(\Delta)$. The set $\mathcal{Z}(\mathcal{S}_f)$ of all semigroups $\mathcal{S} = \{G_t\}_{t\geq 0}$ such that

$$F_t \circ G_s = G_s \circ F_t, \quad t, s \geq 0,$$

is called the **centralizer of** $\mathcal{S}_f$.

It is clear that for each $f \in \mathcal{G}(\Delta)$ the centralizer $\mathcal{Z}(\mathcal{S}_f)$ contains $\mathcal{S}_{\alpha f}$ for all $\alpha \geq 0$.

Therefore we will say that **the centralizer of** $\mathcal{S}_f$ **is trivial** when the inclusion $\mathcal{S} \in \mathcal{Z}(\mathcal{S}_f)$ implies that $\mathcal{S} = \mathcal{S}_{\alpha f}$ for some $\alpha \geq 0$.

Proposition 6.3. *Let f be the generator of a semigroup $\mathcal{S}_f = \{F_t\}_{t\geq 0}$, and let $\tau = 1$ be the Denjoy–Wolff point of $\mathcal{S}_f$. Then if one of the following conditions holds, the centralizer $\mathcal{Z}(\mathcal{S}_f)$ is trivial:*

(i) *$\mathcal{S}_f$ is a hyperbolic type semigroup ($f'(1) > 0$) which is not a group;*

(ii) *f admits the expansion*

$$f(z) = a(z-1)^3 + o\left((z-1)^3\right) \quad \text{with} \quad a \neq 0$$

when $z \to 1$ in each non-tangential approach region at τ.

The first statement is based on the following simple lemma:

Lemma 6.3. *Let f and g be generators of two nontrivial (neither f nor g are identically zero) commuting semigroups $\mathcal{S}_f = \{F_t\}_{t\geq 0}$ and $\mathcal{S}_g = \{G_t\}_{t\geq 0}$, respectively. Then $\mathcal{S}_f$ is of hyperbolic type if and only if $\mathcal{S}_g$ is. In this case $f = \alpha g$ with real α. Moreover, $\alpha < 0$ implies that $\mathcal{S}_f$ and $\mathcal{S}_g$ are both groups of hyperbolic automorphisms having 'opposite' fixed points, i.e., the attractive point for $\mathcal{S}_f$ is the repelling point for $\mathcal{S}_g$ and conversely.*

Proof. Since $\mathcal{S}_f$ and $\mathcal{S}_g$ are commuting, by Theorem 6.14, there exists $\alpha \in \mathbb{C}$ such that $f = \alpha g$. In our settings, α is not zero. Suppose that $\tau = 1$ is the Denjoy–Wolff point of $\mathcal{S}_f$. Then $f(1) = 0$ and therefore also $g(1) = 0$. Now, since $f'(1) > 0$ then $g'(1) = \frac{1}{\alpha} f'(1)$ exists finitely, and it must be a real number by Theorem 2.11. So must be α.

Now let us assume that α is negative. Then $g'(1) = \frac{1}{\alpha} f'(1) < 0$. Hence the semigroup $\mathcal{S}_g$ generated by g must have the Denjoy–Wolff point $\sigma \in \overline{\Delta}$ different from $\tau = 1$.

It is clear that σ cannot be inside Δ since otherwise it must be a common fixed point of both semigroups $\mathcal{S}_f$ and $\mathcal{S}_g$ because of the commuting property.

So, $\sigma \in \partial\Delta$ and $g'(\sigma) \geq 0$ (see Theorem 2.10), then $f(\sigma) = 0$ and $f'(\sigma) \leq 0$. It follows by Corollary 2.2 that

$$0 < f'(\tau) \leq -f'(\sigma), \tag{6.3.24}$$

and the equality is possible if and only if f is the generator of a group of hyperbolic automorphisms. From the same Corollary 2.2, we have the reversed inequality for g

$$0 \leq g'(\sigma) \leq -g'(\tau)$$

that means

$$0 \leq \frac{1}{\alpha} f'(\sigma) \leq -\frac{1}{\alpha} f'(\tau).$$

Comparing this inequality with (6.3.24) gives us that $f'(\tau) = -f'(\sigma) > 0$ and $g'(\tau) = -g'(\sigma) < 0$, which means that both f and g generate groups of hyperbolic automorphisms with opposite fixed points. □

Remark 6.9. The last assertion of this lemma follows also by a result of Behan (see [14]). Indeed, let $\alpha < 0$. Then the equality $f(z) = \alpha g(z)$ implies that $g'(\tau)$ exists and is a real negative number. So, the Denjoy–Wolff point τ of the semigroup $\mathcal{S}_f$ cannot be the Denjoy–Wolff point of the semigroup $\mathcal{S}_g$. Hence, by [14], we conclude that $\mathcal{S}_f$ and $\mathcal{S}_g$ are groups of hyperbolic automorphisms.

Proof of Proposition 6.3. The statement (i) is a direct consequence of the previous lemma. To prove the second statement, we note that the number a is a nonnegative real number (see Theorem 7.18 below). On the other hand, since $\mathcal{S}_f$ and $\mathcal{S}_g$ commute, by Theorem 6.14 there is a number $\alpha \in \mathbb{C}$ such that $f = \alpha g$.

Therefore, since $\alpha \neq 0$ also g admits the expansion

$$g(z) = \frac{a}{\alpha}(z-1)^3 + o\left((z-1)^3\right),$$

and we have that also $\frac{a}{\alpha} \geq 0$. This implies that α is a nonnegative real number. □

A natural question which arises in the context of the above results is:

◊ *If two elements F_p and G_q of semigroups $\mathcal{S}_f = \{F_t\}_{t\geq 0}$ and $\mathcal{S}_g = \{G_t\}_{t\geq 0}$ commute for some positive p and q, do these semigroup $\mathcal{S}_f$ and $\mathcal{S}_g$ commute?*

We study this problem for dilation, hyperbolic, and parabolic cases separately. Our exposition is based on the paper [51].

6.3.2 Dilation type

In the following theorems, the condition $F_1 \circ G_1 = G_1 \circ F_1$ can be replaced by the condition $F_p \circ G_q = G_q \circ F_p$ for some $p, q > 0$.

Theorem 6.15. *Let $\mathcal{S}_1 = \{F_t\}_{t\geq 0}$ and $\mathcal{S}_2 = \{G_t\}_{t\geq 0}$ be two continuous semigroups on Δ generated by f and g, respectively, and assume that $F_1 \circ G_1 = G_1 \circ F_1$. Suppose that f has an interior null point $\tau \in \Delta$. If $\mathcal{S}_1$ and $\mathcal{S}_2$ are not groups of automorphisms of Δ, then they commute.*

Proof. Since τ is an interior null point of the generator f, it is the unique interior fixed point of the semigroup $\mathcal{S}_1$ (see [3]). The commutativity of F_1 and G_1 implies that τ is a fixed point of G_1 and, consequently, τ is a fixed point of G_t for each $t > 0$.

By our assumption, $\mathcal{S}_1$ and $\mathcal{S}_2$ are not groups of automorphisms of Δ. By the Schwarz–Pick lemma and the univalence of F_t and G_t on Δ, we have $0 < |F_t'(\tau)| < 1$ and $0 < |G_t'(\tau)| < 1$ for all $t > 0$.

Therefore a result in [39] and the semigroup property implies that $G_1 \circ F_t = F_t \circ G_1$ for all $t \geq 0$. Similarly, $G_s \circ F_t = F_t \circ G_s$ for all $s, t \geq 0$. □

Surprisingly, the case where $\mathcal{S}_1$ contains elliptic automorphisms is more complicated. First we prove that a semigroup commuting with a group of elliptic automorphisms has a specific form.

Proposition 6.4. *Let $\mathcal{S}_1 = \{F_t\}_{t\geq 0}$ be a nontrivial group of elliptic automorphisms of Δ with a common fixed point at $\tau \in \Delta$, and let $\mathcal{S}_2 = \{G_t\}_{t\geq 0}$ be a semigroup of self-mappings of Δ. Then $\mathcal{S}_1$ and $\mathcal{S}_2$ commute if and only if $\mathcal{S}_2$ is a semigroup of linear fractional transformations of the form*

$$G_t(z) = m_\tau(e^{-at} \cdot m_\tau(z)) \tag{6.3.25}$$

for some $a \in \mathbb{C}$, where $m_\tau(z) = \frac{\tau - z}{1-\overline{\tau}z}$.

Note that the function G_t defined by equality (6.3.25) is a self-mapping of Δ if and only if $\operatorname{Re} a \geq 0$.

Proof. Let $\mathcal{S}_2$ be of the form (6.3.25). Since both $\mathcal{S}_1$ and $\mathcal{S}_2$ are actually linear semigroups up to conjugation with m_τ, they must commute.

Conversely, suppose that $F_t \circ G_s = G_s \circ F_t$ for all $s, t \geq 0$. Let

$$\widetilde{F}_t(z) = e^{i\varphi t}z, \ \widetilde{G}_t = m_\tau \circ G_t \circ m_\tau.$$

Then $\{\widetilde{F}_t\}_{t\geq 0}$ is a group of automorphisms of Δ with a fixed point at zero, and $\{\widetilde{G}_t\}_{t\geq 0}$ is a semigroup of self-mappings of Δ with a fixed point at zero. It is obvious that the semigroups $\{\widetilde{F}_t\}_{t\geq 0}$ and $\{\widetilde{G}_t\}_{t\geq 0}$ commute. Consequently, their generators $\widetilde{g}(z)$ and $\widetilde{f}(z) = -i\varphi z$ are proportional (see [53]). So $\widetilde{g}(z) = az$ for some $a \in \mathbb{C}$. Therefore $\widetilde{G}(z) = e^{-at}z$ and $G_t(z) = m_\tau(e^{-at}m_\tau(z))$. □

We will see below that if $\mathcal{S}_1$ is a group of elliptic automorphisms the commutativity of F_1 and G_1 does not imply that the semigroups $\mathcal{S}_1$ and $\mathcal{S}_2$ commute. Nevertheless, in this case, one can still obtain some additional information about the semigroup $\mathcal{S}_2$. The following assertions explain our claim.

Theorem 6.16. *Let F be an elliptic automorphism of Δ and let $\mathcal{S}_2 = \{G_t\}_{t\geq 0}$ be a semigroup of self-mappings of Δ which are not automorphisms. Then the commutativity of F and G_1 implies that $F \circ G_t = G_t \circ F$ for all $t \geq 0$.*

Proof. Let $\tau \in \Delta$ be the common fixed point of F and G_t, $t \geq 0$. Then the function F is of the form $F(z) = m_\tau(e^{i\varphi}m_\tau(z))$, $\varphi \in \mathbb{R}$, $z \in \Delta$, where $m_\tau(z) = \frac{\tau - z}{1-\overline{\tau}z}$.

Let $\widetilde{F}(z) := e^{i\varphi}z$ and $\widetilde{G}_t(z) = m_\tau(G_t(m_\tau(z)))$. Then $\{\widetilde{G}_t\}_{t\geq 0}$ is a semigroup of self-mappings of Δ which are not automorphisms with its common fixed point at zero.

It is obvious that for each $t > 0$, F and G_t commute if and only if $\widetilde{F}$ and $\widetilde{G}_t$ commute. Hence, by our assumption, $\widetilde{F} \circ \widetilde{G}_1 = \widetilde{G}_1 \circ \widetilde{F}$ or, which is one and the same, $e^{i\varphi}\widetilde{G}_1(z) = \widetilde{G}_1(e^{i\varphi}z)$. It follows that for all $n \in \mathbb{N}$, $\widetilde{F} \circ \widetilde{G}_n = \widetilde{G}_n \circ \widetilde{F}$, where $\widetilde{G}_n$ are the iterates of $\widetilde{G}_1$, i.e., $\widetilde{G}_n = \widetilde{G}_1 \circ \widetilde{G}_{n-1}$.

Since $\widetilde{G}_1$ is a self-mapping of Δ (which is not an automorphism) with a fixed point at the origin, there exists a unique univalent solution h of Schröder's functional equation

$$h(\widetilde{G}_1(z)) = \alpha h(z), \quad \text{with} \quad \alpha = \widetilde{G}_1'(0),$$

normalized by $h(0) = 0,\ h'(0) = 1$ (see Theorem 5.2). This solution is given by

$$h(z) = \lim_{n\to\infty} \frac{\widetilde{G}_n(z)}{\alpha^n}.$$

Moreover, for all real positive t (see, for instance, [49]),

$$h(\widetilde{G}_t(z)) = \alpha^t h(z)\,.$$

Therefore,

$$\begin{aligned} h(\widetilde{F}(\widetilde{G}_t(z))) &= h(e^{i\varphi}\widetilde{G}_t(z)) = \lim_{n\to\infty} \frac{\widetilde{G}_n(e^{i\varphi}\widetilde{G}_t(z))}{\alpha^n} \\ &= \lim_{n\to\infty} \frac{e^{i\varphi}\widetilde{G}_n(\widetilde{G}_t(z))}{\alpha^n} = e^{i\varphi}h(\widetilde{G}_t(z)) = e^{i\varphi}\alpha^t h(z) = \alpha^t \lim_{n\to\infty} \frac{e^{i\varphi}\widetilde{G}_n(z)}{\alpha^n} \\ &= \alpha^t \lim_{n\to\infty} \frac{\widetilde{G}_n(e^{i\varphi}z)}{\alpha^n} = \alpha^t h(e^{i\varphi}z) = h(\widetilde{G}_t(e^{i\varphi}z)) = h(\widetilde{G}_t(\widetilde{F}(z))) \end{aligned}$$

and, by the univalence of h, we get $\widetilde{F}\circ\widetilde{G}_t = \widetilde{G}_t\circ\widetilde{F}$ for all $t\geq 0$. Consequently, F and G_t commute for all $t \geq 0$ as asserted. □

Corollary 6.5. *Let $\mathcal{S}_1 = \{F_t\}_{t\geq 0}$ be a group of elliptic automorphisms of Δ, i.e., $F_t(z) = m_\tau(e^{i\varphi t}m_\tau(z))$, $\varphi\in\mathbb{R}$, $\tau\in\Delta$, and let $\mathcal{S}_2 = \{G_t\}_{t\geq 0}$ be a semigroup of self-mappings of Δ. Suppose that $\frac{\varphi}{\pi}$ is an irrational number and F_1 and $G_1 \neq I$ commute. Then $G_t(z) = m_\tau(e^{-at}m_\tau(z))$, $a\in\mathbb{C}$, and, consequently, the semigroups $\mathcal{S}_1$ and $\mathcal{S}_2$ commute.*

Proof. Once again, we define the functions $\widetilde{F}_t = e^{i\varphi t}z$ and $\widetilde{G}_t = m_\tau\circ G_t\circ m_\tau$. The commutativity of F_1 and G_1 implies that $\widetilde{F}_1\circ\widetilde{G}_1 = \widetilde{G}_1\circ\widetilde{F}_1$ and, by Theorem 6.16, $\widetilde{F}_1\circ\widetilde{G}_t = \widetilde{G}_t\circ\widetilde{F}_1$ for all $t\geq 0$. Therefore $\widetilde{G}_t(e^{in\varphi}z) = e^{in\varphi}\widetilde{G}_t(z)$ for all $n\in\mathbb{N}$. Since the set $\{e^{in\varphi}\}_{n\in\mathbb{N}}$ is dense in the unit circle, $\widetilde{G}_t(\lambda z) = \lambda\widetilde{G}_t(z)$ for all λ with $|\lambda| = 1$ and $z\in\Delta$, by the continuity of $\widetilde{G}_t$ on Δ.

Fix $0\neq z\in\Delta$ and $t>0$, and consider the analytic function $q(\lambda)$ on the closed unit disk defined by

$$q(\lambda) = \begin{cases} \dfrac{\widetilde{G}_t(\lambda z)}{\lambda}, & \lambda\neq 0, \\[2ex] \lim\limits_{\lambda\to 0} \dfrac{\widetilde{G}_t(\lambda z)}{\lambda} = z\,\dfrac{\partial}{\partial w}\widetilde{G}_t(w)\Big|_{w=0}, & \lambda = 0. \end{cases} \tag{6.3.26}$$

This function is constant on the unit circle: $q(\lambda) = \widetilde{G}_t(z)$. Moreover, $q(\lambda)\neq 0$ for all $\lambda\in\Delta$. Therefore, $q(\lambda) = \widetilde{G}_t(z)$ for all $\lambda\in\overline{\Delta}$. So for each $z\neq 0$ and $t>0$, $\widetilde{G}_t(\lambda z) = \lambda\widetilde{G}_t(z)$. Consequently, this equality holds for all $z\in\Delta$. Hence $\widetilde{G}_t$ is a linear function for each $t>0$, i.e., $\widetilde{G}_t(z) = e^{-at}z$ for some $a\in\mathbb{C}$, $\operatorname{Re} a\geq 0$, and the assertion follows. □

In contrast with this corollary, if $\frac{\varphi}{\pi}$ is a rational number, the semigroups $\mathcal{S}_1$ and $\mathcal{S}_2$ do not necessarily commute. The following example gives a large class of semigroups $\mathcal{S}_2 = \{G_t\}_{t\geq 0}$ such that $F_1 \circ G_t = G_t \circ F_1$ for all $t \geq 0$, but the semigroups $\mathcal{S}_1$ and $\mathcal{S}_2$ do not commute.

Example 6.2. Let $\mathcal{S}_1 = \{F_t\}_{t\geq 0}$, where $F_t(z) = e^{i\frac{2\pi}{n}t}z$, $n \in \mathbb{N}$, and let $\mathcal{S}_2 = \{G_t\}_{t\geq 0}$ be the semigroup generated by $g(z) = zp(z^n)$, where $\operatorname{Re} p(z) \geq 0$ for all $z \in \Delta$. Then $F_1 \circ G_t = G_t \circ F_1$ for all $t \geq 0$.

Indeed, let $u = u(t,z) := G_t(z)$. Then u is the unique solution of the Cauchy problem

$$\begin{cases} \dfrac{\partial u}{\partial t} + up(u^n) = 0, \\ u(0,z) = z, \quad z \in \Delta, \end{cases} \tag{6.3.27}$$

and, consequently,

$$\int_z^{G_t(z)} \frac{d\varsigma}{\varsigma p(\varsigma^n)} = -t \quad \text{for all} \quad z \in \Delta. \tag{6.3.28}$$

Substituting $e^{i\frac{2\pi}{n}} z$ instead of z, we get

$$\int_{e^{i\frac{2\pi}{n}} z}^{G_t(e^{i\frac{2\pi}{n}} z)} \frac{d\varsigma}{\varsigma p(\varsigma^n)} = -t.$$

Now substitute $\varsigma = e^{i\frac{2\pi}{n}} w$:

$$\int_z^{G_t(e^{i\frac{2\pi}{n}} z)e^{-i\frac{2\pi}{n}}} \frac{dw}{wp(w^n e^{i2\pi})} = \int_z^{G_t(e^{i\frac{2\pi}{n}} z)e^{-i\frac{2\pi}{n}}} \frac{dw}{wp(w^n)} = -t \tag{6.3.29}$$

for all $z \in \Delta$. Equalities (6.3.28) and (6.3.29) imply that

$$\int_{G_t(z)}^{G_t(e^{i\frac{2\pi}{n}} z)e^{-i\frac{2\pi}{n}}} \frac{dw}{wp(w^n)} = 0, \quad z \in \Delta. \tag{6.3.30}$$

By the uniqueness of the solution to the Cauchy problem (6.3.27), the equation

$$\int_z^u \frac{dw}{wp(w^n)} = -s, \quad s \geq 0, \quad z \in \Delta$$

has the unique solution $u = G_s(z)$ for each $s \geq 0$. Thus, it follows from (6.3.30) that $G_t(e^{i\frac{2\pi}{n}} z)e^{-i\frac{2\pi}{n}} = G_0(G_t(z)) = G_t(z)$. Hence, $G_t(e^{i\frac{2\pi}{n}} z) = e^{i\frac{2\pi}{n}} G_t(z)$. Therefore, F_1 commutes with G_t for all $t \geq 0$. At the same time, if p is not a constant function, the semigroups do not commute because their generators are not proportional (see Theorem 6.14).

Remark 6.10. *If at least one element of a semigroup $\mathcal{S}_2 = \{G_t\}_{t\geq 0}$, say G_1, has the form $G_1(z) = z\phi_1(z^n)$, $\phi_1 \in \mathrm{Hol}(\Delta)$, then all the elements have the same form: $G_t(z) = z\phi_t(z^n)$ for some functions $\phi_t \in \mathrm{Hol}(\Delta)$; so the semigroup generator can be represented as $g(z) = zp(z^n)$ with $\mathrm{Re}\, p(z) \geq 0$.*

Indeed, the representation $G_1(z) = z\phi_1(z^n)$ is equivalent to the commutativity of G_1 with $F_1(z) = e^{i\frac{2\pi}{n}}z$. By Theorem 6.16, each mapping G_t, $t \geq 0$, must commute with F_1. Hence $G_t(z) = z\phi_t(z^n)$, $\phi_t \in \mathrm{Hol}(\Delta)$. Differentiating G_t at $t = 0^+$, we arrive at our claim.

6.3.3 Hyperbolic type

We start this section with a result of Heins [80].

Lemma 6.4. *Let F be a hyperbolic automorphism of Δ, and let $G \in \mathrm{Hol}(\Delta)$, $G \neq I$, commute with F. Then G is also a hyperbolic automorphism of Δ.*

This result can be complemented by the following assertion which is of independent interest.

Proposition 6.5. *Let F and G be two commuting holomorphic self-mappings of Δ and assume that G is not the identity. If F is of hyperbolic type, then G is of hyperbolic type too.*

Proof. If F is a hyperbolic automorphism of Δ, then by Lemma 6.4, G is a hyperbolic automorphism of Δ.

Let F be a holomorphic self-mapping of Δ which is not an automorphism of Δ. In this case, the mappings F and G have a common Denjoy–Wolff point $\tau \in \partial\Delta$ (see [14]). We have to show that G is of hyperbolic type, i.e., $0 < G'(\tau) < 1$. Suppose, to the contrary, that G is of parabolic type, i.e., $G'(\tau) = 1$. Then, by a result in [39], G must be a parabolic automorphism.

Let $\varphi := C \circ F \circ C^{-1}$ and $\psi := C \circ G \circ C^{-1}$, where $C(z) = \dfrac{\tau + z}{\tau - z}$. Then f and g are two commuting holomorphic self-mappings of the right half-plane $\Pi_+ = \{z \in \mathbb{C} : \mathrm{Re}\, z > 0\}$ with their common Denjoy–Wolff point at infinity. Moreover, ψ is a parabolic automorphism of Π_+ while φ is a hyperbolic self-mapping of Π_+. Consequently, φ and ψ are of the forms (see [127]):

$$\varphi(w) = cw + \Gamma_F(w) \quad \text{with} \quad c = \frac{1}{F'(\tau)} > 1 \quad \text{and} \quad \angle \lim_{w\to\infty} \frac{\Gamma_F(w)}{w} = 0,$$

and

$$\psi(w) = w + ib \quad \text{with} \quad b \in \mathbb{R} \setminus \{0\} \quad \text{and} \quad w \in \Pi_+.$$

By a simple calculation and the commutativity of φ and ψ_n, we infer from the above representations that

$$\varphi(w + nib) = \varphi(w) + nib, \quad w \in \Pi_+. \tag{6.3.31}$$

Hence,

$$\frac{\varphi(w+nib)}{w+nib} = \frac{\varphi(w)}{w+nib} + \frac{nib}{w+nib}, \quad w \in \Pi_+.$$

Letting $n \to \infty$, we obtain that for each $w \in \Pi_+$, the limit $\lim\limits_{n\to\infty} \frac{\varphi(w+nib)}{w+nib}$ exists and equals 1.

Fix $w_0 \in \Pi_+$. Consider the curve $\ell := \{w_0 + it : t \in \mathbb{R},\ \operatorname{sgn} t = \operatorname{sgn} b\}$. We intend to show that the limit $\lim\limits_{\ell \ni z \to \infty} \frac{\varphi(z)}{z}$ exists and equals 1.

To this end, fix an arbitrary $\varepsilon > 0$ and take $N \in \mathbb{N}$ such that

$$N > \frac{1}{|b|}\left(\frac{|\varphi(w) - w|}{\varepsilon} + |w|\right) \quad \text{and} \quad N > \frac{|w|}{|b|}$$

for all $w \in [w_0, w_0 + ib]$.

Then $\left|\frac{\varphi(z)}{z} - 1\right| < \varepsilon$ for all $z \in \ell$ with $\operatorname{sgn} b \cdot \operatorname{Im} z > \operatorname{sgn} b(\operatorname{Im} w_0 + Nb)$.

Indeed, if $\operatorname{sgn} b \cdot \operatorname{Im} z > \operatorname{sgn} b(\operatorname{Im} w_0 + Nb)$, then $z = \alpha + ikb$ for some $\alpha \in [w_0, w_0 + ib]$ and $k \geq N$.

Hence, $k|b| \geq |\alpha|$ and $k > \frac{1}{|b|}\left(\frac{|\varphi(\alpha)-\alpha|}{\varepsilon} + |\alpha|\right)$. Consequently, $|\alpha + ikb| > k|b| - |\alpha| > \frac{|\varphi(\alpha)-\alpha|}{\varepsilon}$.

Now using (6.3.31), we obtain that

$$\left|\frac{\varphi(z)}{z} - 1\right| = \left|\frac{\varphi(\alpha + kib)}{\alpha + kib} - 1\right| = \left|\frac{\varphi(\alpha) - \alpha}{\alpha + ikb}\right| < \varepsilon.$$

Thus, $\lim\limits_{\ell \ni z \to \infty} \frac{\varphi(z)}{z} = 1$. It now follows from Lindelöf's theorem 1.6 that $\angle \lim\limits_{z\to\infty} \frac{\varphi(z)}{z} = 1$, which contradicts our assumption. Therefore the mapping G is indeed of hyperbolic type. □

Theorem 6.17. *Let $\mathcal{S}_1 = \{F_t\}_{t\geq 0}$ and $\mathcal{S}_2 = \{G_t\}_{t\geq 0}$ be continuous semigroups on Δ generated by f and g, respectively, and assume that $F_1 \circ G_1 = G_1 \circ F_1$ and $G_1 \neq I$. Suppose that f has a boundary null point $\tau \in \partial\Delta$, such that $f'(\tau) := \angle \lim\limits_{z\to\tau} f'(z) > 0$, i.e., the semigroup $\mathcal{S}_1$ is of hyperbolic type. Then the semigroups $\mathcal{S}_1$ and $\mathcal{S}_2$ commute.*

Proof. By our assumption, τ is the Denjoy–Wolff point of the semigroup $\mathcal{S}_1$.

First we suppose that $\mathcal{S}_1$ consists of automorphisms of Δ. Since $f'(\tau) > 0$, $\mathcal{S}_1$ consists of hyperbolic automorphisms of Δ and its generator f is of the form

$$f(z) = \frac{a_1}{\tau - \varsigma}(z - \tau)(z - \varsigma),$$

where a_1 is a positive real number and ς is the second common fixed point of the semigroup $\mathcal{S}_1$ (see [16]).

Again Lemma 6.4 and the commutativity of F_1 and G_1 imply that G_1 (hence, G_t, $t \geq 0$) are hyperbolic automorphisms of Δ. Moreover, $\mathcal{S}_2$ has the same fixed points τ and ς; consequently, its generator g is of the form

$$g(z) = \frac{a_2}{\varsigma - \tau}(z-\tau)(z-\varsigma),$$

where a_2 is a non-zero real number. Hence, $g(z) = -\frac{a_2}{a_1} f(z)$, and by Theorem 6.14, the semigroups commute.

Suppose now that the semigroup $\mathcal{S}_1$ consists of self-mappings of Δ that are not automorphisms. By a result in [14], τ is the common Denjoy–Wolff point of $\mathcal{S}_1$ and $\mathcal{S}_2$.

Then by Lemma 6.4 and Proposition 6.5, $\mathcal{S}_2$ consists also of hyperbolic mappings which are not automorphisms. Now our theorem is seen to be a consequence of a result of Cowen (see [39]). □

Remark 6.11. Note that if $\mathcal{S}_1$ and $\mathcal{S}_2$ are commuting semigroups of hyperbolic type generated by f and g, respectively, then $f(z) = kg(z)$, where $k = \frac{f'(\tau)}{g'(\tau)}$ is a real constant. (This constant is positive whenever the semigroups are not groups.)

Therefore $\mathcal{S}_1$ and $\mathcal{S}_2$ coincide up to rescaling. In particular, if in Theorem 6.17 the derivatives $F_1'(\tau) = e^{-f'(\tau)}$ and $G_1'(\tau) = e^{-g'(\tau)}$ are equal, then $F_t(z) = G_t(z)$ for all $t \geq 0$ and $z \in \Delta$.

6.3.4 Parabolic type

We start with the following auxiliary result:

Lemma 6.5. *Let $F, G \in \mathrm{Hol}(\Delta)$ be two commuting univalent parabolic mappings and let $\tau = 1$ be the Denjoy–Wolff point of F. If one of the following conditions*

(i) $F, G \in C^2(1)$, $F''(1) \neq 0$, $G''(1) \neq 0$;

(ii) $F, G \in C_A^2(1)$, $G''(1) \neq 0$, $\operatorname{Re} F''(1) > 0$;

(iii) $F, G \in C^3(1)$, $F''(1) = G''(1) = 0$, $F'''(1) \neq 0$, $G'''(1) \neq 0$

holds, then there exists a univalent function $\sigma \in \mathrm{Hol}(\Delta, \mathbb{C})$ such that

$$\sigma \circ F = \sigma + 1 \tag{6.3.32}$$

and

$$\sigma \circ G = \sigma + \lambda \quad \text{with} \quad \lambda \in \mathbb{C},\ \lambda \neq 0. \tag{6.3.33}$$

Proof. Let $\varphi = C \circ F \circ C^{-1}$, $\psi = C \circ G \circ C^{-1}$, $\varphi, \psi \in \mathrm{Hol}(\Pi_+)$, where $\Pi_+ = \{z \in \mathbb{C} : \operatorname{Re} z > 0\}$ and C is the Cayley transformation given by $C(z) = \dfrac{1+z}{1-z}$. Then φ and ψ are commuting parabolic maps in $\mathrm{Hol}(\Pi_+)$ having ∞ as their common Denjoy–Wolff point.

Denote $w_0 := 1$, $w_n^0 := \varphi_n(1)$, $n = 1, 2, \ldots$,

$$w_n := \varphi_n(w), \quad w_n \in \Pi_+,$$

and

$$h_n(w) := \frac{w_n - w_n^0}{w_{n+1}^0 - w_n^0}, \qquad w \in \Pi_+.$$

Then $h_n \in \text{Hol}(\Pi_+, \mathbb{C})$, and the sequence $\{h_n\}_{n=1}^{\infty}$ converges in the compact open topology to a holomorphic function $h \in \text{Hol}(\Pi_+, \mathbb{C})$ such that $h \circ \varphi = h + 1$ and the function $\sigma := h \circ C$ solves equation (6.3.32) (see [35]). Since F is univalent in Δ, the solution σ of Abel's equation (6.3.32) is also univalent in Δ.

Suppose that (i) holds. Then the following expansions of φ and ψ at ∞ are satisfied (see [18]):

$$\varphi(w) = w + F''(1) + \gamma_\varphi(w), \quad \lim_{w\to\infty} \gamma_\varphi(w) = 0 \tag{6.3.34}$$

and

$$\psi(w) = w + G''(1) + \gamma_\psi(w), \quad \lim_{w\to\infty} \gamma_\psi(w) = 0. \tag{6.3.35}$$

Hence,

$$\begin{aligned} h_n(\psi(w)) &= \frac{\varphi_n(\psi(w)) - w_n^0}{w_{n+1}^0 - w_n^0} = \frac{\psi(\varphi_n(w)) - w_n^0}{w_{n+1}^0 - w_n^0} \\ &= \frac{w_n + G''(1) + \gamma_\psi(w_n) - w_n^0}{w_{n+1}^0 - w_n^0} = \frac{w_n - w_n^0}{w_{n+1}^0 - w_n^0} + \frac{G''(1) + \gamma_\psi(w_n)}{w_{n+1}^0 - w_n^0} \\ &= h_n(w) + \frac{G''(1) + \gamma_\psi(w_n)}{F''(1) + \gamma_\varphi(w_n)} \cdot \frac{F''(1) + \gamma_\varphi(w_n)}{w_{n+1}^0 - w_n^0}. \end{aligned}$$

Letting $n \to \infty$, we obtain

$$h(\psi(w)) - h(w) = \frac{G''(1)}{F''(1)} \cdot \lim_{n\to\infty} \frac{F''(1) + \gamma_\varphi(w_n)}{w_{n+1}^0 - w_n^0}. \tag{6.3.36}$$

Repeating this calculation with φ instead of ψ, we find that

$$h(\varphi(w)) = h(w) + \lim_{n\to\infty} \frac{F''(1) + \gamma_\varphi(w_n)}{w_{n+1}^0 - w_n^0}.$$

At the same time, $h \circ \varphi = h + 1$. Hence $\lim\limits_{n\to\infty} \dfrac{F''(1) + \gamma_\varphi(w_n)}{w_{n+1}^0 - w_n^0} = 1$.

Rewrite (6.3.36) as follows:

$$h(\psi(w)) - h(w) = \lambda, \quad \text{where} \quad \lambda = \frac{G''(1)}{F''(1)} \neq 0 \quad \text{and} \quad w \in \Pi_+.$$

Substituting $h = \sigma \circ C^{-1}$ and $\psi = C \circ G \circ C^{-1}$ in the last equality, we get (6.3.33).

If (ii) holds, then Theorem 14 in [36] implies that for each $z \in \Delta$, the sequence $\{F_n(z)\}_{n=1}^{\infty}$ converges to 1 (and, consequently, $\{w_n\}$ converges to ∞) non-tangentially. So, in this case, one can repeat the proof of item (i), replacing the unrestricted limits in (6.3.34) and (6.3.35) by the angular limits.

Suppose now that (iii) holds. Then the following expansions of φ and ψ at ∞ hold (see [18]):

$$\varphi(w) = w - \frac{2}{3}\frac{F'''(1)}{w+1} + \Gamma_\varphi(w), \quad \lim_{w\to\infty} \Gamma_\varphi(w)w = 0 \tag{6.3.37}$$

and

$$\psi(w) = w - \frac{2}{3}\frac{G'''(1)}{w+1} + \Gamma_\psi(w), \quad \lim_{w\to\infty} \Gamma_\psi(w)w = 0. \tag{6.3.38}$$

Therefore

$$\begin{aligned} h_n(\varphi(w)) &= \frac{\varphi(w_n) - w_n^0}{w_{n+1}^0 - w_n^0} = \frac{w_n - w_n^0}{w_{n+1}^0 - w_n^0} + \frac{-\frac{2}{3}\frac{F'''(1)}{w_n+1} + \Gamma_\varphi(w_n)}{w_{n+1}^0 - w_n^0} \\ &= h_n(w) + \frac{-\frac{2}{3}\frac{F'''(1)}{w_n+1} + \Gamma_\varphi(w_n)}{w_{n+1}^0 - w_n^0}. \end{aligned}$$

Letting $n \to \infty$, we obtain

$$h(\varphi(w)) = h(w) + \lim_{n\to\infty} \frac{-\frac{2}{3}\frac{F'''(1)}{w_n+1} + \Gamma_\varphi(w_n)}{w_{n+1}^0 - w_n^0}.$$

On the other hand, $h(\varphi(w)) = h(w) + 1$. Hence,

$$\lim_{n\to\infty} \frac{-\frac{2}{3}\frac{F'''(1)}{w_n+1} + \Gamma_\varphi(w_n)}{w_{n+1}^0 - w_n^0} = 1. \tag{6.3.39}$$

Now using (6.3.38), we find

$$\begin{aligned} h_n(\psi(w)) &= \frac{\psi(w_n) - w_n^0}{w_{n+1}^0 - w_n^0} = \frac{w_n - \frac{2}{3}\frac{G'''(1)}{w_n+1} + \Gamma_\psi(w_n) - w_n^0}{w_{n+1}^0 - w_n^0} \\ &= h_n(w) + \frac{-\frac{2}{3}G'''(1) + \Gamma_\psi(w_n)(w_n+1)}{-\frac{2}{3}F'''(1) + \Gamma_\varphi(w_n)(w_n+1)} \cdot \frac{-\frac{2}{3}\frac{F'''(1)}{w_n+1} + \Gamma_\varphi(w_n)}{w_{n+1}^0 - w_n^0}. \end{aligned}$$

Letting $n \to \infty$ and using (6.3.39), we get

$$h(\psi(w)) - h(w) = \lambda, \quad w \in \Pi_+, \quad \text{where} \quad \lambda = \frac{G'''(1)}{F'''(1)} \neq 0.$$

Consequently, $\sigma \circ G - \sigma = \lambda$. $\square$

Following [35], we say that the function σ mentioned in the lemma is *the Kœnigs intertwining function* associated with F.

Remark 6.12. The function σ in Lemma 6.5 is completely determined by the function F. It does not depend on G. So if the conditions of the lemma hold for the same function F and another function $G_1 \in \mathrm{Hol}(\Delta)$, then we have the equality

$$\sigma \circ G_1 = \sigma + \lambda_1$$

with the same function σ and a constant $\lambda_1 \neq 0$.

Theorem 6.18. *Let $\mathcal{S}_1 = \{F_t\}_{t\geq 0}$ and $\mathcal{S}_2 = \{G_t\}_{t\geq 0}$ be two nontrivial continuous semigroups on Δ generated by f and g, respectively, and let $F_1 \circ G_1 = G_1 \circ F_1$.*

Suppose that $\tau = 1$ is the boundary null point of f such that $f'(1) = 0$. If $\mathcal{S}_1, \mathcal{S}_2 \subset C^0(1)$ and one of the following conditions

(i) *the semigroups $\mathcal{S}_1$ and $\mathcal{S}_2$ are of non-automorphic type;*

(ii) *the unrestricted limits $\alpha := \lim\limits_{z\to 1} f''(z)$ and $\widetilde{\alpha} := \lim\limits_{z\to 1} g''(z)$ exist and are different from zero;*

holds, then the semigroups commute.

Proof. Since $\tau = 1$ is a boundary null point of f and $f'(1) = 0$, it is the common Denjoy–Wolff point of the semigroup $\mathcal{S}_1$. The commutativity of F_1 and G_1 implies that $\tau = 1$ is the Denjoy–Wolff point of G_1 (see [14]) and, consequently, it is also the common Denjoy–Wolff point of the semigroup $\mathcal{S}_2$.

Suppose that condition (i) holds. Consider the holomorphic function σ defined by

$$\sigma'(z) = \frac{-1}{f(z)}, \quad \sigma(0) = 0. \tag{6.3.40}$$

It follows by the Berkson–Porta formula (see Theorem 2.7) that f can be presented in the form $f(z) = -(1-z)^2 p(z)$, where $\mathrm{Re}\, p(z) \geq 0$. Thus

$$\mathrm{Re}\, \frac{\sigma'(z)}{q'(z)} \geq 0,$$

where $q(z) = \dfrac{z}{1-z}$ is a univalent convex function. Then the function σ is close-to-convex, hence univalent in the open unit disk Δ (see, for example, Theorem 2.17 in [43]).

In addition, it can be shown (see, for example, [136] and [49]) by using (6.3.40) that the function σ satisfies the following functional equation:

$$\sigma\left(F_t(z)\right) = \sigma(z) + t. \tag{6.3.41}$$

Define now a univalent function $\widetilde{\sigma}$ in the unit disk by $\widetilde{\sigma} = \sigma \circ G_1$. It follows from (6.3.41) with $t = 1$ and the commutativity of F_1 and G_1 that

$$\widetilde{\sigma}\left(F_1(z)\right) = \widetilde{\sigma}(z) + 1.$$

Then by Theorem 3.1 in [35], there exists a constant $\lambda \in \mathbb{C}$ such that $\widetilde{\sigma} = \sigma + \lambda$, i.e.,

$$\sigma\left(G_1(z)\right) = \sigma(z) + \lambda. \tag{6.3.42}$$

Note that if $\lambda = 0$, the univalence of σ implies that $G_1(z) \equiv z$. In this case all the functions G_t, $t \geq 0$, coincide with the identity mapping. Hence the semigroups $\mathcal{S}_1$ and $\mathcal{S}_2$ commute. Therefore we can suppose that $\lambda \neq 0$.

Consider now the holomorphic function σ_1 defined by

$$\sigma_1'(z) = \frac{-1}{g(z)}, \quad \sigma_1(0) = 0. \tag{6.3.43}$$

As above, σ_1 is univalent and satisfies the functional equation

$$\sigma_1\left(G_t(z)\right) = \sigma_1(z) + t.$$

In particular, σ_1 is the Kœnigs intertwining function associated with G_1,

$$\sigma_1\left(G_1(z)\right) = \sigma_1(z) + 1. \tag{6.3.44}$$

Comparing equations (6.3.42) and (6.3.44) and using again Theorem 3.1 in [35], we obtain that $\sigma = \lambda\sigma_1 + \mu$ for some complex number μ. Differentiating the last equality, we obtain by (6.3.40) and (6.3.43), that $g(z) = \lambda f(z)$. So, by Theorem 6.14, the semigroups $\mathcal{S}_1$ and $\mathcal{S}_2$ commute.

Suppose now that condition (ii) holds. It can be shown that for each $t \geq 0$,

$$\lim_{z \to 1} F_t''(z) = -\alpha t$$

and

$$\lim_{z \to 1} G_t''(z) = -\widetilde{\alpha} t.$$

We have already seen in the proof of Lemma 6.5 that

$$\sigma(G_1(z)) - \sigma(z) = \frac{G_1''(1)}{F_1''(1)}. \tag{6.3.45}$$

Since $\operatorname{Re} F_1''(1) = 0$ and $\operatorname{Re} G_1''(1) = 0$ (see Theorem 4.4 in [18]), it follows that $\frac{G_1''(1)}{F_1''(1)} \in \mathbb{R} \setminus \{0\}$. Moreover,

$$F_t''(1) = -\alpha t \quad \text{and} \quad G_t''(1) = -\widetilde{\alpha} t, \quad t > 0,$$

where $\alpha = f''(1) \neq 0$ and $\widetilde{\alpha} = g''(1) \neq 0$. So equality (6.3.45) has the form

$$\sigma(G_1(z)) - \sigma(z) = p, \quad \text{where} \quad p := \frac{\widetilde{\alpha}}{\alpha}. \tag{6.3.46}$$

On the other hand,

$$\sigma(F_t(z)) - \sigma(z) = \frac{F_t''(1)}{F_1''(1)} = \frac{\alpha t}{\alpha} = t \quad \text{for all} \quad t \geq 0. \tag{6.3.47}$$

First we suppose that $p > 0$. From (6.3.46) and (6.3.47) we have $\sigma(G_1(z)) = \sigma(F_p(z))$, $z \in \Delta$, and by the univalence of σ on Δ, $G_1(z) = F_p(z)$ for all $z \in \Delta$. Hence, $G_1 \circ F_t = F_t \circ G_1$ for all $t \geq 0$.

Fix $t > 0$ and repeat these considerations with G_1, F_t, G_s and $\widetilde{\sigma}$ instead of F_1, G_1, F_t and σ, respectively. Namely,

$$\widetilde{\sigma}(F_t(z)) - \widetilde{\sigma}(z) = \frac{F_t''(1)}{G_1''(1)} = \frac{\alpha t}{\widetilde{\alpha}} > 0$$

and

$$\widetilde{\sigma}(G_s(z)) - \widetilde{\sigma}(z) = \frac{G_s''(1)}{G_1''(1)} = s \quad \text{for all} \quad s > 0.$$

Let $\widetilde{s} := \frac{\alpha t}{\widetilde{\alpha}} > 0$. Then $\widetilde{\sigma}(F_t(z)) = \widetilde{\sigma}(G_{\widetilde{s}}(z))$, $z \in \Delta$. By the univalence of $\widetilde{\sigma}$ on Δ we have $F_t(z) = G_{\widetilde{s}}(z)$. Therefore $G_s \circ F_t = F_t \circ G_s$ for all $s > 0$. Since $t > 0$ is arbitrary, it follows that the semigroups $\mathcal{S}_1 = \{F_t\}_{t \geq 0}$ and $\mathcal{S}_2 = \{G_s\}_{s \geq 0}$ commute.

Let now $p < 0$. Then by (6.3.47), $\sigma(F_{-p}(z)) - \sigma(z) = -p$ for all $z \in \Delta$. Hence, by (6.3.46),

$$\sigma(F_{-p}(G_1(z))) - \sigma(G_1(z)) = \sigma(z) - \sigma(G_1(z)), \quad z \in \Delta,$$

and, therefore,

$$\sigma(F_{-p}(G_1(z))) = \sigma(z), \quad z \in \Delta.$$

By the univalence of σ on Δ, $F_{-p}(G_1(z)) = z$. Consequently, $F_{-p} = G_1^{-1}$ on $G_1(\Delta)$. Since $F_{-p} \in \mathrm{Hol}(\Delta)$, G_1^{-1} is well defined on Δ and so G_1, as well as F_{-p}, are automorphisms of Δ. Therefore, $\{F_t\}_{t \geq 0}$ is a semigroup of automorphisms (see, for example, [3]). Consequently, it can be extended to a group $\mathcal{S}_F = \{F_t\}_{t \in \mathbb{R}}$ and $G_1 = F_p^{-1} = F_{-p} \in \mathcal{S}_F$. In particular, $G_1 \circ F_t = F_t \circ G_1$ for all $t \geq 0$.

Fix $t > 0$. In a similar way, using the commutativity of F_t and G_1, one can show that the semigroup $\{G_s\}_{s \geq 0}$ can be extended to a group $\mathcal{S}_G = \{G_s\}_{s \in \mathbb{R}}$ and that $F_t \circ G_s = G_s \circ F_t$ for all $s, t \in \mathbb{R}$. □

Recently, Levenshtein and Reich [100] showed that condition (ii) in Theorem 6.18 may be weakened.

For discrete time semigroups, different interesting results on fixed points of commuting mappings based on the construction of Poggi-Corradini [107] are presented in [19].

Chapter 7

Asymptotic Behavior of One-parameter Semigroups

In this chapter, we continue an advanced study of the asymptotic behavior of discrete and continuous semigroups using some geometric properties of associated Kœnigs functions.

Our main goal is to establish rates of convergence of continuous semigroups to their attractive fixed point and to study angular characteristics of their trajectories at the Denjoy–Wolff point and other boundary regular fixed points (when they exist). It turns out that the three mutually exclusive classes of semigroups, although sharing some common features, also exhibit strikingly different phenomena.

For example, semigroups of parabolic type, which in general possess the most complicated behavior, are the simplest in the sense of their angular asymptotic characteristics.

In particular, all such semigroups which have the same second derivative at their common Denjoy–Wolff point share the same asymptote, irrespective of their initial points.

In the hyperbolic case, the situation is completely different. Although in this case each semigroup trajectory has an asymptote at its Denjoy–Wolff point, this asymptote does depend on the initial point of the trajectory. Moreover, if two such semigroups behave similarly as $t \to \infty$, they must coincide up to rescaling (see Section 7.2.2 below).

In general, the trajectories of a semigroup of dilation type have no asymptotes (except in the case when the trajectories themselves are straight lines). Nevertheless, one can use the angular characteristics of such trajectories to establish an asymptotic rigidity result in this case too (see Section 7.1.2 below).

7.1 Dilation case

7.1.1 General remarks and rates of convergence

The classical Schwarz–Pick Lemma implies that if $\zeta \in \Delta$ is an interior fixed point of $F \in \mathrm{Hol}(\Delta)$, i.e.,

$$F(\zeta) = \zeta, \tag{7.1.1}$$

then F leaves each pseudohyperbolic disk $D_r(\zeta)$ centered at ζ invariant.

In other words, for each $r \in (0,1)$,

$$F\left(D_r(\zeta)\right) \subseteq D_r(\zeta), \tag{7.1.2}$$

where

$$D_r(\zeta) = \left\{ z \in \Delta : \left| \frac{z-\zeta}{1-\bar{\zeta}z} \right| < r \right\} = \left\{ z \in \Delta : \frac{|1-z\bar{\zeta}|^2}{1-|z|^2} < K \right\}, \tag{7.1.3}$$

$K = \left(1-|\zeta|^2\right)\left(1-r^2\right)^{-1}$.

In turn, this fact and Vitali's Theorem show that a holomorphic self-mapping of Δ different from the identity has at most one interior fixed point in Δ (see, for example, Proposition 1.3.2 in [130]). However, simple examples show that holomorphic self-mappings of Δ may have many fixed points on the circle $\partial\Delta$. We begin with the dilation case when $F \in \mathrm{Hol}(\Delta)$ has an interior fixed point $\zeta \in \Delta$.

An additional consequence of the Schwarz–Pick Lemma is that

$$|F'(\zeta)| \leq 1. \tag{7.1.4}$$

Moreover, equality in (7.1.2) or (7.1.4) holds if and only if F is an (elliptic) automorphism of Δ.

On the other hand, if the strong inequality $|F'(\zeta)| < 1$ holds, then it follows by the Banach Contraction Principle (locally applying to F) and Compact Convergence Principle that for each $z \in \Delta$, the sequence $\left\{F^{(n)}(z)\right\}_{n=0}^{\infty}$ (the orbit) converges to ζ as n goes to infinity. This is a part of the Denjoy–Wolff Theorem (see Section 1.3), which is helpful in the study of the asymptotic behavior of discrete-time semigroups defined by iterates of a holomorphic self-mapping of Δ.

To summarize we formulate the following assertion.

Proposition 7.1. *Let $F \in \mathrm{Hol}(\Delta)$ have a fixed point $\zeta \in \Delta$. Then*

(i) *for each $r \in (0,1)$ and $n = 0,1,2,\dots$, the following invariance condition holds:*

$$F^{(n)}(D_r(\zeta)) \subseteq D_r(\zeta);$$

(ii) *if F is not the identity, then the point $\zeta \in \Delta$ is a unique fixed point of F in Δ. Moreover, the following are equivalent:*

(a) *For each* $z \in \Delta$*, the sequence* $\left\{F^{(n)}(z)\right\}_{n=0}^{\infty}$ *of iterates converges to* ζ *as* n *goes to infinity.*

(b) *The mapping* F *is not an (elliptic) automorphism of* Δ.

(c) $|F'(\zeta)| < 1$.

Combining this statement with Brouwer's Fixed Point Principle , one gets the following sufficient condition of existence and uniqueness of an interior fixed point for holomorphic self-mappings of the unit disk.

Corollary 7.1. *Suppose that* $F \in \mathrm{Hol}(\Delta)$ *maps* Δ *strictly inside, i.e., for some* $r \in (0,1)$,

$$|F(z)| \le r \tag{7.1.5}$$

for all $z \in \Delta$*. Then* F *has a unique fixed point* $\zeta \in \Delta$, $|\zeta| \le r$*, and for each* $z \in \Delta$ *the orbit* $\left\{F^{(n)}(z)\right\}_{n=0}^{\infty}$ *converges to* ζ *as* n *goes to infinity.*

In fact, (7.1.5) means that $F \in \mathrm{Hol}(\Delta)$ is a strict contraction in the Poincaré hyperbolic metric on Δ. So, Corollary 7.1 is also a consequence of the Banach Fixed Point Principle. At the same time, if condition (7.1.5) does not hold, the problem of finding a global rate of convergence of $\{F_n\}_{n=1}^{\infty}$ to the interior fixed point $\zeta \in \Delta$ seems to be open.

This problem, however, can be successfully solved for continuous semigroups (see Proposition 7.4 below).

To proceed with continuous semigroups we want to trace a connection between the iteration theory of functions in one complex variable and the asymptotic behavior of solutions of ordinary differential equations governed by evolution problems. Therefore, our terminology is related to both these topics.

Definition 7.1. *A point* $\zeta \in \Delta$ *is said to be a stationary point of a continuous semigroup* $\mathcal{S} = \{F_t\}_{t>0} \subset \mathrm{Hol}(\Delta)$ *if*

$$F_t(\zeta) = \zeta \tag{7.1.6}$$

for all $t > 0$.

In other words, $\zeta \in \Delta$ is a stationary point of $\mathcal{S}$ if it is a common fixed point of all $F_t \in \mathcal{S}$.

Note that the family $\mathcal{S} = \{F_t\}_{t\ge 0}$ is commuting, that is $F_t \circ F_s = F_s \circ F_t = F_{t+s}$ for all $t, s \ge 0$. Hence, it follows by Shields' theorem [129] that if each F_t had been continuously extended to $\partial\Delta$, the boundary of Δ, then the stationary point set of $\mathcal{S}$ would not be empty.

As a matter of fact, it is enough to require the existence of an interior fixed point only for one $t > 0$ to ensure the existence of such a point for the whole semigroup. Indeed, if for at least one $t > 0$ the mapping $F_t \in \mathcal{S}$ has an interior fixed point $\zeta \in \Delta$, then it is a unique fixed point for F_t, and for each $s \geq 0$ we have:

$$F_s(\zeta) = F_s(F_t(\zeta)) = F_t(F_s(\zeta)) = \zeta,$$

i.e., ζ is also a fixed point of each element $F_s \in \mathcal{S},\ s \geq 0$. Hence, this fixed point is a unique stationary point of $\mathcal{S}$.

Using the chain rule and the Schwarz–Pick Lemma, it is easy to show that if $\zeta \in \Delta$ is a stationary point for $\mathcal{S} = \{F_t\}_{t>0}$, then $(F_t)'(\zeta) = e^{-kt}$ with some complex k such that $\operatorname{Re} k \geq 0$.

Naturally, the strategy now is to study the convergence of a semigroup to its stationary point. The foregoing fact is the first step in the study of the asymptotic behavior of a continuous semigroup in Δ.

Proposition 7.2 ([1, 91]). *Let $\mathcal{S} = \{F_t\}_{t>0} \subset \operatorname{Hol}(\Delta)$ be a semigroup on Δ. Then this semigroup converges uniformly on compact subsets of Δ to a holomorphic mapping $F \in \operatorname{Hol}(\Delta, \mathbb{C})$ if and only if, for at least one t_0, the sequence $\{F_{t_0 n}\}_{n=0}^{\infty}$ converges uniformly on compact subsets of Δ. Moreover, if F_{t_0} is not the identity, then F is a constant with modulus less than or equal to 1.*

The fact that we have formulated above implies immediately a continuous analog of the Denjoy–Wolff Theorem [17, 1, 91] (see Section 2.3).

♦ *Let $\mathcal{S} = \{F_t\}_{t\geq 0} \subset \operatorname{Hol}(\Delta)$ be a flow on Δ. If for at least one t_0 the mapping F_{t_0} is not the identity and is not an elliptic automorphism of Δ, then the net $\{F_t\}_{t\geq 0}$ converges to a constant $\zeta \in \overline{\Delta}$ as $t \to \infty$ uniformly on each compact subset of Δ.*

Since every continuous semigroup $\mathcal{S} = \{F_t\}_{t\geq 0}$ of holomorphic self-mappings of Δ is differentiable in $t \geq 0$, it is natural to describe its asymptotic behavior in terms of the generator $f = \lim\limits_{t\to 0} \frac{1}{t}\left(I - F_t\right)$. This becomes more desirable when such a semigroup is not given explicitly, but is defined as the solution of the Cauchy problem:

$$\begin{cases} \dfrac{\partial u(t,z)}{\partial t} + f\left(u(t,z)\right) = 0 \\ u(0,z) = z \in \Delta. \end{cases} \tag{7.1.7}$$

Here we set $F_t(z) = u(t,z)$.

Note also that if f is holomorphic in a neighborhood of the point $\zeta \in \overline{\Delta}$, then it follows by the uniqueness of the solution of the Cauchý problem that $f(\zeta) = 0$ if and only if ζ is a stationary point of $\mathcal{S} = \{F_t\}_{t\geq 0}$. In particular, an interior null point of a generator is a stationary point of the generated semigroup. However, this fact is no longer true for a boundary null point even when f is continuous (but not differentiable) on $\overline{\Delta}$.

The following assertion is a nonlinear analog of Lyapunov's Stability Theorem.

Proposition 7.3 (see [130]). *Let $f \in \mathrm{Hol}(\Delta, \mathbb{C})$ be the generator of a one-parameter semigroup $\mathcal{S} = \{F_t\}_{t\geq 0}$ with $f(\tau) = 0$ for some $\tau \in \Delta$. Then*

(i) $\operatorname{Re} f'(\tau) \geq 0$;

(ii) $\operatorname{Re} f'(\tau) > 0$ *if and only if* $\mathcal{S} = \{F_t\}_{t\geq 0}$ *converges to* τ *uniformly on compact subsets of* Δ.

Note that if $f \not\equiv 0$ and $\operatorname{Re} f'(\tau) = 0$, then the semigroup $\mathcal{S} = \{F_t\}_{t\geq 0}$ is actually a group of elliptic automorphisms (see [130]). Combining this fact with the Berkson–Porta Theorem 2.7, we conclude that *if $f \in \mathcal{G}(\Delta)$ is not a generator of the form*

$$f(z) = (z - \tau)(1 - \bar{\tau} z)a$$

with $\tau \in \Delta$ and $\operatorname{Re} a = 0$, then the semigroup $\mathcal{S} = \{F_t\}_{t\geq 0}$ generated by f does not consist of elliptic automorphisms and converges uniformly on each compact subset of Δ to a point in $\overline{\Delta}$. If this point belongs to Δ, then it is a unique uniformly attractive stationary point of the semigroup, and the question of finding a rate of convergence arises naturally.

We consider here the case when $f(0) = 0$, i.e., the origin is a stationary point of the semigroup $S = \{F_t\}_{t\geq 0}$.

Proposition 7.4 (see [77], [114] and [130]). *Let $f \in \mathcal{G}$ be a holomorphic generator with $f(0) = 0$ and $k = \operatorname{Re} f'(0) > 0$, and let $\mathcal{S} = \{F_t\}_{t\geq 0}$ be the semigroup generated by f. Then there exists $c \in [0, 1]$ such that for all $z \in \Delta$ and $t \geq 0$ the following estimates hold:*

(a) $|z| \exp(-k \dfrac{1 + c|z|}{1 - c|z|} t) \leq |F_t(z)| \leq |z| \cdot \exp\left(-k \dfrac{1 - c|z|}{1 + c|z|} t\right)$;

(b) $\exp(-kt) \dfrac{|z|}{(1 + c|z|)^2} \leq \dfrac{|F_t(z)|}{(1 - c|F_t(z)|)^2} \leq \exp(-kt) \dfrac{|z|}{(1 - c|z|)^2}$.

Note that estimate (a) with $c = 1$ is due to Gurganus [77], while estimate (b) was established by Poreda [114].

A more general case, when $f \in \mathcal{G}$ with $f(\tau) = 0$ for some $\tau \in \Delta$ which is not necessarily zero, can be obtained easily from the above theorem by using the Möbius transformation

$$M_\tau(z) = \frac{\tau - z}{1 - \bar{\tau} z}, \quad z \in \Delta.$$

Some other rates of convergence in terms of the hyperbolic Poincaré metric on Δ can be found in [57] and [58].

7.1.2 Argument rigidity principle

Now we turn again to a continuous time semigroup of dilation type generated by $f \in \mathcal{G}[\tau]$,$\tau \in \Delta$, $f(\tau) = 0$ and $\operatorname{Re} f'(\tau) > 0$. In this situation, the trajectories $\{F_t(z)\}_{t\geq 0}$ are either straight lines (if $\operatorname{Im} f'(\tau) = 0$) or spirals around τ (if

$\operatorname{Im} f'(\tau) \neq 0$). In the latter case one cannot consider the angular characteristic of the trajectories as in the hyperbolic or parabolic case (see Sections 7.2.2 and 7.3.3 below). Nevertheless, we are able to treat two (possibly different) semigroups in the sense of their similar asymptotic behavior according to the following notation.

Definition 7.2. *Let $\mathcal{S}_f = \{F_t\}_{t\geq 0}$ and $\mathcal{S}_g = \{G_t\}_{t\geq 0}$ be two semigroups generated by f and g respectively, such that they have the same Denjoy–Wolff point $\tau \in \overline{\Delta}$. We say that these semigroups have similar asymptotic behavior if*

$$\lim_{t\to\infty}\left(\arg\frac{\tau - F_t(z)}{\tau - G_t(z)}\right) = 0. \tag{7.1.8}$$

Theorem 7.1 ([60]). *Let f and g be generators of semigroups $\mathcal{S}_f = \{F_t\}_{t\geq 0}$ and $\mathcal{S}_g = \{G_t\}_{t\geq 0}$, respectively. Suppose that $f(\tau) = g(\tau) = 0$ for some $\tau \in \Delta$ and $\operatorname{Re} f'(\tau)\cdot\operatorname{Re} g'(\tau) > 0$. In other words, $\mathcal{S}_f$ and $\mathcal{S}_g$ have the same (interior) Denjoy–Wolff point $\tau \in \Delta$. Then they have similar asymptotic behavior, i.e.,*

$$\lim_{\tau\to\infty}\left(\arg\frac{\tau - F_t(z)}{\tau - G_t(z)}\right) = 0 \tag{7.1.9}$$

if and only if

$$f(z) = kg(z) \tag{7.1.10}$$

for some $k \in \mathbb{C}$ and

$$\operatorname{Im} f'(\tau) = \operatorname{Im} g'(\tau). \tag{7.1.11}$$

Moreover, in this case, the semigroups $\mathcal{S}_f$ and $\mathcal{S}_g$ must commute.

Proof. Set $\tau = 0$. Again we consider Kœnigs type functions (see Remark 5.4)

$$h_f(z) = \lim_{t\to\infty} e^{t\beta_f} F_t(z) \tag{7.1.12}$$

and

$$h_g(z) = \lim_{t\to\infty} e^{t\beta_g} G_t(z), \tag{7.1.13}$$

where $\beta_f = f'(0)$ and $\beta_g = g'(0)$ ($\operatorname{Re}\beta_f \cdot \operatorname{Re}\beta_g > 0$).

Then, by (7.1.9), (7.1.12) and (7.1.13), we get that

$$\lim_{t\to\infty} \arg\frac{F_t(z)}{G_t(z)} = \arg\frac{h_f(z)}{h_g(z)} + \lim_{t\to\infty} \arg e^{(\beta_g-\beta_f)t} = 0. \tag{7.1.14}$$

Setting $\mu := \beta_g - \beta_f$, we have also that

$$\arg e^{\mu t} = t \operatorname{Im}\mu. \tag{7.1.15}$$

Therefore, we obtain from (7.1.14) and (7.1.15) that (7.1.9) holds if and only if

$$\arg h_f(z) = \arg h_g(z) \tag{7.1.16}$$

and

$$\operatorname{Im}\beta_f = \operatorname{Im}\beta_g. \tag{7.1.17}$$

Since $h_f(0) = h_g(0) = 0$ and $h'_f(0) = h'_g(0) = 0$, we have from (7.1.16) that

$$h_f(z) = h_g(z) =: h(z), \quad z \in \Delta, \tag{7.1.18}$$

and, consequently,

$$f(z) = kg(z),$$

where $k = \frac{\beta_f}{\beta_g}$.

Now Theorem 6.14 implies that the semigroups $\mathcal{S}_f$ and $\mathcal{S}_g$ commute.

The reverse consideration completes our proof. □

Remark 7.1. Note that, in contrast to the hyperbolic case, the trajectories of a dilation type of semigroups that satisfy condition (7.1.9) are not necessarily the same. Moreover, they coincide if and only if

$$\operatorname{Im}\beta_f = \operatorname{Im}\beta_g = 0 \quad \text{or} \quad \beta_f = \beta_g.$$

7.2 Hyperbolic case

7.2.1 Criteria for the exponential convergence

Let $f \in \mathcal{G}[1]$, i.e., f is a semigroup generator with $\lim\limits_{r\to 1^-} f(r) = 0$ and $\beta = f'(1) := \lim\limits_{r\to 1^-} \frac{f(r)}{r-1}$ exists (see (2.2.9)), and let $\mathcal{S} = \{F_t\}_{t\geq 0}$ be a semigroup generated by f. It follows by Theorems 2.12 and 1.10 that

$$\frac{|1 - F_t(z)|^2}{1 - |F_t(z)|^2} \leq e^{-t\beta} \frac{|1-z|^2}{1-|z|^2}. \tag{7.2.1}$$

In addition, $\lim\limits_{t\to\infty} F_t(z) = 1$ if and only if $\beta \geq 0$ (see Theorem 2.10). Moreover, if $\beta > 0$, this estimate establishes the exponential rate of convergence of $\{F_t\}_{t\geq 0}$ to the point $\tau = 1$ in terms of a non-Euclidean distance defined by

$$d(z, \tau) = \frac{|\tau - z|^2}{1 - |z|^2},$$

where $\tau \in \partial\Delta$, $z \in \Delta$ (see details in Theorem 2.10).

The natural problem that has arisen here (see, for example, [10, 79]) is

◊ *Whether the norm convergence is exactly of exponential type, i.e.,*

$$F_t(z) - 1 \sim e^{-\beta t} \ ?$$

In other words, the question is

◊ *Whether the following limit*

$$K(z) := \lim_{t\to\infty} \exp(t\beta)\,(1 - F_t(z))$$

exists and is not zero?

The following well-known fact is a very important result in the study of the probability asymptotics of branching processes (see, for example, [126]):

♦ *Let F_t, $t \geq 0$, be the generating function of a subcritical continuous branching process. Then*

$$1 - F_t(0) = Ke^{-\beta t}(1 + o(1))$$

if and only if the (real) integral

$$\int_0^1 \frac{\beta x + f(1-x)}{xf(1-x)}\,dx$$

converges, where f is the infinitesimal generator of the semigroup $\mathcal{S} = \{F_t\}_{t\geq 0}$ of the form (5.2.6). *In this case, the integral is equal to $-\log K$.*

As a matter of fact, we will see below that this result does not depend on the probability nature of the semigroup rather than on the boundary behavior of the Kœnigs–Valiron function h defined by (5.4.19), which in the probability sense is closely related to the solution $\hat{g}$ of the so-called Yaglom equation

$$\hat{g}(F_t(z)) = m_t\hat{g}(z) + 1 - m_t. \tag{7.2.2}$$

Note that, for discrete-time semigroups this problem has been studied in [143] and [110]. It turns out that, for continuous-time semigroups the answer can be given in terms of the (K-V)–function

$$h(z) = h_0(z) = \lim_{t\to\infty} \frac{1 - F_t(z)}{1 - F_t(0)},$$

defined in Chapter 5.

First, we observe that, actually, in Theorem 5.6 we proved the following fact:

♦ *Let $f \in \mathcal{G}[1]$ have a positive angular derivative $\beta := f'(1) > 0$ and generate a semigroup of holomorphic self-mappings $\{F_t\}_{t\geq 0}$. Then the limit*

$$h(z) := \lim_{t\to\infty} \frac{1 - F_t(z)}{1 - F_t(0)} \tag{7.2.3}$$

exists and is a function of class $\mathrm{Star}_1[1]$.

Moreover, this function satisfies the differential equation

$$\beta h(z) = h'(z)f(z), \tag{7.2.4}$$

which is equivalent to Schröder's functional equation

$$h(F_t(z)) = \lambda_t h(z), \quad t \geq 0, \tag{7.2.5}$$

with $\lambda_t = e^{-t\beta}$ (see Theorem 5.1).

We now need the following consequence of Proposition 4.13 in [113].

Lemma 7.1. *Let ϕ be a univalent function on the unit disk Δ, and let $g \in \mathrm{Hol}(\Delta, \mathbb{C})$ satisfy $g(\Delta) \subset \phi(\Delta)$. Given a boundary point $\zeta \in \partial\Delta$, suppose that the limits $\angle\lim\limits_{z\to\zeta} g(z)$ and $\lim\limits_{z\to\zeta} \phi(z)$ exist finitely and coincide. Suppose also that $\lim\limits_{z\to\zeta} \phi'(z) = s \neq 0, \infty$. If there is a sequence $\{z_n\} \subset \Delta$ non-tangentially convergent to ζ such that*

$$g'(z_n) \to \alpha \in \mathbb{C},$$

then

$$\angle\lim_{z\to\zeta} g'(z) = \alpha.$$

Proof. First we define

$$g_1(z) := \left(\phi^{-1} \circ g\right)(z).$$

This function is a well-defined self-mapping of Δ because of $g(\Delta) \subset \phi(\Delta)$. Moreover, by our supposition, the point ζ is a boundary fixed point of g_1. Then, by Proposition 4.13 in [113], the limit

$$\angle\lim_{z\to\zeta} g_1'(z) = A \leq \infty$$

exists. Now let $\gamma : [0, \infty) \mapsto \Delta$ be a continuous curve in Δ such that $\gamma(t) \to \zeta$ non-tangentially as $t \to \infty$ and $\gamma(n) = z_n$. We already know that

$$\lim_{t\to\infty} g_1'(\gamma(t)) = A.$$

Hence,

$$\lim_{n\to\infty} g_1'(z_n) = \lim_{n\to\infty} g_1'(\gamma(n)) = A.$$

By definition, $g = \phi \circ g_1$, and so $g'(z) = \phi'(g_1(z)) \cdot g_1'(z)$. Therefore,

$$\alpha = \lim_{n\to\infty} g'(z_n) = \lim_{n\to\infty} \Big((\phi' \circ g_1)(z_n) \cdot g_1'(z_n) \Big) = sA.$$

Then $A \neq \infty$, and

$$\angle\lim_{z\to\zeta} g'(z) = \angle\lim_{z\to\zeta} \Big((\phi' \circ g_1)(z) \cdot g_1'(z) \Big) = sA = \alpha.$$

The lemma is proved. □

Definition 7.3. *Let f be a continuous function on the open unit disk Δ. We say that the integral $\int_0^1 f(z)dz$ converges non-tangentially if the function*

$$\varphi(z) = \int_0^z f(z)dz$$

has a non-tangential limit at the boundary point $z = 1$.

Theorem 7.2 (cf., [66]). *Let $f \in \mathcal{G}_h[1]$, i.e.,*

$$f(z) = -(1-z)^2 p(z) + \beta/2(z^2 - 1), \tag{7.2.6}$$

where $\beta (= f'(1)) > 0$ and $p \in \mathrm{Hol}(\Delta, \mathbb{C})$ with $\mathrm{Re}\, p > 0$ and $\angle \lim_{z\to 1} (1-z)p(z) = 0$. Let $\{F_t\}_{t\geq 0}$ be the semigroup generated by f. Then

(I) *the limit*

$$K(z) := \lim_{t\to\infty} \exp(t\beta)\,(1 - F_t(z))$$

exists.

(II) *the limit function K is either identically zero or a univalent function with $\mathrm{Re}\, K > 0$.*

(III) *$K(z) \not\equiv 0$ if and only if one of the following assertions holds:*

(a) *the integral*

$$\int_0^1 \frac{\beta(1-z) + f(z)}{(1-z)f(z)}\, dz$$

converges non-tangentially;

(b) *the integral*

$$\int_0^1 p(z)\, dz$$

converges non-tangentially;

(c) *the (K-V)-function h defined by (7.2.3) is conformal at the point $z = 1$.*

In this case $K(z) = K(0)h(z)$.

Proof. Any element F_t, $t \geq 0$, of the semigroup is a univalent self-mapping of the unit disk Δ. Therefore, each function

$$K_t(z) := \exp(t\beta)\,(1 - F_t(z))$$

is univalent and satisfies $\operatorname{Re} K_t > 0$. So, the family $\{K_t\}_{t\geq 0}$ is normal. In addition, it follows by (7.2.1) that this family is compact:

$$|K_t(z)| \leq \frac{|1-z|^2}{1-|z|^2}\frac{1-|F_t(z)|^2}{|1-F_t(z)|}$$
$$\leq \frac{|1-z|^2}{1-|z|^2}\frac{1-|F_t(z)|}{|1-F_t(z)|}(1+|F_t(z)|) \leq 2\frac{|1-z|^2}{1-|z|^2}.$$

Hence, there exists a convergent sequence $\{K_{t_n}\}_{n=1}^{\infty}$.

Since $\{F_t\}_{t\geq 0}$ converges to $\tau = 1$ as $t \to \infty$ non-tangentially, we see that

$$\frac{K_t'(z)}{K_t(z)} = \frac{-F_t'(z)}{1-F_t(z)} = \frac{1}{f(z)}\frac{f(F_t(z))}{F_t(z)-1} \to \frac{\beta}{f(z)} \quad \text{as } t \to \infty.$$

Thus, each limit function K satisfies the following equation:

$$\beta K(z) = K'(z) f(z). \tag{7.2.7}$$

Since (7.2.7) is a homogeneous linear differential equation which coincides with (7.2.4), it follows that K is either equal to zero identically, or $K(z) = \alpha h(z)$ for some $\alpha \neq 0$.

Suppose that there exists a sequence $\{K_{t_n}\}_{n=1}^{\infty}$ converging to a non-zero function

$$K(z) = \lim_{n\to\infty} K_{t_n}(z) = \lim_{n\to\infty} \exp(t_n\beta)\,(1 - F_{t_n}(z)) = \alpha h(z).$$

In this case, K is a univalent function on the unit disk Δ. We now show that the angular derivative $K'(1)$ exists and equals -1. We already know that any limit function K satisfies the differential equation (7.2.7). Then, by Theorem 5.1, K is a solution of Schröder's equation (7.2.5). Therefore, denoting $w_n = F_{t_n}(z)$ for fixed $z \in \Delta$ and using Schröder's equation (7.2.5), differential equation (7.2.7) and representation (7.2.6), one can calculate

$$\begin{aligned}
0 &= \lim_{n\to\infty}\left(e^{t_n\beta}(1-w_n) - e^{t_n\beta}K(w_n)\right)\\
&= \lim_{n\to\infty} e^{t_n\beta}\left[(1-w_n) - \frac{K'(w_n)f(w_n)}{\beta}\right]\\
&= \lim_{n\to\infty} e^{t_n\beta}\left[(1-w_n) + \frac{K'(w_n)\left[(1-w_n)^2p(w_n) - \beta/2(w_n^2-1)\right]}{\beta}\right]\\
&= \lim_{n\to\infty} e^{t_n\beta}(1-w_n)\left[1 + \frac{K'(w_n)(1-w_n)p(w_n)}{\beta} + \frac{K'(w_n)(w_n+1)}{2}\right]\\
&= K(z)\lim_{n\to\infty}\left[1 + K'(w_n) + K'(w_n)\frac{(1-w_n)p(w_n)}{\beta}\right].
\end{aligned}$$

So,

$$\lim_{n\to\infty} K'(F_{t_n}(z)) = -1.$$

By Lemma 7.1, we conclude that the angular derivative $K'(1)$ exists and equals -1. This implies that if there exists a sequence $\{K_{t_n}\}_{n=1}^{\infty}$ convergent to a non-zero function K, then the function $h(z) = K(z)/\alpha$ has the angular derivative $h'(1) = \frac{-1}{\alpha} \neq 0$, i.e., h is conformal at the point $z = 1$. Moreover, such a non-zero limit function is unique: $K(z) = -h(z)/h'(1)$ and

$$h(z) = \frac{1}{\alpha}\lim_{n\to\infty} K_{t_n}(z) = \frac{1}{\alpha}\lim_{n\to\infty} \exp\left(t_n\beta\right)\left(1 - F_{t_n}(z)\right).$$

Further, by (7.2.4) and (7.2.6), we get

$$h(z) = -h'(z)(1-z)\left[\frac{(1-z)p(z)}{\beta} + \frac{1+z}{2}\right].$$

Combining these equalities with (7.2.5), we simply calculate

$$\begin{aligned}
&\lim_{t\to\infty} \frac{1}{\alpha} K_t(z) - h(z) = \lim_{t\to\infty}\left[\frac{1}{\alpha}K_t(z) - e^{t\beta}h(F_t(z))\right]\\
&= \lim_{t\to\infty} K_t(z)\left[\frac{1}{\alpha} + h'(F_t(z))\left(\frac{(1-F_t(z))p(F_t(z))}{\beta} + \frac{1+F_t(z)}{2}\right)\right]\\
&= \lim_{t\to\infty} K_t(z)\left(\frac{1}{\alpha} + h'(F_t(z))\right) = 0,
\end{aligned}$$

because of $h'(1) = -\frac{1}{\alpha}$ and $\{K_t\}_{t\geq 0}$ is a compact family.

So, we have proved (I), (II) and (III c).

To proceed, we note that by (7.2.6),

$$\frac{\beta(1-z) + f(z)}{(1-z)f(z)} = \frac{p(z) - \frac{\beta}{2}}{(1-z)p(z) + \frac{\beta}{2}(1+z)}.$$

Since the denominator in the last fraction does not vanish in Δ and tends to $\beta \neq 0$ as z goes to 1 non-tangentially, we conclude that conditions (a) and (b) in assertion (III) of the theorem are equivalent.

Since

$$\beta h(z) = h'(z)f(z),$$

we have

$$\frac{\beta(1-z) + f(z)}{(1-z)f(z)} = \frac{h'(z)}{h(z)} + \frac{1}{1-z} \quad \text{for all } z \in \Delta,$$

and thus

$$\int_0^z \frac{\beta(1-u) + f(u)}{(1-u)f(u)}du = \log\left(\frac{h(z)}{1-z}\right). \tag{7.2.8}$$

Therefore, condition (a) holds if and only if the function h is conformal at the boundary point $z = 1$, i.e., condition (c) holds. The proof of the theorem is complete. □

Another proof that $K \not\equiv 0$ if and only if h is conformal at the boundary point $z = 1$ was given in [131].

Based on this fact, one can prove the equivalence of assertions (a) and (c) of the theorem by the following simple argument.

Using the identity

$$\frac{\beta}{f(z)} = \left(\frac{\beta(z-1) - f(z)}{(z-1)f(z)}\right) - \frac{1}{1-z},$$

one can solve the Cauchy problem (2.1.3) as follows:

$$\beta t = \log(1-z) - \log(1 - F_t(z)) - \int_z^{F_t(z)} \left(\frac{\beta(1-u) + f(u)}{(1-u)f(u)}\right) du,$$

and, consequently,

$$\log\Big(\exp(\beta t)\,(1 - F_t(0))\Big) = -\int_0^{F_t(0)} \left(\frac{\beta(1-u) + f(u)}{(1-u)f(u)}\right) du. \tag{7.2.9}$$

It is known that for any point $z \in \Delta$, the curve $\{w = F_t(z),\ t \geq t_0\}$ lies in a Stolz angle with the vertex at $\tau = 1$. Therefore, if condition (a) holds, relation (7.2.9) implies that $K(0) \neq 0$. By the proof of the first part of the theorem, K is a univalent function with $\operatorname{Re} K(z) > 0,\ z \in \Delta$.

Conversely, suppose that K is a univalent function. It follows by (7.2.8) and (7.2.9) that

$$\exp(\beta t)\,(1 - F_t(0)) = \frac{1 - F_t(0)}{h(F_t(0))}.$$

Then

$$K(0) = \lim_{t\to\infty} \frac{1 - F_t(0)}{h(F_t(0))}.$$

By Proposition 4.9 in [113], the angular derivative

$$h'(1) = \angle \lim_{z\to 1} \frac{h(z)}{z-1}$$

exists and is different from 0, i.e., h is conformal at the point $z = 1$. □

Remark 7.2. It is clear that the function $\hat{g}(z) = 1-h(z)$ is a solution of the Yaglom equation (7.2.2) with $m_t = e^{-\beta t}$. In addition, it is known that $\hat{g}(z) = \sum_{k=0}^{\infty} b_k z^k$, where $b_k = \lim_{t\to\infty} P(Z(t) = k|\, Z(t) > 0)$. Therefore, our theorem can be easily reformulated in terms of the generating function $\hat{g}$.

Also inequality (7.2.1) evolves a similar problem on exponential convergence in terms of the non-Euclidean distance $d(z,\tau) = \dfrac{|\tau - z|^2}{1-|z|^2}$. Namely, the question is:

◊ *Whether the limit*

$$\lim_{t\to\infty} e^{t\beta} d(F_t(z),\tau) =: m(z)$$

exists and is not zero?

The following considerations show that, in fact, the non-tangential convergence of a hyperbolic type semigroup implies that

$$d(F_t(z),\tau) \sim |\tau - F_t(z)|.$$

Recall that for $\kappa > 1$ the set

$$\Gamma(\zeta,\kappa) = \{z \in \Delta : \ |z-\zeta| < \kappa(1-|z|)\}$$

is a non-tangential approach region at $\zeta \in \partial\Delta$.

We already know that each hyperbolic type semigroup $\mathcal{S} = \{F_t\}_{t\geq 0}$ converges to its boundary Denjoy–Wolff point non-tangentially (see the proof of Theorem 5.6). In other words, for each $z \in \Delta$ there is $\kappa = \kappa(z) \in (1,\infty)$ such that

$$|1 - F_t(z)| < \kappa(z)\,(1 - |F_t(z)|) \quad \text{for all } t \geq 0, \tag{7.2.10}$$

or, which is same, for each $z \in \Delta$ the function

$$\ell_t(z) = \frac{1-|F_t(z)|}{|1-F_t(z)|} \tag{7.2.11}$$

is bounded away from zero on $[0,\infty)$, i.e.,

$$1 \geq \ell_t(z) > \frac{1}{\kappa(z)} > 0.$$

At the same time,

$$d(F_t(z),1) = \frac{|1-F_t(z)|}{1+|F_t(z)|} \cdot \frac{1}{\ell_t(z)}.$$

So, the question is:

◊ *Whether the limit* $\ell(z) := \lim_{t\to\infty} \ell_t(z)$ *exists?*

We now show that the answer is affirmative.

To do this, we will use a modified Kœnigs–Valiron function defined as follows. Recall that by Theorems 5.5 and 5.6, for fixed $w \in \Delta$ the limit

$$\psi_w(z) = \lim_{t\to\infty} \frac{1 - F_t(z)}{|1 - F_t(w)|} \tag{7.2.12}$$

exists and is a univalent starlike function with respect to $\psi_w(1)=0$ with $\operatorname{Re}\psi_w(z) > 0$, $z \in \Delta$. This function satisfies Schröder's functional equation

$$\psi_w(F_t(z) = e^{-t\beta}\psi_w(z).$$

Lemma 7.2. *Let* $\mathcal{S} = \{F_t\}_{t\geq 0} \subset \mathrm{Hol}(\Delta)$ *be a semigroup of hyperbolic type and let* ψ_w *be defined by* (7.2.12). *Then*

$$\ell(w) := \lim_{t\to\infty} \ell_t(w) = \operatorname{Re}\psi_w(w). \tag{7.2.13}$$

Proof. Fix $w \in \Delta$ and consider the net of univalent functions

$$g_t(z) = \frac{1 - F_t(z)}{1 + F_t(z)} \cdot \left|\frac{1 + F_t(w)}{1 - F_t(w)}\right|.$$

It is clear that

$$\lim_{t\to\infty} g_t(z) = \psi_w(z).$$

Hence $\operatorname{Re} g_t(z) \to \operatorname{Re}\psi_w(z)$ as $t \to \infty$. On the other hand,

$$\begin{aligned}\operatorname{Re} g_t(w) &= \frac{1 - |F_t(w)|^2}{|1 + F_t(w)|^2} \cdot \left|\frac{1 + F_t(w)}{1 - F_t(w)}\right| \\ &= \frac{1 - |F_t(w)|}{|1 - F_t(w)|} \cdot \frac{1 + |F_t(w)|}{|1 + F_t(w)|} \to \ell(w),\end{aligned}$$

and we are done. □

This lemma and Theorem 7.2 imply the following assertion.

Corollary 7.2. *Let* $f \in \mathcal{G}_h[1]$, *and let* $\mathcal{S} = \{F_t\}_{t\geq 0} \subset \mathrm{Hol}(\Delta)$ *be the semigroup generated by* f. *Then the limit*

$$m(z) := \lim_{t\to\infty} e^{t\beta} d(F_t(z), 1)$$

exists. Moreover,

$$m(z) = \frac{|K(z)|}{2\ell(z)},$$

where $K(z)$ *is defined in Theorem* 7.2.

Finally, regarding the asymptotic behavior of a semigroup generated by $g \in \mathcal{G}[1,-1]$, we note that using Theorem 2.12, one can now derive the following rates of convergence.

Corollary 7.3. *Let $g \in \mathcal{G}[1,-1]$ with $\beta_+ = g'(1) \geq 0$ and $g'(-1) = \beta_- < 0$. Then for $z \in \Delta$*

$$e^{t\beta_-}\left[\varphi_-(z)\right]^{-1} \leq \varphi_+(F_t(z)) \leq e^{-t\beta_+}\varphi_+(z),$$

where

$$\varphi_+(z) = \frac{|1-z|^2}{1-|z|^2} \quad \text{and} \quad \varphi_-(z) = \frac{|1+z|^2}{1-|z|^2}.$$

In particular, if $z = x \in (-1,1)$ then

$$e^{t\beta_-}\frac{1-x}{1+x} \leq \frac{|1-F_t(x)|^2}{1-|F_t(x)|^2} \leq e^{-t\beta_+}\frac{1-x}{1+x}.$$

7.2.2 Angular similarity principle

In contrast with the dilation case, for a semigroup with the boundary Denjoy–Wolff point $\tau \in \partial\Delta$, a natural question is:

$\Diamond$ *Does the limit* $\lim\limits_{t\to\infty} \arg(\tau - F_t(z))$ *exist?*

For semigroups of hyperbolic type, the answer is affirmative. Moreover, we will see that a kind of rigidity results for semigroups of holomorphic mappings for the hyperbolic case can be established also in terms of their asymptotic angular behavior.

It turns out, that in contrast to the parabolic case, two semigroups of hyperbolic or dilation types having similar asymptotic behavior at their common Denjoy–Wolff point coincide if and only if they have the same derivatives of the first order at this point. In addition, we note that for hyperbolic type semigroups (in contrast to the parabolic case) the limiting tangent of their trajectories $\{F_t(z)\}_{t\geq 0}$ does depend on the choice of the initial point $z \in \Delta$. Moreover, each other semigroup having the same limiting tangents (depending on $z \in \Delta$) must be the same semigroup up to a rescaling.

We begin with the following angular asymptotic formula which is important by itself:

Theorem 7.3 (see [60]). *Let $\mathcal{S} = \{F_t\}_{t\geq 0}$ be a semigroup of hyperbolic type with the Denjoy–Wolff point $\tau = 1$ and let f be its generator. Then*

$$\lim_{z\to 1} \arg(1 - F_t(z)) = f'(1)\left[\operatorname{Im}\left(\int_0^z \frac{d\zeta}{f(\zeta)}\right) - \lim_{r\to 1}\operatorname{Im}\left(\int_0^r \frac{d\zeta}{f(\zeta)}\right)\right]. \tag{7.2.14}$$

Proof. Consider a net of holomorphic functions $h_t : \Delta \mapsto \mathbb{C}$ defined as

$$h_t(z) = \alpha_t(1 - F_t(z)), \tag{7.2.15}$$

where $\{\alpha_t\}_{t\geq 0}$ is a net of complex numbers, such that the limit

$$\lim_{t\to\infty} \alpha_t(1 - F_t(0)) = \alpha \tag{7.2.16}$$

exists and is different from zero. Then

$$\frac{\partial h_t(z)}{\partial z} = -\alpha_t \frac{\partial F_t(z)}{\partial z} = \alpha_t \cdot \frac{f(F_t(z))}{f(z)}. \tag{7.2.17}$$

Since each trajectory $\{F_t(z)\}_{t\geq 0}$, $z \in \Delta$, converges to $z = 1$ non-tangentially, we get from (7.2.15) and (7.2.17) that

$$\lim_{t\to\infty} \frac{\partial h_t(z)}{\partial z} : h_t(z) = \lim_{t\to\infty} \frac{f(F_t(z))}{1 - F_t(z)} \cdot \frac{1}{f(z)} = \frac{f'(1)}{f(z)}. \tag{7.2.18}$$

Furthermore, since $\operatorname{Re}(1 - F_t(z)) > 0$, for all $t \geq 0$ and $z \in \Delta$, we have by (7.2.16) that the family $\{h_t\}_{t\geq 0}$ is normal. In addition, (7.2.18) implies that each of its limit functions satisfies the differential equation

$$f'(1) \cdot h(z) = h'(z) \cdot f(z) \tag{7.2.19}$$

and the conditions

$$h(1) = 0 \quad \text{and} \quad h(0) = \alpha. \tag{7.2.20}$$

Therefore, the net $\{h_t\}_{t\geq 0}$ is convergent locally uniformly by the uniqueness of the solution of (7.2.19) with conditions (7.2.20).

Further, we will prove our theorem by using a different choice of the net $\{\alpha_t\}_{t\geq 0}$.

First we set

$$\alpha_t = \frac{1}{1 - F_t(0)}. \tag{7.2.21}$$

Then h defined as

$$h(z) = \lim_{t\to\infty} \frac{1 - F_t(z)}{1 - F_t(0)} \tag{7.2.22}$$

exists and satisfies equation (7.2.19) with $h(0) = 1 - h(1) = 0$.

In addition, for all $s \geq 0$, we have by (7.2.19) and the Cauchy problem that

$$h(F_s(z)) = e^{-sf'(1)} \cdot h(z). \tag{7.2.23}$$

Now let us set $\widetilde{\alpha}_t$ in (7.2.15) in place of α_t defined as

$$\widetilde{\alpha}_t = \frac{1}{|1 - F_t(0)|}. \tag{7.2.24}$$

Show that

$$\lim_{t\to\infty} \widetilde{\alpha}_t(1 - F_t(0)) := \widetilde{\alpha} \tag{7.2.25}$$

exists and is different from zero. Indeed, it follows by Valiron's theorem [143] that there exists the limit

$$\lim_{n\to\infty} \frac{1-F_n(z)}{|1-F_n(0)|} = \widetilde{h}(z) \neq 0. \tag{7.2.26}$$

Then we have

$$h(z) = \lim_{n\to\infty} \frac{1-F_n(z)}{1-F_n(0)} = \lim_{n\to\infty} \frac{1-F_n(z)}{|1-F_n(0)|} \cdot \frac{|1-F_n(0)|}{1-F_n(0)} = \frac{\widetilde{h}(z)}{\widetilde{h}(0)}. \tag{7.2.27}$$

It follows now by (7.2.23) that, for all $s \geq 0$,

$$\widetilde{h}(F_s(z)) = e^{-sf'(1)}\widetilde{h}(z). \tag{7.2.28}$$

Fix any $s \geq 0$ and consider the functions

$$\widetilde{h}_{n+s}(z) = \frac{1-F_{n+s}(z)}{|1-F_{n+s}(0)|}.$$

We have

$$\begin{aligned}
\lim_{n\to\infty} \widetilde{h}_{n+s}(z) &= \lim_{n\to\infty} \frac{1-F_n(F_s(z))}{|1-F_{n+s}(0)|} \\
&= \lim_{n\to\infty} \frac{1-F_n(F_s(z))}{|1-F_n(0)|} \cdot \frac{|1-F_n(0)|}{|1-F_n(F_s(0))|} \\
&= \widetilde{h}(F_s(z)) \cdot \frac{1}{\left|\widetilde{h}(F_s(0))\right|} = \frac{\widetilde{h}(z)}{\left|\widetilde{h}(0)\right|} = \widetilde{h}(z),
\end{aligned}$$

because of (7.2.28) and $\left|\widetilde{h}(0)\right| = 1$.

Thus, this limit does not depend on s and is equal to $\widetilde{h}(z)$. By using the compactness argument, we have that

$$\lim_{t\to\infty} \widetilde{\alpha}_t(1-F_t(0)) = \widetilde{h}(0) \neq 0$$

exists and is a constant of modulus 1. To end our proof, we note that since each nontrivial solution $h \in \mathrm{Hol}(\Delta, \mathbb{C})$ of (7.2.19) with $h(1) = 0$ is of the form $h(z) = c\widetilde{h}(z)$, $c \neq 0$, we get by (7.2.24) and (7.2.25) that $\lim_{t\to\infty} \arg(1-F_t(z)) = \arg \widetilde{h}(z) = \arg h(z) - \arg c$. Finally, observe that both h and $\widetilde{h}$ are starlike functions with respect to a boundary point $z = 1$. Moreover, since they satisfy the so-called Visser–Ostrovski condition (see [113]),

$$\angle \lim_{z\to 1} \frac{(z-1)h'(z)}{h(z)} = \angle \lim_{z\to 1} \frac{(z-1)\widetilde{h}'(z)}{\widetilde{h}(z)} = \angle \lim_{z\to 1} \frac{z-1}{f(z)} f'(1) = 1,$$

we have that the smallest wedges W and $\widetilde{W}$, respectively, containing $h(\Delta)$ and $\widetilde{h}(\Delta)$ are half-planes passing through the origin. Since $\operatorname{Re} \widetilde{h}(z) \geq 0,\ z \in \Delta$, we have that $\widetilde{W} = \Pi_+ = \{z \in \mathbb{C} : \operatorname{Re} z > 0\}$. It follows now by a result in [63] that $W = e^{i\varphi}\widetilde{W}$ where $\varphi = \lim_{r\to 1^-} \arg h(r) = -\arg c$. The theorem is proved. $\square$

Remark 7.3. In fact, we have shown in our theorem that

$$\lim_{t\to\infty} \arg(1 - F_t(z)) = \arg \widetilde{h}(z),$$

where $\widetilde{h}$ is the solution of equation (7.2.19) such that $\widetilde{h}(\Delta)$ lies in the right-half plane $\Pi_+ = \{z \in \mathbb{C} : \operatorname{Re} z \geq 0\}$. In addition, it was shown in [63] that $\widetilde{h}(\Delta)$ is a starlike domain with respect to the boundary point $\widetilde{h}(1) = 0$, having a corner of opening at this point equal to π. Therefore, for each $\alpha \in \left(-\frac{\pi}{2}, \frac{\pi}{2}\right)$, there is a point $z \in \Delta$, such that $\lim_{z\to 1} \arg(1 - F_t(z)) = \alpha$. Thus, our Theorem 7.3 actually completes Theorem 8 in [32].

Theorem 7.4. *Let f and g be generators of hyperbolic type semigroups $\mathcal{S}_f = \{F_t\}_{t\geq 0}$ and $\mathcal{S}_g = \{G_t\}_{t\geq 0}$, respectively. Suppose that $\mathcal{S}_f$ and $\mathcal{S}_g$ have the same Denjoy–Wolff point $\tau \in \partial\Delta$. Then*

$$\lim_{t\to\infty} \left(\arg \frac{\tau - F_t(z)}{\tau - G_t(z)} \right) = 0$$

if and only if

$$f(z) = kg(z) \tag{7.2.29}$$

for some $k > 0$, i.e., the semigroups $\mathcal{S}_f$ and $\mathcal{S}_g$ coincide up to a rescaling.

Proof. Set $\tau = 1$. Let h_f and h_g be Kœnigs type functions for $\mathcal{S}_f$ and $\mathcal{S}_g$ defined by

$$h_f(z) = \lim_{t\to\infty} \frac{1 - F_t(z)}{|1 - F_t(0)|} \tag{7.2.30}$$

and

$$h_g(z) = \lim_{t\to\infty} \frac{1 - G_t(z)}{|1 - G_t(0)|}, \tag{7.2.31}$$

respectively. Then we have by (7.1.8), (7.2.30) and (7.2.31) that

$$\begin{aligned}
&\lim_{t\to\infty} \left(\arg \frac{1 - F_t(z)}{1 - G_t(z)} \right) \\
&= \lim_{t\to\infty} \left(\arg \left(\frac{1 - F_t(z)}{|1 - F_t(0)|} \cdot \frac{|1 - G_t(0)|}{1 - G_t(z)} \cdot \left| \frac{1 - F_t(0)}{1 - G_t(0)} \right| \right) \right) \\
&= \arg h_f(z) - \arg h_g(z) = 0.
\end{aligned}$$

Since both $h_f(z)$ and $h_g(z)$ do not vanish in Δ, the latter equality means that

$$h_f(z) = \alpha(z) h_g(z),$$

where $\alpha(z) \in \mathbb{R}$. However, α is holomorphic on Δ, and hence, is constant: $\alpha(z) \equiv \alpha,\ z \in \Delta$. At the same time, we know from (7.2.30) and (7.2.31) that

$$|h_f(0)| = |h_g(0)| = 1.$$

Therefore, $|\alpha| = 1$. In addition, (7.2.30) and (7.2.31) imply that both $\operatorname{Re} h_f(z)$ and $\operatorname{Re} h_g(z)$ are nonnegative. Consequently, α must be 1, so we have

$$h_f(z) = h_g(z) =: h(z), \quad z \in \Delta. \tag{7.2.32}$$

Denote $\beta_f = f'(1)$ and $\beta_g = g'(1)$. As we already mentioned, both β_f and β_g are positive real numbers. In this setting, we have that

$$\beta_f h(z) = h'(z) \cdot f(z)$$

and

$$\beta_g h(z) = h'(z) \cdot g(z).$$

These equalities imply (7.2.29) with $k = \frac{\beta_f}{\beta_g}$. In other words,

$$F_t(z) = h^{-1}\left(e^{-t\beta_f} h(z)\right) = h^{-1}\left(e^{-s\beta_g} h(z)\right) = G_s(z),$$

where $s = t \cdot k$. This completes our proof. □

As we have already mentioned, an important question related to rigidity is how to realize that a given semigroup consists of linear-fractional transformations of the unit disk. The following assertion answers this question in terms of the asymptotic behavior of semigroups.

Corollary 7.4. *Let $\mathcal{S}_f = \{F_t\}_{t\geq 0}$ be a semigroup of hyperbolic type generated by $f \in \operatorname{Hol}(\Delta, \mathbb{C})$. Assume that $\tau = 1$ is the Denjoy–Wolff point of $\mathcal{S}_f$ and $f \in C_A^2(1)$, and let $g \in \mathcal{G}$ be defined by the quadratic part of f:*

$$g(z) = f'(1)(z-1) + \frac{1}{2} f''(1)(z-1)^2,$$

i.e., g is a generator of the semigroup (of hyperbolic type) $\mathcal{S}_g = \{G_t\}_{t\geq 0}$ of linear-fractional transformations which map Δ into itself. Then $f = g$ if and only if

$$\lim_{t\to\infty} \arg(1 - F_t(z)) = -\arg\left(\frac{f'(1)}{1-z} - \frac{1}{2} f''(1)\right). \tag{7.2.33}$$

Proof. Note that the semigroup $\mathcal{S}_g = \{G_t\}_{t\geq 0}$ of linear-fractional transformations generated by $g,\ g(z) = f'(1)(z-1) + \frac{1}{2} f''(1)(z-1)^2$, can be written as

$$G_t(z) = C^{-1}\left(e^{t\beta} C(z) + A(e^{t\beta} - 1)\right),$$

where $\beta = f'(1)$, $A = 1 - \frac{1}{\beta}f''(1)$ and $C(z) = \frac{1+z}{1-z}$. Then we get by direct calculation that

$$\lim_{t\to\infty} \arg(1 - G_t(z)) = -\arg\left(\frac{1+z}{1-z} + A\right)$$
$$= \arg\frac{1-z}{\beta - \frac{1}{2}f''(1)(1-z)} = -\arg\left(\frac{\beta}{1-z} - \frac{1}{2}f''(1)\right).$$

Applying now Theorem 7.4, we complete our proof. □

Corollary 7.5. *Let f and g be generators of semigroups $\mathcal{S}_f$ and $\mathcal{S}_g$, respectively. Suppose that $\mathcal{S}_f$ and $\mathcal{S}_g$ have the same Denjoy–Wolff point $\tau \in \overline{\Delta}$ and* $\operatorname{Re} f'(\tau) \cdot \operatorname{Re} g'(\tau) > 0$ *(i.e., both semigroups are either of hyperbolic or dilation type). Then $f = g$ if and only if* $\operatorname{Re} f'(\tau) = \operatorname{Re} g'(\tau)$ *and*

$$\lim_{t\to\infty}\left(\arg\frac{\tau - F_t(z)}{\tau - G_t(z)}\right) = 0.$$

7.3 Parabolic case

Let now $\mathcal{S} = \{F_t\}$ be a discrete ($t \in \mathbb{N}$) or continuous ($t \in \mathbb{R}^+$) semigroup of parabolic type. This means that for at least one $t_0 > 0$ (hence for all t) $\dfrac{dF_{t_0}(1)}{dz} = 1$. In this situation, the study of the asymptotic behavior of the semigroup is much more delicate and complicated.

As we have already mentioned, the class of parabolic self-mappings falls into two subclasses of automorphic (of positive hyperbolic step) and non-automorphic (of zero hyperbolic step) types. The problem is:

◊ *Whether these classes have a certain specification regarding the asymptotic behavior of discrete or continuous semigroups?*

7.3.1 Discrete case

This section is devoted to a brief discussion of some recent results for discrete-time semigroups defined by iterations of a single self-mapping. The first assertion we mention in this direction is given by Bourdon and Shapiro (see [18, p. 97]).

Theorem 7.5. *Suppose that F is of parabolic type, and that for some $z \in \Delta$ the discrete-time semigroup of iterates $\{F^n(z)\}_{n=0}^{\infty}$ converges to $\tau = 1$ non-tangentially. Then F is of non-automorphic type.*

In a sense, a converse assertion is due to Poggi-Corradini [109]. To formulate it, we again consider, in parallel with $F \in \operatorname{Hol}(\Delta)$, its associated iteration function $\Phi \in \operatorname{Hol}(\Pi_+)$ given by $\Phi(z) = C(F(C^{-1}(z)))$. We use the following notion suggested by Contreras, Díaz-Madrigal and Pommerenke in [36].

Definition 7.4. *Given a mapping* $F \in \mathrm{Hol}(\Delta)$ *(respectively, its associated iteration mapping* $\Phi = C \circ F \circ C^{-1}$*), we say that its orbit* $\{F^n\}_{n=1}^{\infty}$ *(respectively,* $\{\Phi^n\}_{n=1}^{\infty}$*) is of* **finite shift** *if for some* $w \in \Pi_+$ *(hence, for each* $w \in \Pi_+$*),* $L(w) := \lim\limits_{n\to\infty} \operatorname{Re}\Phi^n(w) < \infty$*. Otherwise, if* $L(w) = \infty$*, the orbit is said to be of* **infinite shift**.

Thus, the property of $\{F_n\}_{n=1}^{\infty}$ to be of finite shift is equivalent to the fact that it converges to the Denjoy–Wolff point strongly tangentially. At the same time, there are examples of semigroups (even continuous time) which converge to their Denjoy–Wolff points tangentially, but not strongly tangentially, i.e., their orbits are of infinite shift.

If the orbit $\{F^n\}_{n=1}^{\infty}$ is of finite shift, then for each pair z_1, z_2 in Δ,

$$\lim_{n\to\infty} \frac{d(F^n(z_1))}{d(F^n(z_2))} < \infty, \quad \text{where } d(z) = \frac{|1-z|^2}{1-|z|^2}.$$

Theorem 7.6 ([109] (see also, [36])). *Let* $F \in \mathrm{Hol}(\Delta)$ *be a parabolic self-mapping of* Δ *with the boundary Denjoy–Wolff point* $\tau = 1$*. If its orbits* $\{F^n(z)\}_{n=0}^{\infty}$*,* $z \in \Delta$*, are of finite shift (strongly tangentially convergent), then* F *is of automorphic type (of positive hyperbolic step).*

Of course, the property of F to be of non-automorphic type in general does not imply the non-tangential convergence.

So, the question now is:

◊ *Whether the property of* F *to be of automorphic type is equivalent to tangential (strongly tangential) convergence?*

The answer is affirmative under some smoothness conditions.

To be more concrete, we recall that for $\tau \in \partial\Delta$ *a function* $f \in \mathrm{Hol}(\Delta, \mathbb{C})$ *is of class* $C_A^{p+\varepsilon}(\tau)$*, where* $p \in \mathbb{N} \cup \{0\}$ *and* $\varepsilon \in [0, 1)$*, if it admits the expansion*

$$f(z) = \sum_{j=0}^{p} \frac{f^j(\tau)(z-\tau)^j}{j!} + \gamma(z), \quad z \in \Delta, \tag{7.3.1}$$

where $\angle \lim\limits_{z\to\tau} \frac{\gamma(z)}{(z-\tau)^{p+\varepsilon}} = 0$*; and we say that* $f \in C^{p+\varepsilon}(\tau)$ *when this expansion holds as* $z \to \tau$ *unrestrictedly.* Of course, if $f \in C^{p+\varepsilon}(\tau)$, we have that $f \in C_A^{p+\varepsilon}(\tau)$.

Theorem 7.7 ([36]). *Let* $F \in \mathrm{Hol}(\Delta)$ *be a parabolic self-mapping of* Δ *of automorphic type. The following are equivalent:*

(1) *The mapping* F *belongs to* $C_A^2(1)$.

(2) *The mapping* F *is of finite shift, i.e., the sequence* $\{F^n\}_{n=1}^{\infty}$ *converges to* $\tau = 1$ *strongly tangentially.*

Combining this result with Theorem 7.6 above, we get the following characteristic condition:

Theorem 7.8. *Let $F \in \mathrm{Hol}(\Delta)$ be a parabolic-type self-mapping of Δ. The following are equivalent:*

(i) *The mapping F is of finite shift, i.e., $\{F^n\}_{n=1}^{\infty}$ converges to $\tau = 1$ strongly tangentially;*

(ii) *The mapping F belongs to $C_A^2(1)$ and is of automorphic type (of positive hyperbolic step).*

Actually, for mappings smooth enough at the boundary Denjoy–Wolff point τ, the property of the mapping F to be of automorphic (or non-automorphic) type can be recognized by using the value $F''(\tau)$.

Theorem 7.9 ([18] p. 52). *Let $F \in \mathrm{Hol}(\Delta)$ have the Denjoy–Wolff point $\tau = 1$. Suppose that $F \in C^2(1)$ and $F'(1) = 1$, i.e., F is of parabolic type. Then*

(a) $\mathrm{Re}\, F''(1) \geq 0$;

(b) *if either $F''(1) = 0$ or $\mathrm{Re}\, F''(1) > 0$, then F is of non-automorphic type;*

(c) *conversely, if $F \in C^{3+\varepsilon}(1)$, $\varepsilon > 0$, and $F''(1)$ is pure imaginary, i.e., $F''(1) \neq 0$ and $\mathrm{Re}\, F''(1) = 0$, then F is of automorphic type.*

Corollary 7.6. *If $F \in C^{3+\varepsilon}(1)$, $\varepsilon > 0$, then F is of automorphic type if and only if $F''(1)$ is purely imaginary.*

In the case when $F''(1) \neq 0$, Bourdon and Shapiro have presented the following criterion of non-tangential convergence for each orbit $\{F^n(z)\}_{n=1}^{\infty}$ of a self-mapping F (see [18, p. 54]).

Theorem 7.10. *Let $F \in \mathrm{Hol}(\Delta)$ have the Denjoy–Wolff point $\tau = 1$. Suppose that $F \in C^2(1)$ and F is of parabolic type. Assume that $F''(1) \neq 0$. Then for each $z \in \Delta$ the orbit $\{F^n(z)\}_{n=1}^{\infty}$ converges to $\tau = 1$ non-tangentially if and only if* $\mathrm{Re}\, F''(1) > 0$.

This theorem may be partially completed by the following assertion, which is a direct consequence of a result in Tauraso [139, p. 950].

Theorem 7.11. *Let $F \in C^3(1)$ be of parabolic type, and let $F''(1) = 0$. Then $\{F^n(0)\}_{n=1}^{\infty}$ is of infinite shift. Moreover, there is a subsequence $n_k \in \mathbb{N}$, such that $\{F^{n_k}(0)\}_{k=1}^{\infty}$ converges to $\tau = 1$ non-tangentially.*

On the other hand, a sufficient condition for the orbit of F to be strongly tangentially convergent was given earlier in [18, p. 65].

Theorem 7.12. *Let $F \in C^{3+\varepsilon}(1)$ be a self-mapping of Δ of parabolic type, with Denjoy–Wolff point $\tau = 1$. Suppose that $F''(1)$ is pure imaginary. Then for each z the sequence $\{F^n(z)\}_{n=1}^{\infty}$ converges to $\tau = 1$ strongly tangentially, i.e., it lies outside some disk in Δ that is tangent to $\partial\Delta$ at 1.*

Note also that by Theorems 7.9 and 7.10, if $F \in C^2(1)$ and $\{F^n\}_{n=1}^{\infty}$ converges non-tangentially to $\tau = 1$, then F must be of non-automorphic type. As a matter of fact, Theorem 7.5 states that this assertion holds without any smoothness requirement.

Table 7.1 presents a scheme that summarizes the relation between different subclasses of parabolic type mappings and the asymptotic behavior of their orbits.

7.3.2 Continuous case

Since each continuous (in $t \geq 0$) semigroup $\mathcal{S} = \{F_t\}_{t\geq 0}$ is differentiable, its asymptotic behavior is much more "ordered" than that of discrete semigroups. It seems that a trajectory of the solution of the Cauchy problem may behave like a "snake". In fact, however, we will see that some smoothness of a generator at the Denjoy–Wolff point implies that the generated semigroup's asymptotic behavior is rigid.

Let $\mathcal{S} = \{F_t\}_{t\geq 0}$ be a continuous semigroup of holomorphic self-mappings of Δ of parabolic type having the Denjoy–Wolff point at $\tau = 1$. Thus, for its generator f, we have

$$f(1) = 0 \quad \text{and} \quad f'(1) = 0.$$

So, f is of the form $f(z) = -(1-z)^2 p(z)$ with $\operatorname{Re} p \geq 0$ and $\angle \lim\limits_{z\to 1}(1-z)p(z) = 0$.

To study the asymptotic behavior of the semigroup $\mathcal{S} = \{F_t\}_{t\geq 0}$, consider two positive functions $\varepsilon_t(z)$ and $\delta_t(z)$ defined as follows:

$$\varepsilon_t(z) = \frac{|1 - F_t(z)|^2}{1 - |F_t(z)|^2}, \quad z \in \Delta, \quad t \geq 0,$$

and

$$\delta_t(z) = \frac{|1 - F_t(z)|}{1 - |F_t(z)|}, \quad z \in \Delta, \quad t \geq 0.$$

Definition 7.5. *We say that a semigroup $\mathcal{S} = \{F_t\}_{t\geq 0}$ is strongly tangentially convergent if, for each $z \in \Delta$,*

$$\varepsilon(z) := \lim_{t\to\infty} \varepsilon_t(z) > 0.$$

Note that by the Julia–Wolff–Carathéodory Theorem (see Theorem 2.10), the function $\varepsilon_t(z)$ is nonincreasing in t for each $z \in \Delta$; hence, the limit above exists.

We recall also that *a semigroup $\mathcal{S} = \{F_t\}_{t\geq 0}$ is said to be non-tangentially convergent if for each $z \in \Delta$ there exists $1 < \delta < \infty$ such that*

$$\sup \delta_t(z) \leq \delta.$$

Table 7.1: Parabolic-type mappings

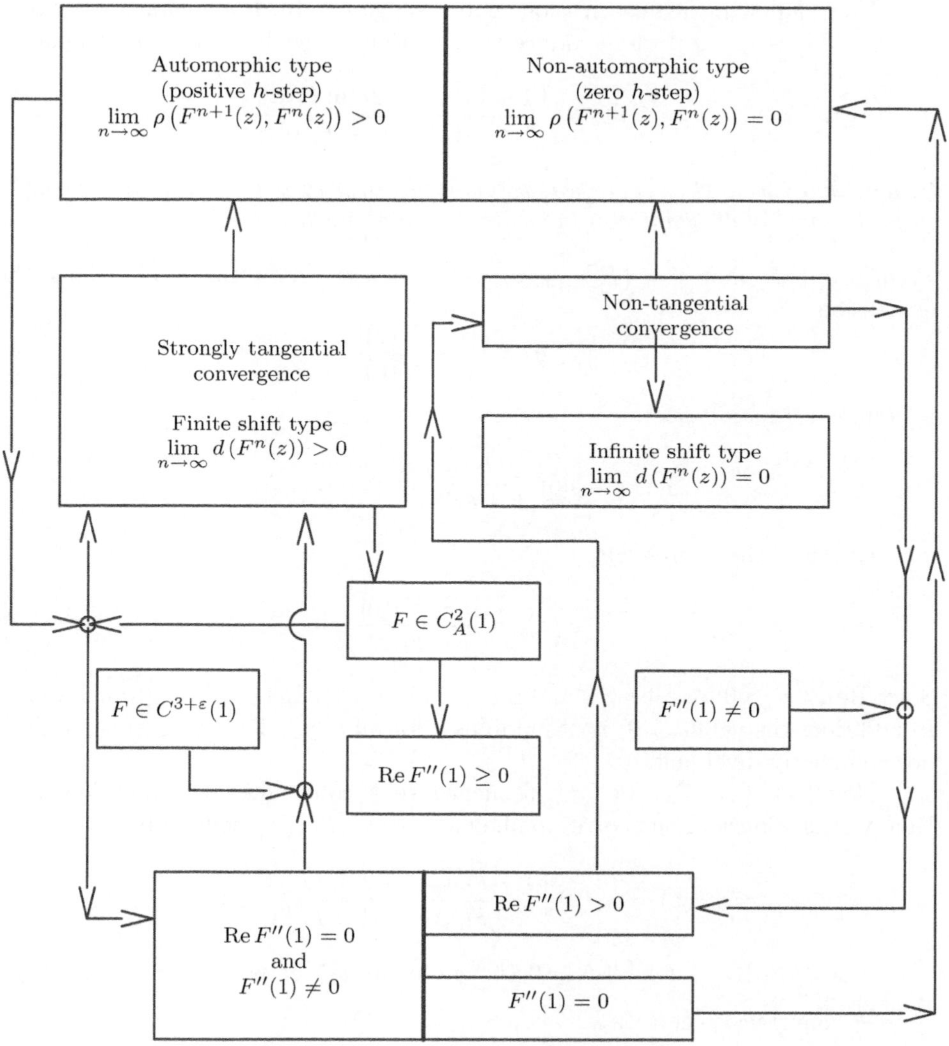

In fact, this definition means that for each $z \in \Delta$, there is a non-tangential approach region $\Gamma(1,\kappa)$, $\kappa = \kappa(z) > 1$, such that the whole trajectory $\{F_t(z),\ t \geq 0\}$ lies in $\Gamma(1,\kappa)$.

Now, following a linearization model suggested by Pommerenke (see Section 5.5, Theorem 5.8), we introduce the function h_t by the following formula:

$$h_t(z) = \varepsilon_t(0)\left[\frac{1+F_t(z)}{1-F_t(z)} - \frac{2i\,\mathrm{Im}\,F_t(0)}{|1-F_t(0)|^2}\right].$$

Proposition 7.5. *If $\mathcal{S} = \{F_t\}_{t\geq 0}$ is a semigroup of parabolic type, which converges to its Denjoy–Wolff point non-tangentially, then* $\lim_{t\to\infty} h_t(z) = 1$.

Proof. Suppose that $\mathcal{S} = \{F_t\}_{t\geq 0}$ converges to $\tau = 1$ non-tangentially. Then the expression

$$\delta_t(0) := \frac{|1-F_t(0)|}{1-|F_t(0)|}$$

is bounded. In addition,

$$f'(1) = \lim_{t\to\infty} \frac{f(F_t(z))}{1-F_t(z)} = 0, \qquad z \in \Delta.$$

Consider the expression

$$\omega_t(z) := \frac{|1-F_t(0)|}{1-F_t(z)}.$$

Since $\mathrm{Re}\,\omega_t(z) > 0$ for all $t \geq 0$ and $z \in \Delta$, we see that $\{\omega_t\}_{t\geq 0}$ is a normal family. In addition, the value $\omega_t(0)$ is of modulus 1 for all $t \geq 0$. Therefore, this family is not compactly divergent.

Now, let $\{\omega_{t_k}\}_{k=1}^{\infty}$ be any sequence that converges to a function $\omega \in \mathrm{Hol}(\Delta, \Pi_+)$. Consider the corresponding sequence $\{h_k\}_{k=1}^{\infty}$ defined by

$$h_k(z) = \varepsilon_{t_k}(0)\left[\frac{1+F_{t_k}(z)}{1-F_{t_k}(z)} - \frac{2i\,\mathrm{Im}\,F_{t_k}(0)}{|1-F_{t_k}(0)|^2}\right].$$

Since $h_t(0) = 1$ for all $t \geq 0$ (hence for all t_k), it is enough to show that $\dfrac{dh_k(z)}{dz}$ converges to zero as $k \to \infty$.

Indeed, by direct calculation we get

$$\begin{aligned}\frac{dh_k(z)}{dz} &= 2\,\frac{|1-F_{t_k}(0)|^2}{1-|F_{t_k}(0)|^2}\cdot\frac{f(F_{t_k}(z))}{(1-F_{t_k}(z))^2}\frac{1}{f(z)}\\ &= \frac{2}{f(z)}\frac{|1-F_{t_k}(0)|}{1-|F_{t_k}(0)|^2}\cdot\frac{|1-F_{t_k}(0)|}{1-F_{t_k}(z)}\cdot\frac{f(F_{t_k}(z))}{1-F_{t_k}(z)},\end{aligned}$$

which converges to zero, and we are done. □

Corollary 7.7. *Let $\mathcal{S} = \{F_t\}_{t\geq 0}$ be a one-parameter continuous semigroup with the boundary Denjoy–Wolff point. If $\mathcal{S} = \{F_t\}_{t\geq 0}$ converges to this point nontangentially, then all its elements F_t, $t \geq 0$, are either of hyperbolic or of nonautomorphic parabolic type.*

As we have mentioned, under some smoothness conditions on f at the Denjoy–Wolff point, one can provide more complete information regarding the behavior of the semigroup at this point.

Assume now that $f(z) = -(1-z)^2 p(z)$ and

$$a = -\lim_{z\to 1} p(z) \tag{7.3.2}$$

exists finitely. This is equivalent to the fact that $\mathcal{S}$ is of parabolic type and $f \in C^2(1)$ with $a = \frac{1}{2}f''(1)$.

The following assertion is needed for our characterization of the asymptotic behavior of semigroups of parabolic type:

Lemma 7.3. *Let $\mathcal{S} = \{F_t\}_{t\geq 0} \subset \mathrm{Hol}(\Delta)$ be a semigroup of parabolic type with a boundary Denjoy–Wolff point $\tau \in \partial\Delta$. Suppose that its generator $f \in \mathcal{G}$ is of class $C^2(\tau)$. If $f''(\tau) := \lim_{z\to\tau} f''(z)$ exists finitely, then for each $t \geq 0$, $F_t''(\tau) := \angle\lim_{z\to\tau} F''(z)$ also exists and*

$$F_t''(\tau) = -2at, \tag{7.3.3}$$

where $a = \dfrac{1}{2}f''(\tau)$.

Proof. Recall that the semigroup elements solve the Cauchy problem (2.1.3). Differentiating the equality

$$\frac{\partial F_t(z)}{\partial t} + f(F_t(z)) = 0,\ \ z \in \Delta,\ t \geq 0, \tag{7.3.4}$$

two times with respect to $z \in \Delta$, we get

$$\frac{\partial}{\partial t}\left(\frac{\partial^2 F_t(z)}{\partial z^2}\right) + f''(F_t(z))\left(\frac{\partial F_t(z)}{\partial z}\right)^2 + f'(F_t(z))\frac{\partial^2 F_t(z)}{\partial z^2} = 0 \tag{7.3.5}$$

for all $z \in \Delta$ and $t \geq 0$.

Define the functions $p(z,t) := f'(F_t(z))$, $q(z,t) := -f''(F_t(z))\left(\frac{\partial F_t(z)}{\partial z}\right)^2$ and $u(z,t) := \dfrac{\partial^2 F_t(z)}{\partial z^2}$, $z \in \Delta$, $t \geq 0$. It is clear that $u(z,0) = 0$. Rewriting (7.3.5) in the form

$$\frac{\partial u(z,t)}{\partial t} + p(z,t)u(z,t) = q(z,t), \quad z \in \Delta, \quad t \geq 0,$$

we find

$$u(z,t) = e^{-\int_0^t p(z,s)ds}\cdot \int_0^t q(z,s)e^{\int_0^s p(z,\varsigma)d\varsigma}ds.$$

Now we fix t and let z tend to τ in the right-hand side of this equality. Since $\mathcal{S}$ consists of mappings of parabolic type, $F_t(\tau) = \tau$ and $F_t'(\tau) = 1$ for all $t \geq 0$. We conclude that $\lim\limits_{z\to\tau} p(z,t) = 0$ and $\lim\limits_{z\to\tau} q(z,t) = -2a$ for each $t > 0$. Hence,

$$\lim_{z\to\tau}\left(e^{-\int_0^t p(z,t)ds} \cdot \int_0^t q(z,s)e^{\int_0^s p(z,\varsigma)d\varsigma}ds\right)$$

$$= e^{-\int_0^t \lim\limits_{z\to\tau} p(z,s)ds} \cdot \int_0^t \lim_{z\to\tau} q(z,s) \cdot e^{\int_0^s \lim\limits_{z\to\tau} p(z,\varsigma)d\varsigma} ds = -2at.$$

Therefore,

$$\lim_{z\to\tau} \frac{\partial^2 F_t(z)}{\partial z^2} = -2at, \quad 0 \leq t < \infty. \qquad \square$$

The next theorem shows that if $\mathcal{S} = \{F_t\}_{t\geq 0}$ is strongly tangentially convergent, then the number a must be purely imaginary. Moreover, in the case when $a = 0$, each element of $\mathcal{S}$ is of non-automorphic type.

Theorem 7.13 ([133]). *Let $f \in \mathcal{G}$ admit the representation*

$$f(z) = a(z-1)^2 + \gamma(z),$$

where $\lim\limits_{z\to 1} \frac{\gamma(z)}{(z-1)^2} = 0$, and let $\mathcal{S} = \{F_t\}_{t\geq 0}$ be the semigroup (of parabolic type) generated by f. Then:

1. *The limit*

$$h(z) = \lim_{t\to\infty} h_t(z),$$

where

$$h_t(z) = \varepsilon_t(0)\left[\frac{1+F_t(z)}{1-F_t(z)} - \frac{2i\operatorname{Im} F_t(0)}{|1-F_t(0)|^2}\right]$$

exists finitely, uniformly on each compact subset of Δ.

2. *The following assertions are equivalent:*

   (i) *$a \neq 0$ and $\mathcal{S}$ is strongly tangential convergent to $\tau = 1$, i.e., $\varepsilon = \lim\limits_{t\to\infty} \varepsilon_t(0) > 0$;*

   (ii) *the mapping F_1 is of automorphic type;*

   (iii) *for each $t > 0$ the mapping F_t is of automorphic type;*

   (iv) *the limit function $h(z) = \lim\limits_{t\to\infty} h_t(z)$ is not constant on Δ (i.e., $h(z) \neq 1$).*

   Moreover, in this case $\operatorname{Re} a = 0$, and h is a univalent function on Δ mapping Δ into Π_+ which satisfies Abel's equation

$$h(F_t(z)) = h(z) + ibt$$

   with $b = -\varepsilon \operatorname{Im} f''(1)$.

Proof. Differentiating $h_t(z)$ at $z \in \Delta$ we have

$$\frac{\partial h_t(z)}{\partial z} = \varepsilon_t(0) \cdot \frac{2}{(1 - F_t(z))^2} \frac{f(F_t(z))}{f(z)}.$$

As we have already mentioned, it follows by the Julia–Wolff–Carathéodory Theorem (see, Section 2.3) that, for each $z \in \Delta$ the semigroup $F_t(z)$ converges to $\tau = 1$ as $t \to \infty$, and the function

$$\varepsilon_t(z) = \frac{|1 - F_t(z)|^2}{1 - |F_t(z)|^2}$$

is nonincreasing. In particular, the limit $\varepsilon = \lim\limits_{t\to\infty} \varepsilon_t(z)$ exists and is a nonnegative number.

Also, by our assumptions, we have

$$\lim_{t\to\infty} \frac{f(F_t(z))}{(1 - F_t(z))^2} = a.$$

Hence, for each $z \in \Delta$, we get

$$\lim_{t\to\infty} \frac{\partial h_t(z)}{\partial z} = 2\varepsilon a \cdot \frac{1}{f(z)}.$$

Now observe that for each $t \geq 0$ the image $h_t(\Delta)$ lies in the right half-plane. Hence, the family $\{h_t(z)\}$ is a normal family of univalent functions on Δ. Yet for all $t \geq 0$, we have also $h_t(0) = 1$.

Hence, by the Hurwitz Theorem each limit function $h(z)$ of the net $\{h_t(z)\}_{t\geq 0}$ is either a univalent function on Δ or identically equal to 1.

At the same time, the latter formula implies that

$$h'(z) = \frac{2\varepsilon a}{f(z)}, \tag{7.3.6}$$

which means that all limit functions of this net must coincide. So, assertion 1 is proved.

Now we prove assertion 2.

First we note that the above considerations and formula (7.3.6) immediately imply that (iv) is equivalent to the fact that neither a nor ε are not zero. So, (iv) implies (i). In turn, (i) implies (ii) and (iii) by Theorem 7.8.

Now the above Lemma 7.3 and Theorem 7.9 (b) imply that if (ii) (or (iii)) holds, then $a \neq 0$ and $\operatorname{Re} a = 0$. Hence (i) and (ii) imply (iv).

This proves the equivalence of (i)–(iv).

To prove our last statement, we just integrate formula (7.3.6), and we get

$$h(z) = 2\varepsilon a \int_0^z \frac{dw}{f(w)} + 1.$$

On the other hand, we know that

$$\begin{cases} \dfrac{dF_t(z)}{dt} = -f(F_t(z)), \\ F_0(z) = z; \end{cases}$$

hence,

$$\int_z^{F_t(z)} \frac{dw}{f(w)} = -t.$$

Now it follows that

$$h(F_t(z)) = 2\varepsilon a \int_0^{F_t(z)} \frac{dw}{f(w)} + 1$$

$$= 2\varepsilon a \left[\int_0^z \frac{dw}{f(w)} + \int_z^{F_t(z)} \frac{dw}{f(w)} \right] + 1 = h(z) - 2\varepsilon a t, \quad t \geq 0.$$

Thus, for each $t \geq 0$ the mapping F_t satisfies Abel's functional equation

$$h(F_t(z)) = h(z) - 2\varepsilon a t.$$

On the other hand, by Pommerenke's Theorem 5.8, we see that $F = F_1$ satisfies Abel's equation

$$h(F(z)) = h(z) + ib$$

for some $b \in \mathbb{R}$, $b \neq 0$. Comparing these equations for $t = 1$ completes our proof. □

Remark 7.4. Note that once the existence of the limit $\lim_{t\to\infty} h_t(z) = h(z)$ has been shown, it is enough to require that $\varepsilon_t(0)$ in Definition 7.5 is bounded away from zero to provide the strong inequality $\varepsilon(z) = \lim_{t\to\infty} \varepsilon_t(z) > 0$ for all $z \in \Delta$. Indeed, in this case, we have

$$\varepsilon_t(0) = \varepsilon_t(z) \operatorname{Re} h_t(z).$$

Remark 7.5. If $f \in \mathcal{G} \cap C^2(1)$, then it follows by (7.3.2) that

$$\operatorname{Re} f''(1) \leq 0.$$

In turn, the above theorem implies that if

$$\operatorname{Re} f''(1) < 0,$$

then all elements of $\mathcal{S} = \{F_t\}_{t\geq 0}$ are of non-automorphic type and

$$\varepsilon(z) = \lim_{t\to\infty} \varepsilon_t(z) = 0 \quad \text{for all} \quad z \in \Delta,$$

i.e., the semigroup $\mathcal{S}$ does not converge to $\tau = 1$ strongly tangentially.

The question now is:

◊ *Whether in this case there is at least one element* $t_0 \in \mathbb{R}^+$, $t_0 \neq 0$, *such that the orbit* $\{F_{t_0 n}\}_{n=1}^{\infty}$ *converges to* $\tau = 1$ *non-tangentially?*

The answer is affirmative. Indeed, applying Theorem 14 in [36] and Theorem 7.10 as in the above theorem and by using some simple compactness argument, one can easily obtain the following assertion:

Theorem 7.14. *Let* $f \in \mathcal{G}(\Delta)$ *admit the representation*

$$f(z) = a(z-1)^2 + \gamma(z), \quad z \in \Delta, \tag{7.3.7}$$

where $\lim\limits_{z \to 1} \frac{\gamma(z)}{(z-1)^2} = 0$, *and let* $\mathcal{S} = \{F_t\}_{t \geq 0}$ *be the semigroup generated by* f. *Then the following assertions are equivalent:*

(i) $\operatorname{Re} a \neq 0$;

(ii) $a \neq 0$ *and there is* $t_0 \in \mathbb{R}^+$ *such that* $\{F_{nt_0}\}_{n=0}^{\infty}$ *converges to* $\tau = 1$ *non-tangentially;*

(iii) $a \neq 0$ *and for each* $t \in \mathbb{R}^+$ *the orbit* $\{F_{nt}\}_{n=0}^{\infty}$ *converges to* $\tau = 1$ *non-tangentially;*

(iv) $a \neq 0$ *and the semigroup* $\mathcal{S} = \{F_t\}_{t \geq 0}$ *converges to* $\tau = 1$ *non-tangentially;*

(v) *the limit function*

$$h(z) = \lim_{t \to \infty} \left(\frac{1 + F_t(z)}{1 - F_t(z)} - \frac{1 + F_t(0)}{1 - F_t(0)} \right) \tag{7.3.8}$$

is univalent on Δ *and satisfies Abel's equation*

$$h(F(z)) = h(z) + b \tag{7.3.9}$$

with $\operatorname{Re} b > 0$.

Moreover, in this case, $\operatorname{Re} a < 0$, $b = -2a$, *and each element* F_t, $t \in \mathbb{R}^+$, *of* $\mathcal{S}$ *is of non-automorphic type.*

Remark 7.6. Note that the condition $a \neq 0$ itself implies that the limit in (7.3.8) is a non-constant univalent function on Δ. However, if $\mathcal{S}$ consists of automorphic type mappings, this function satisfies Abel's equation (7.3.9) with $\operatorname{Re} b = -2 \operatorname{Re} a = 0$.

In contrast to the limit scheme in Theorem 7.13, the converse assertion does not hold in general.

At the same time, if $f \in C^{3+\epsilon}(1)$ and a is purely imaginary, then the semigroup $\mathcal{S} = \{F_t\}_{t \geq 0}$ consists of automorphic type mappings (see Theorem 7.9); hence, it converges to $\tau = 1$ strongly tangentially.

7.3.3 Universal asymptotes

A detailed description of the asymptotic behavior of one-parameter semigroups of holomorphic self-mappings in the sense of angular characteristics of their trajectories was given in [32]. In particular, it was proven there (see Theorem 2.9) that if $\mathcal{S} = \{F_t\}_{t\geq 0}$ is of parabolic type, then angular characteristics of their trajectories do not depend on the initial point $z \in \Delta$. Moreover, if for at least one $z \in \partial\Delta$ the trajectory $\{F_t\}_{t\geq 0}$ converges to the Denjoy–Wolff point $\tau \in \partial\Delta$ tangentially, then all the trajectories converge tangentially to this point.

An open question mentioned there is:

$\diamondsuit$ *What happens in the general case (when $\{F_t(z)\}_{t\geq 0}$ converge to $\tau \in \partial\Delta$ not necessarily tangentially)?* In particular, *does the limit*

$$\lim_{t\to\infty} \arg(1 - \overline{\tau} F_t(z))$$

exist?

We will answer this question below for the case when the generator f of the semigroup is twice differentiable at the boundary Denjoy–Wolff point $\tau \in \partial\Delta$.

We just note that Theorem 2.9 in [32] implies that

$\blacklozenge$ *If for at least one point $z \in \Delta$ the limit $\lim\limits_{t\to\infty} \arg\left(1 - \overline{\tau} F_t(z)\right) = \theta$ exists, then it exists for all $z \in \Delta$ and is equal to θ.*

We show inter alia that if $f \in \mathcal{G}(\Delta) \cap C^2(\tau)$, then the angle θ is uniquely determined by the value $f''(\tau)$; so even different semigroups of such type have the same asymptote for their trajectories whenever the second derivatives of their generators coincide.

- We say that a function $k \in \operatorname{Hol}(\Delta, \mathbb{C})$ is of class $P^{(n)}[\tau]$, $n = 2, 3, \dots$, $\tau \in \partial\Delta$, if k has a pole of order $(n-1)$ at τ in the sense that $k(z) = \dfrac{h(z)}{(z-\tau)^{n-1}}$ for some $h \in \operatorname{Hol}\left(\Delta \cup \{\tau\}, \mathbb{C}\right)$ with $h(\tau) \neq 0$.

Let $\mathcal{S} = \{F_t\}_{t\geq 0}$ be a semigroup of holomorphic self-mappings of Δ with the Denjoy–Wolff point $\tau = 1$, and let $k \in P^{(n)}[1]$.

For a pair $(z, w) \in \Delta \times \Delta = \Delta^2$, define a family $\sigma_t : \Delta^2 \times \mathbb{R}^+ \mapsto \mathbb{C}$ of holomorphic functions by the formula

$$\sigma_t(z, w) = k(F_t(z)) - k(F_t(w)), \quad (z, w) \in \Delta^2,\ t \geq 0. \tag{7.3.10}$$

It is clear that $\sigma_t(z, w) = -\sigma_t(w, z)$ and

$$\sigma_t(w, w) = 0, \quad t \geq 0. \tag{7.3.11}$$

Theorem 7.15 (see [60]). *Let $\mathcal{S} = \{F_t\}_{t\geq 0}$ be a semigroup of parabolic type generated by $f \in \mathcal{G}_p[1]$.*

(A) *Fix $w \in \Delta$ and assume that for some $n \geq 2$ and $k \in P^{(n)}[1]$ the limit*

$$\lim_{t\to\infty} \sigma_t(z,w) =: \sigma(z,w) \tag{7.3.12}$$

exists locally uniformly on Δ. The following assertions hold:

(i) *$f \in C_A^n(1)$ with $f^{(m)}(1) = 0$, $m = 1, 2, \ldots, n-1$, and*

$$(-1)^n \operatorname{Re} f^{(n)}(1) \leq 0. \tag{7.3.13}$$

(ii) *$f^{(n)}(1) \neq 0$ if and only if $\sigma(\cdot, w)$ is locally univalent on Δ.*

(iii) *If f is not identically zero, then $n < 4$, i.e., n is equal to either 2 or 3.*

(iv) *The limit σ in (7.3.12) exists for any choice of the function $k \in P^{(n)}[1]$ and satisfies Abel's functional equation*

$$\sigma(F_t(z), w) = \sigma(z,w) - \frac{1}{n!} f^{(n)}(1) b \cdot t \tag{7.3.14}$$

where $b = \lim_{z\to 1} (1-n)(z-1)^{n-1} k(z) \neq 0$.

(v) *For all $z \in \Delta$, we have*

$$\lim_{t\to\infty} \frac{1}{t(1-F_t(z))^{n-1}} = (-1)^{n-1}(n-1)\frac{f^{(n)}(1)}{n!}, \tag{7.3.15}$$

$$\lim_{t\to\infty} \frac{1}{t^{\frac{1}{n-1}}\, d(F_t(z), 1)} = 2\operatorname{Re}\left((-1)^{n-1}(n-1)\frac{f^{(n)}(1)}{n!}\right)^{\frac{1}{n-1}}, \tag{7.3.16}$$

where $d(z,\tau)$, $z \in \Delta$, $\tau \in \overline{\Delta}$, is a non-Euclidean distance on Δ defined by $d(z,\tau) = \dfrac{|z-\tau|^2}{1-|z|^2}$, and

$$\lim_{t\to\infty} \arg(1 - F_t(z)) = \begin{cases} -\arg(-f''(1)) & n = 2, \\ 0 & n = 3, \\ \arg(1-z) & n = 4, 5, \ldots. \end{cases} \tag{7.3.17}$$

(B) *Conversely. If for some $n = 2, 3, \ldots$, $f \in C^n(1)$ (or $f \in C_A^n(1)$ and $\mathcal{S}$ converges to $\tau = 1$ non-tangentially), then for some $m \leq n$ the limit in (7.3.12) exists for any choice of the function $k \in P^{(n)}[1]$.*

Proof. Step 1. Suppose that for some $w \in \Delta$ and $n = 2, 3, \ldots$, there is $k \in P^{(n)}[1]$, such that the net $\{\sigma_t(\cdot, w)\}_{t\geq 0}$ defined by (7.3.10) converges to $\sigma(\cdot, w)$ as t goes to ∞. Then we have that

$$\frac{\partial \sigma(z,w)}{\partial z} = \lim_{t\to\infty} k'(F_t(z)) \cdot \frac{\partial F_t(z)}{\partial z}. \tag{7.3.18}$$

Since k has a pole of order $n-1$ at $\tau = 1$, the function $k'(z)$ can be written in the form

$$k'(z) = \frac{h(z)}{(z-1)^n}, \tag{7.3.19}$$

where $h \in \mathrm{Hol}(\Delta \cup \{1\}, \mathbb{C})$ and $h(1) \neq 0$. Also we know from the Cauchy problem that

$$\frac{\partial F_t(z)}{\partial z} = \frac{f(F_t(z))}{f(z)}, \quad z \in \Delta. \tag{7.3.20}$$

So, we have by (7.3.18)–(7.3.20) that

$$\begin{aligned}\frac{\partial \sigma(z,w)}{\partial z} &= \frac{1}{f(z)} \cdot \lim_{t\to\infty} h(F_t(z)) \cdot \frac{f(F_t(z))}{(F_t(z)-1)^n} \\ &= \frac{b}{f(z)} \cdot \lim_{t\to\infty} \frac{f(F_t(z))}{(F_t(z)-1)^n},\end{aligned} \tag{7.3.21}$$

where

$$b = h(1) = \lim_{z\to 1}(1-n)(z-1)^{n-1}k(z) \neq 0. \tag{7.3.22}$$

Thus, the limit

$$\lim_{t\to\infty} \frac{f(F_t(z))}{(F_t(z)-1)^n} = \frac{1}{b}\frac{\partial \sigma(z,w)}{\partial z} \cdot f(z) \tag{7.3.23}$$

exists finitely. Now we claim that this limit does not depend on $z \in \Delta$ and is equal to

$$\frac{1}{n!} f^{(n)}(1) := \frac{1}{n!} \angle \lim_{z\to 1} f^{(n)}(z).$$

Indeed, it follows by the Berkson–Porta formula that $f(z) = -(z-1)^2 p(z)$, where $\operatorname{Re} p(z) \geq 0, \; z \in \Delta$. Then we have that

$$\lim_{t\to\infty} \frac{p(F_t(z))}{(F_t(z)-1)^{n-2}} = \left(-\lim_{t\to\infty} \frac{f(F_t(z))}{(F_t(z)-1)^n}\right) = -\frac{1}{b}\frac{\partial \sigma(z,w)}{\partial z} \cdot f(z) \tag{7.3.24}$$

exists finitely.

Applying now the Lindelöf Theorem 1.6, one can show that

$$f^{(m)}(1) := m! \angle \lim_{z\to 1} \frac{f(z)}{(z-1)^m}$$

for all $m = 1, 2, \ldots, n$ exists; and $f^{(m)}(1) = 0$ for $m = 1, 2, \ldots, n-1$, while

$$f^{(n)}(1) = n! \frac{1}{b}\frac{\partial \sigma(z,w)}{\partial z} \cdot f(z). \tag{7.3.25}$$

Moreover, Proposition 6.2 also implies that if f (hence p) is not identically zero, the number n must be less or equal to 3. For $n = 2$ we have

$$f''(1) = 2\angle \lim_{t\to\infty} \frac{f(z)}{(z-1)^2} = -\angle \lim_{t\to\infty} p(z).$$

So, in this case $\operatorname{Re} f''(1) \leq 0$. For $n = 3$ we get that

$$f'''(1) = 3!\angle \lim_{t\to\infty} \frac{f(z)}{(z-1)^3} = -\angle \lim_{t\to\infty} \frac{p(z)}{z-1}.$$

So, $f'''(1)$ is, in fact, a real positive number. (Actually, we will show below that this fact holds in a more general case even $f''(1) \neq 0$). Finally, observe that the limit (7.3.24) is not zero if and only if $\dfrac{\partial\sigma(z,w)}{\partial z}$ is not zero. This complete the proof of assertions (i)–(iii).

Step 2. As soon as we know that for some $n = 1, 2, \ldots,$ the limit

$$\lim_{t\to\infty} \frac{f(F_t(z))}{(F_t(z)-1)^n} = \frac{1}{n!} f^{(n)}(1) \tag{7.3.26}$$

exists, repeating Step 1 for any other choice of the function $k \in P^{(n)}[1]$ we see that each limit function of the net (7.3.10) must satisfy the differential equation (7.3.25) with the initial condition (7.3.11). Hence they must coincide. Noting that by the Cauchy problem

$$\int_0^{F_t(z)} \frac{d\xi}{f(\xi)} = \int_0^z \frac{d\xi}{f(\xi)} + \int_z^{F_t(z)} \frac{d\xi}{f(\xi)} = \int_0^z \frac{d\xi}{f(\xi)} - t,$$

we get the equality

$$\sigma(F_t(z), w) - \sigma(0, w) = \sigma(z, w) - \sigma(0, w) - ct,$$

where $c = \frac{1}{n!} f^{(n)}(1)b$, which proves assertion (iv) as well as assertion (B).

Step 3. Now by using again (7.3.26) and the Cauchy problem we have that the average

$$\frac{1}{t}\int_0^t \frac{dF_s(z)}{(F_s(z)-1)^n} = -\frac{1}{t}\int_0^t \frac{f(F_s(z))\,ds}{(F_s(z)-1)^n} \to -\frac{1}{n!} f^{(n)}(1).$$

Integrating the left-hand side of this formula we get (7.3.15). Rewriting this formula in the form

$$\lim_{t\to\infty} \frac{1+F_t(z)}{t^{\frac{1}{n-1}}(1-F_t(z))} = 2\left((-1)^{n-1}(n-1)\frac{f^{(n)}(1)}{n!}\right)^{\frac{1}{n-1}}$$

and taking the real part in both sides, we get (7.3.16). Formula (7.3.17) is also a consequence of (7.3.15). The theorem is proved. □

Remark 7.7. Observe that even for a different choice of functions $k \in P^{(n)}[1]$ the functions σ obtained by (7.3.12) coincide up to a complex factor. Namely, if k_1 and k_2 belong to $P^{(n)}[1]$, then the corresponding functions σ_1 and σ_2 are connected by the formula $\sigma_2 = \sigma_1 \lim\limits_{z\to 1} \frac{k_2(z)}{k_1(z)}$. If $n = 2$ then an appropriate choice of k may be made by the Cayley transform $k(z) = C(z) = \frac{1+z}{1-z}$. In this case the limit scheme (7.3.10) is a continuous analog of the discrete scheme used in [18]. If $n = 3$, then an appropriate choice of k may be made, for example, by the function $k(z) = (C(z))^2$ or by the Koebe function $k(z) = \frac{z}{(z-1)^2}$.

Since our theorem is too long because of its generality and actually is relevant only for the cases $n = 2$ or $n = 3$ we formulate the corresponding consequence separately.

Theorem 7.16. *Let $\mathcal{S} = \{F_t\}_{t\geq 0}$ be a parabolic type semigroup generated by $f \in \mathcal{G} \cap C^2_A(1)$ with the Denjoy–Wolff point $\tau = 1$, i.e., $f(1) = f'(1) = 0$. Assume that $a := \frac{f''(1)}{2} \neq 0$. Then:*

(i) $\operatorname{Re} a \leq 0$.

(ii) *If for at least one $z \in \Delta$ the trajectory $\{F_t(z)\}_{t\geq 0}$ converges to the point $\tau = 1$ non-tangentially, then $\operatorname{Re} a < 0$. Moreover, in this case for all $z \in \Delta$ we have*

$$\lim_{t\to\infty} \arg(1 - F_t(z)) = -\arg(-a), \tag{7.3.27}$$

i.e., all trajectories $\{F_t(z)\}_{t\geq 0}$ converge to $\tau = 1$ non-tangentially. In other words, there is one and the same straight line ℓ passing through $\tau = 1$, which does not depend on the initial point $z \in \Delta$, and such that these trajectories are tangential to ℓ at the point $\tau = 1$.

Conversely. If $\operatorname{Re} a < 0$ and $f \in C^2(1)$, then all the trajectories $\{F_t(z)\}_{t\geq 0}$, $z \in \Delta$, converge to $\tau = 1$ non-tangentially.

Corollary 7.8. *Suppose that conditions of Theorem 7.16 hold and $f \in C^2(1)$. Then $\operatorname{Re} a = 0$ if and only if each trajectory $\{F_t(z)\}_{t\geq 0}$, $z \in \Delta$, converges to $\tau = 1$ tangentially.*

Moreover, in this case for all $z \in \Delta$,

$$\lim_{t\to\infty} \arg(1 - F_t(z)) = \frac{\pi}{2}(\operatorname{sgn} \operatorname{Im} f''(1)).$$

Thus, if $f \in \operatorname{Hol}(\Delta, \mathbb{C})$ is a generator of a parabolic type semigroup with the Denjoy–Wolff point $\tau \in \partial\Delta$ and $f''(\tau) \neq 0$, then either all trajectories converge to τ tangentially, or all trajectories converge to τ non-tangentially in such a way that each domain bounded by two trajectories has a cusp at the point τ.

In fact, these assertions are a consequence of the following comparative result on the asymptotic behavior of two related semigroups.

Theorem 7.17. *Let $f \in \mathcal{G} \cap C_A^2(1)$ be of the form*

$$f(z) = a(z-1)^2 + \gamma(z), \tag{7.3.28}$$

where

$$\angle \lim_{z \to 1} \frac{\gamma(z)}{(z-1)^2} = 0$$

and $a \neq 0$. The following assertions hold:

(i) *The function $g \in \operatorname{Hol}(\Delta, \mathbb{C})$ defined by the first term of (7.3.28),*

$$g(z) = a(z-1)^2, \tag{7.3.29}$$

is a generator of a parabolic type semigroup $\mathcal{S}_g = \{G_t\}_{t \geq 0}$ of linear-fractional transformations, hence $\operatorname{Re} a \leq 0$;

(ii) *If the semigroup $\mathcal{S}_f = \{F_t\}$ generated by f converges to $\tau = 1$ non-tangentially, or $f \in C^2(1)$ then*

$$\lim_{t \to \infty} \frac{1 - F_t(z)}{1 - G_t(z)} = 1. \tag{7.3.30}$$

Proof. Assertion (i) of Theorem 7.17 as well as assertion (i) of Theorem 7.16 are direct consequences of Theorem 6.10 and the Berkson–Porta formula (2.1.2).

Further, it follows by the Cauchy problem, the Berkson–Porta representation and (7.3.29) that semigroups $\mathcal{S}_f = \{F_t\}_{t \geq 0}$ and $\mathcal{S}_g = \{G_t\}_{t \geq 0}$ satisfy the following equations:

$$\frac{dF_t(z)}{(1 - F_t(z))^2} - p(F_t(z))dt = 0$$

and

$$\frac{dG_t(z)}{(1 - G_t(z))^2} + adt = 0,$$

respectively. Integrating these equations we get

$$\frac{1}{1 - F_t(z)} = \int_0^t p(F_s(z))ds + \frac{1}{1-z}$$

and

$$\frac{1}{1 - G_t(z)} = -at + \frac{1}{1-z}. \tag{7.3.31}$$

Then

$$\frac{1 - F_t(z)}{1 - G_t(z)} = \frac{-a + \frac{1}{t}\frac{1}{1-z}}{\frac{1}{t}\int_0^t p(F_s(z))ds + \frac{1}{t}\frac{1}{1-z}}.$$

It is clear that if for at least one $z \in \Delta$ the trajectory $\{F_t(z)\}_{t\geq 0}$ converges to the point $\tau = 1$ non-tangentially, or $f \in C^2(1)$, then $p(F_s(z))$ converges to $-a$, as $s \to \infty$. Therefore, for such $z \in \Delta$ we have that

$$\lim_{t\to\infty} \frac{1}{t} \int_0^t p(F_s(z))ds = -a \tag{7.3.32}$$

and

$$\lim_{t\to\infty} \frac{1 - F_t(z)}{1 - G_t(z)} = 1. \tag{7.3.33}$$

This proves assertion (ii) of Theorem 7.17. In turn, (7.3.33) implies that in both cases $\{F_t(z)\}_{t\geq 0}$ converges to $\tau = 1$ non-tangentially, or for $f \in C^2(1)$),

$$\lim_{t\to\infty} \arg(1 - F_t(z)) = \lim_{t\to\infty} \arg(1 - G_t(z)) = -\arg(-a).$$

This completes the proof of assertion (ii) of Theorem 7.16, as well as Corollary 7.8 of this theorem. □

The following assertion establishes asymptotic behavior of parabolic type semigroups in terms of a non-Euclidean distance on Δ defined by $d(z,\tau) = \frac{|z-\tau|^2}{1-|z|^2}$. We show that if $\mathcal{S} = \{F_t\}_{t\geq 0}$ converges to $\tau = 1$ non-tangentially, then

$$d(F_t(z), 1) \sim |F_t(z) - 1| \sim \frac{1}{t}.$$

Corollary 7.9. *Let $\mathcal{S}_f = \{F_t\}_{t\geq 0}$ be a parabolic type semigroup generated by $f \in \mathcal{G} \cap C_A^2(1)$. Assume that $\{F_t\}_{t\geq 0}$ converges to $\tau = 1$ non-tangentially. Then*

$$\lim_{t\to\infty} t\,(1 - F_t(z)) = -\frac{2}{f''(1)} \tag{7.3.34}$$

and

$$\lim_{t\to\infty} t\frac{|1 - F_t(z)|^2}{1 - |F_t(z)|^2} = -\frac{1}{\operatorname{Re} f''(1)} > 0. \tag{7.3.35}$$

Proof. As above denote by $\mathcal{S}_g = \{G_t\}_{t\geq 0}$ the semigroup of linear-fractional transformations generated by $g \in \mathcal{G}$, where $g(z) = \frac{1}{2}f''(1)\cdot(z-1)^2$.

Then (7.3.34) follows by (7.3.31) and (7.3.30).

Consequently, we have that

$$\lim_{t\to\infty} \frac{1}{t}\frac{1 + F_t(z)}{1 - F_t(z)} = -f''(1).$$

Taking here the real part in both sides we get the equality

$$\lim_{t\to\infty} \frac{1}{t}\frac{1 - |F_t(z)|^2}{|1 - F_t(z)|^2} = -\operatorname{Re} f''(1)$$

which proves our assertion. □

To study a critical case where $f''(1) = 0$, we will assume that f is three times differentiable at the boundary Denjoy–Wolff point $\tau = 1$. We will see below that in this case all trajectories are either tangential to the real axis or do not converge (i.e., $F_T(z) \equiv z,\ t \geq 0$). Moreover, in this situation (when f is three times differentiable) we can obtain additional information, even $f''(1) \neq 0$, but $\operatorname{Re} f''(1) = 0$.

Theorem 7.18. *Let $f \in \mathcal{G} \cap C^3_A(1)$ and let $f(1) = f'(1) = \operatorname{Re} f''(1) = 0$. The following assertions hold:*

(i) $\operatorname{Im} f'''(1) = 0$ *and* $\operatorname{Re} f'''(1) \geq 0$.

(ii) $\operatorname{Re} f'''(1) = 0$ *if and only if f is a generator of automorphisms of Δ:*

$$f(z) = \frac{1}{2} i \operatorname{Im} f''(1)(z-1)^2.$$

In particular, $f(z) \equiv 0$ (or equivalently, $F_t(z) \equiv z,\ t \geq 0$) if and only if $\operatorname{Im} f''(1) = 0$.

(iii) *If $f \in C^3(1)$ and $f''(1) = 0$, then $\operatorname{Re} f'''(1) \neq 0$ (i.e., $\operatorname{Re} f'''(1) > 0$) if and only if*

$$\lim_{t\to\infty} \arg(1 - F_t(z)) = 0, \tag{7.3.36}$$

that is, all trajectories $\{F_t(z)\}_{t\geq 0}$, $z \in \Delta$, are tangential to the real axis.

Proof. Under our setting we can write

$$f(z) = \frac{1}{2} i \operatorname{Im} f''(1)(z-1)^2 + \frac{1}{3!} f'''(1)(z-1)^3 + \gamma_f(z), \tag{7.3.37}$$

where

$$\angle \lim_{z\to 1} \frac{\gamma_f(z)}{(z-1)^3} = 0. \tag{7.3.38}$$

On the other hand, f admits the Berkson–Porta representation (2.2.7). Then we have from (7.3.37) and (7.3.38) that,

$$p(z) = -\frac{1}{2} i \operatorname{Im} f''(1) - \frac{1}{3!} f'''(1)(z-1) + \gamma_p(z), \tag{7.3.39}$$

where

$$\angle \lim_{z\to 1} \frac{\gamma_p(z)}{z-1} = 0. \tag{7.3.40}$$

Define a positive real part function $p_1 \in \operatorname{Hol}(\Delta, \mathbb{C})$ by

$$p_1(z) := p(z) + \frac{1}{2} i \operatorname{Im} f''(1) = -\frac{1}{3!} f'''(1)(z-1) + \gamma_p(z).$$

Since $p_1(1) = 0$ we have from (7.3.39) and (7.3.40) that

$$p_1'(1) := \angle \lim_{z\to 1} \frac{p_1(z)}{z-1} = -\frac{1}{3!} f'''(1).$$

At the same time we know that $p_1'(1)$ must be a nonpositive real number. Therefore, $\operatorname{Im} f'''(1) = 0$, and

$$p'(1) = -\frac{1}{3!}f'''(1)\left(= -\frac{1}{3!}\operatorname{Re} f'''(1)\right) \leq 0.$$

Assertion (i) is proved.

If now $\operatorname{Re} f'''(1) = 0$, then p_1 must be identically zero, hence f must be of the form $f(z) = \frac{1}{2}i \operatorname{Im} f''(1)(z-1)^2$. It follows directly by solving the Cauchy problem that in this case f generates a group of parabolic automorphisms of Δ. It is clear now that, in particular, f is identically zero if and only if $\operatorname{Im} f''(1) = 0$. This proves assertion (ii).

If $f''(1) = 0$, but $\operatorname{Re} f'''(1)$ is possibly different from zero, then we can calculate (by using again the Cauchy problem) that

$$\frac{dF_t(z)}{(1-F_t(z))^3} = \left[\frac{1}{3!}f'''(1) + \frac{\gamma_f(F_t(z))}{(F_t(z)-1)^3}\right] dt$$

or

$$\frac{1}{t(1-F_t(z))^2} = \frac{1}{3}f'''(1) + \frac{2}{t}\int_0^t \frac{\gamma_f(F_s(z))ds}{(F_s(z)-1)^3} + \frac{1}{t(1-z)^2}.$$

The latter equality implies (7.3.36) if and only if $\operatorname{Re} f'''(1) \neq 0$, i.e., $\operatorname{Re} f'''(1) > 0$, and we are done. □

Exactly as in Corollary 7.9 one can obtain the following additional information on the asymptotic behavior of semigroups.

Corollary 7.10. *Let $\mathcal{S}_f = \{F_t\}_{t\geq 0}$ be a parabolic type semigroup generated by $f \in \mathcal{G} \cap C_A^3(1)$ with $f''(1) = 0$. Then*

$$\lim_{t\to\infty} \frac{1}{\sqrt{t}(1-F_t(z))} = \sqrt{\frac{f'''(1)}{3}} \tag{7.3.41}$$

and

$$\lim_{t\to\infty} \frac{1}{\sqrt{t}\, d(1, F_t(z))} = 2\sqrt{\frac{f'''(1)}{3}}. \tag{7.3.42}$$

Remark 7.8. By Corollary 7.8, if the semigroup generator f belongs to the class $C^2(1)$, then all of the trajectories $\{F_t(z)\}_{t\geq 0}$ converge to the Denjoy–Wolff point $\tau = 1$ tangentially if and only if $\operatorname{Re} f''(1) = 0$. At the same time, there is a more precise characteristic of the asymptotic behavior of parabolic type semigroups. Namely, it may happen that for each $z \in \Delta$ there is a horodisk $D(\tau, k) := \{\zeta \in \Delta : \ d(\tau, \zeta) < k\}$, $k = k(z)$, internally tangent to the unit circle at the point τ, such that the trajectory $\{F_t(z)\}_{t\geq 0}$ lies outside $D(\tau, k)$. In this case we say that the semigroup $\mathcal{S}$ converges to τ **strongly tangentially**. Otherwise, if $d(\tau, F_t(z))$ tends to zero, we say that $\mathcal{S}$ converges to τ **weakly tangentially**.

Transfer a given semigroup in the right half-plane by the formulas

$$\widetilde{F}_t(w) = C \circ F_t \circ C^{-1}(w), \quad \operatorname{Re} w > 0, \tag{7.3.43}$$

where $C(z) = \frac{1+z}{1-z}$ is the Cayley transform. Then Julia's Lemma implies that $\operatorname{Re}\widetilde{F}_t(w)$ is an increasing function in t. The tangential convergence of the semigroup means that the function $\dfrac{\operatorname{Im}\widetilde{F}_t(w)}{\operatorname{Re}\widetilde{F}_t(w)}$ is unbounded when t goes to infinity. Roughly speaking, the semigroup converges tangentially when $|\operatorname{Im}\widetilde{F}_t(w)|$ grows faster than $\operatorname{Re}\widetilde{F}_t(w)$. Moreover, the semigroup $\{F_t\}_{t\geq 0}$ converges strongly tangentially if and only if the function $\operatorname{Re}\widetilde{F}_t(w)$ is bounded for each w, $\operatorname{Re} w > 0$. For this reason, strongly tangentially convergent semigroups were referred to in [36] as semigroups **of finite shift**; and weakly tangentially convergent semigroups — as semigroups **of infinite shift**. An open problem is:

◊ Let a semigroup generator f be of the class $C^2(1)$ and satisfy $f(1) = f'(1) = \operatorname{Re} f''(1) = 0$, $\operatorname{Im} f''(1) \neq 0$ (hence, the generated semigroup converges tangentially). *How to recognize a type of the semigroup convergence?* In particular, if $f \in C^m(1)$, $m > 3$, one can obtain using Theorem 4.15 in [18] that the tangential convergence must be strongly tangential. So, it is natural to ask: *Does there exist a semigroup which converges tangentially but not strongly tangentially?* In the following example we show that the answer is affirmative.

Example 7.1. Consider a semigroup generator f defined by the Berkson–Porta formula $f(z) = -(1-z)^2 p(z)$ with

$$p(z) = \frac{1+i}{2}\sqrt{\frac{1+z}{1-z}}.$$

Let $\mathcal{S} = \{F_t\}_{t\geq 0}$ be the semigroup generated by f. We claim that $\mathcal{S}$ converges not strongly tangentially.

Proof. Let the semigroup $\widetilde{\mathcal{S}} = \{\widetilde{F}_t\}_{t\geq 0}$ be defined by (7.3.43). Differentiating at $t = 0^+$ we find its generator $\widetilde{f}(w) = -(1+i)\sqrt{w}$. Now we solve the Cauchy problem

$$\begin{cases} \dfrac{\partial u(t,w)}{\partial t} - (1+i)\sqrt{u(t,w)} = 0, \\ u(0,w) = w, \quad \operatorname{Re} w > 0, \end{cases}$$

and obtain

$$\widetilde{F}_t(w) = \left(\frac{1+i}{2}t + \sqrt{w}\right)^2 = \frac{it^2}{2} + w + (1+i)t\sqrt{w}.$$

Separating the real and imaginary parts we see that

$$\begin{aligned} \operatorname{Re}\widetilde{F}_t(w) &= \operatorname{Re} w + t\operatorname{Re}\left((1+i)\sqrt{w}\right), \\ \operatorname{Im}\widetilde{F}_t(w) &= \frac{t^2}{2} + \operatorname{Im} w + t\operatorname{Im}\left((1+i)\sqrt{w}\right). \end{aligned}$$

Obviously, $\dfrac{\operatorname{Im}\widetilde{F}_t(w)}{\operatorname{Re}\widetilde{F}_t(w)}$ tends to infinity as $t \to \infty$, i.e., $\widetilde{\mathcal{S}}$ (hence, $\mathcal{S}$) converges tangentially. On the other hand, $\operatorname{Re}\widetilde{F}_t(w)$ tends to infinity, so this tangential convergence is not strongly. □

Concerning individual parabolic type self-mappings of the unit disk (not necessarily embeddable into a one-parameter semigroup), we know by Theorem 4.15 in [18] that if $F \in C^{3+\varepsilon}$ with $F''(1) \neq 0$, then the following assertions are equivalent:

(1) each orbit $\{F^n(z)\}_{n=0}^{\infty}$ converges tangentially;

(2) each orbit $\{F^n(z)\}_{n=0}^{\infty}$ converges strongly tangentially;

(3) $\operatorname{Re} F''(1) = 0$.

On the other hand, by Theorem 6.2 in [18], there are self-mappings not belonging to the class $C^3(1)$ such that their orbits converge non-tangentially.

Comparing these facts with the example above, one can conclude that, to attain strongly tangential convergence, we need a certain smoothness of the semigroup's elements (or its generator). In this context, the following statement seems to be plausible.

Conjecture 7.1. *Let a semigroup generator f be of the class $C^3(1)$ and satisfy $f(1) = f'(1) = \operatorname{Re} f''(1) = 0$, $\operatorname{Im} f''(1) \neq 0$. Then the generated semigroup converges strongly tangentially.*

Chapter 8

Backward Flow Invariant Domains for Semigroups

8.1 Existence

In this section, we study conditions which ensure the existence of backward flow invariant domains for semigroups of holomorphic self-mappings of a simply connected domain $D \subset \mathbb{C}$. More precisely, the problem is the following.

$\lozenge$ *Given a one-parameter continuous semigroup $\mathcal{S} \subset \mathrm{Hol}(D)$, find a simply connected subset $\Omega \subset D$ (if it exists) such that $\mathcal{S} \subset \mathrm{Aut}(\Omega)$.*

Definition 8.1. *Let $\mathcal{S} = \{F_t\}_{t \geq 0}$ be a semiflow on Δ. A domain $\Omega \subset \Delta$ is called a (backward)* **flow-invariant domain** *(shortly,* **FID***) for $\mathcal{S}$ if $\mathcal{S} \subset \mathrm{Aut}(\Omega)$.*

Recall that the set of all semigroup generators (i.e., semi-complete vector fields) on D is denoted by $\mathcal{G}(D)$, and the set of complete vector fields on D is denoted by $\mathrm{aut}(D)$. In these terms, our problem can be rephrased as follows.

$\lozenge$ *Given $f \in \mathcal{G}(D)$, find a domain Ω (if it exists) such that $f \in \mathrm{aut}(\Omega)$.*

Let now $D = \Delta$ be the open unit disk in $\mathbb{C}$. In this case, $\mathcal{G} = \mathcal{G}(\Delta)$ is a real cone in $\mathrm{Hol}(\Delta, \mathbb{C})$, while $\mathrm{aut}(\Delta) \subset \mathcal{G}(\Delta)$ is a real Banach space (see, for example, [118]).

Moreover, by the Berkson–Porta representation formula (2.2.7), a function f belongs to $\mathcal{G}$ if and only if there is a point $\tau \in \overline{\Delta}$ and a function $p \in \mathrm{Hol}(\Delta, \mathbb{C})$ with positive real part ($\mathrm{Re}\, p(z) \geq 0$ everywhere) such that

$$f(z) = (z - \tau)(1 - z\overline{\tau})p(z). \tag{8.1.1}$$

For $\tau \in \overline{\Delta}$, we use the notation $\mathcal{G}^*[\tau]$ for a subcone of $\mathcal{G}[\tau]$ of generators f defined by (8.1.1) for which

$$\mathrm{Re}\, f'(\tau) > 0. \tag{8.1.2}$$

In the case $\tau \in \Delta$, the inclusion $f \in \mathcal{G}^*[\tau]$ means that the semigroup of holomorphic self-mappings of Δ generated by f has the Denjoy–Wolff point at τ and does not consist of elliptic automorphisms. In the case $\tau \in \partial\Delta$, $\mathcal{G}^*[\tau]$ consists of hyperbolic type generators, i.e., coincides with $\mathcal{G}_h[\tau]$.

We solve the problem mentioned above for the class $\mathcal{G}^*[\tau]$ of generators.

Following an analogy with (2.2.13), we also introduce the subcone $\mathcal{G}^*[\tau,\eta]$, where $\tau \in \overline{\Delta}$ and $\eta \in \partial\Delta$, $\eta \neq \tau$, by the formula

$$\mathcal{G}^*[\tau,\eta] := \mathcal{G}^*[\tau] \bigcap \mathcal{G}[\eta]. \tag{8.1.3}$$

So, $f \in \mathcal{G}^*[\tau]$ belongs to $\mathcal{G}^*[\tau,\eta]$ if $f(\eta) = \angle \lim_{z\to\eta} f(z) = 0$ and $\gamma = \angle \lim_{z\to\eta} f'(z)$ exists finitely. It follows by Theorem 2.8 that in this case γ must be a real negative number (below, in Lemma 8.2, we specify a more precise result).

Theorem 8.1 (see [68], cf., [69]). *Let $\mathcal{S} = \{F_t\}_{t\geq 0}$ be a semiflow on Δ generated by $f \in \mathcal{G}^*[\tau]$, for some $\tau \in \overline{\Delta}$ with $f'(\tau) = \beta$, $\operatorname{Re}\beta > 0$. The following assertions are equivalent.*

(i) *$f \in \mathcal{G}^*[\tau,\eta]$ for some $\eta \in \partial\Delta$.*

(ii) *There is a nonempty (backward) flow invariant domain $\Omega \subset \Delta$, so $\mathcal{S} \subset \operatorname{Aut}(\Omega)$.*

(iii) *For some $\alpha > 0$, the differential equation*

$$\alpha\varphi'(z)(z^2-1) = 2f(\varphi(z)) \tag{8.1.4}$$

has a locally univalent solution φ with $|\varphi(z)| < 1$ when $z \in \Delta$. Moreover, in this case φ is univalent and is a Riemann mapping of Δ onto a flow invariant domain Ω.

This theorem can be completed by the following result.

Theorem 8.2. *Let $\mathcal{S} = \{F_t\}_{t\geq 0}$ be a semiflow on Δ generated by $f \in \mathcal{G}^*[\tau]$, for some $\tau \in \overline{\Delta}$ with $f(\tau) = 0$ and $f'(\tau) = \beta$, $\operatorname{Re}\beta > 0$. The following assertions hold.*

(a) *If $f \in \mathcal{G}^*[\tau,\eta]$ for some $\eta \in \partial\Delta$ with $\gamma = \angle \lim_{z\to\eta} f'(z)$, then for each $\alpha \geq -\gamma$, equation (8.1.4) has a univalent solution φ such that $\varphi(1) = \tau$, $\varphi(-1) = \eta$ and $\Omega = \varphi(\Delta)$ is a (backward) flow invariant domain for $\mathcal{S}$. In addition, $\tau = \lim_{t\to\infty} F_t(z) \in \partial\Omega$, $z \in \Omega$, and $\lim_{t\to-\infty} F_t(z) = \eta \in \partial\Delta \cap \partial\Omega$ for each $z \in \Omega$.*

(b) *If $\Omega \subset \Delta$ is a nonempty (backward) flow invariant domain, then it is a Jordan domain such that $\tau \in \partial\Omega$, and there is a point $\eta \in \partial\Omega \cap \partial\Delta$ such that $\lim_{t\to-\infty} F_t(z) = \eta$ whenever $z \in \Omega$, $\angle \lim_{z\to\eta} f(z) = 0$ and $\angle \lim_{z\to\eta} f'(z) =: \gamma$ exists with $\gamma < 0$. In addition, there is a conformal mapping φ of Δ onto Ω which satisfies equation (8.1.4) with some $\alpha \geq -\gamma$.*

(c) *Conversely, if for some* $\alpha > 0$, *the differential equation* (8.1.4) *has a locally univalent solution* $\varphi \in \mathrm{Hol}(\Delta)$, *then it is, in fact, a conformal mapping of* Δ *onto the flow invariant domain* $\Omega = \varphi(\Delta)$ *such that* $\varphi(1) = \tau \in \partial\Omega$ *and* $\varphi(-1) = \eta$ *for some* $\eta \in \partial\Delta \cap \partial\Omega$.

In addition, $f(\eta) = 0$ *and* $f'(\eta) = \gamma$ *with* $0 > \gamma \geq -\alpha$.

The main tool of the proof of our theorems is a linearization method for semigroups which uses Kœnigs type starlike and spirallike functions on Δ (see Section 5).

Recall that by Theorem 2.14, a univalent function h is spirallike (starlike) of the class Spiral$[\tau]$ (of the class Star$[\tau]$) if and only if there is a semigroup generator $f \in \mathcal{G}^*[\tau]$ such that

$$\mu h(z) = h'(z)f(z) \tag{8.1.5}$$

for some μ with $\operatorname{Re}\mu > 0$. Obviously, if $\tau \in \Delta$, the number μ must be equal to $f'(\tau)$. If $\tau \in \partial\Delta$, then by Theorem 9.4, $|\mu - \beta| \leq \beta$. Moreover, by Theorem 5.1, this function h satisfies Schröder's equation

$$h(F_t(z)) = e^{-\mu t}h(z). \tag{8.1.6}$$

We call this function h the **spirallike (starlike) function associated with** f**.**

Since we deal with generators having additional null points on the boundary, we need to recall Definition 4.6 (see also Definition 3.4):

- For $\tau \in \overline{\Delta}$ and $\eta \in \partial\Delta$, $\eta \neq \tau$, we say that a function $h \in$ Spiral$[\tau]$ ($h \in$ Star$[\tau]$) belongs to the subclass Spiral$[\tau, \eta]$ ($S^*[\tau, \eta]$) if the angular limit

$$Q_h(\eta) := \angle \lim_{z \to \eta} \frac{(z - \eta)h'(z)}{h(z)}$$

exists finitely and is different from zero, i.e., h satisfies a generalized Visser–Ostrowski condition at the boundary point η.

To proceed, we note that the inequality $\eta \neq \tau$ implies that, for each $h \in$ Spiral$[\tau, \eta]$,

$$\angle \lim_{z \to \eta} h(z) = \infty.$$

The following fact is an immediate consequence of 8.1.5.

Lemma 8.1. *Let* $h \in$ Spiral$[\tau]$ *and* $f \in \mathcal{G}^*[\tau]$ *be connected by (8.1.5). Then* h *belongs to* Spiral$[\tau, \eta]$ *if and only if* $f \in \mathcal{G}[\tau, \eta]$. *In this case,*

$$Q_h(\eta) = \frac{\mu}{f'(\eta)}.$$

Lemma 8.2. *Let $f \in \mathcal{G}^*[\tau, \eta]$ for some $\tau \in \overline{\Delta}$ (which is the Denjoy–Wolff point for the semiflow $\mathcal{S}$ generated by f) with $\beta = f'(\tau) > 0$ and some $\eta \in \partial\Delta$, such that $f(\eta) := \angle \lim\limits_{z\to\eta} f(z) = 0$ and $\gamma = f'(\eta) = \angle \lim\limits_{z\to\eta} f'(z)$ exists finitely.*

The following assertions hold.

(i) *If $\tau \in \Delta$, then $\gamma < -\frac{1}{2}\operatorname{Re}\beta$.*

(ii) *If $\tau \in \partial\Delta$, then $\gamma \le -\beta < 0$ and the equality $\gamma = -\beta$ holds if and only if $f \subset \operatorname{aut}(\Delta)$ or, which is the same, $\mathcal{S} \subset \operatorname{Aut}(\Delta)$ consists of hyperbolic automorphisms of Δ.*

Proof. (i) Let $\tau \in \Delta$. Then $f \in \mathcal{G}^*[\tau]$ admits the representation

$$f(z) = (z - \tau)(1 - z\bar{\tau})p(z)$$

with $\operatorname{Re} p(z) > 0,\ z \in \Delta$ and

$$\beta\,(= f'(\tau)) = (1 - |\tau|^2)p(\tau).$$

Assume that, for some $\eta \in \partial\Delta$,

$$f(\eta) := \angle \lim_{z\to\eta} f(z) = 0$$

and

$$\gamma = \angle \lim_{z\to\eta} \frac{f(z)}{z-\eta}$$

exists finitely. Then $\angle \lim\limits_{z\to\eta} p(z) = 0$, and

$$\gamma = \eta|\eta - \tau|^2 \cdot p'(\eta),$$

where

$$p'(\eta) = \angle \lim_{z\to\eta} \frac{p(z)}{z-\eta}.$$

To find an estimate for $p'(\eta)$, we introduce a function p_1 of positive real part by the formula

$$p_1(z) = (1 - |\tau|^2)p(m(z)),$$

where

$$m(z) = \frac{\tau - z}{1 - z\bar{\tau}}$$

is the Möbius transformation (involution) taking τ to 0 and 0 to τ. Thus

$$p_1(0) = (1 - |\tau|^2)p(\tau) = \beta;$$

and, setting $\eta_1 = m(\eta)$, we have

$$\begin{aligned} p_1'(\eta_1) &= (1 - |\tau|^2)p'(\eta) \cdot m'(\eta_1) \\ &= \frac{1 - |\tau|^2}{m'(\eta)} \cdot p'(\eta) = -(1 - \eta\bar{\tau})^2 p'(\eta). \end{aligned}$$

On the other hand, introducing the function $q = 1/p_1$ we have

$$\angle \lim_{z\to\eta_1} (z-\eta_1)q(z) = -\eta_1\delta_q(\eta_1),$$

where $\delta_q(\eta_1) = \angle \lim_{z\to\eta_1} (1-z\overline{\eta}_1)q(z)$ is the charge of q at the point η_1. Consequently,

$$\begin{aligned} p_1'(\eta_1) &= \angle \lim_{z\to\eta_1} \frac{p_1(z)}{z-\eta_1} = \angle \lim_{z\to\eta_1} \frac{1}{(z-\eta_1)q(z)} \\ &= \frac{-\overline{\eta_1}}{\delta_q(\eta_1)} = -(1-\eta\overline{\tau})^2 p'(\eta). \end{aligned}$$

Hence

$$p'(\eta) = \frac{\overline{\eta_1}}{(1-\eta\overline{\tau})^2\delta_q(\eta_1)}$$

and

$$\gamma = \frac{\eta|\eta-\tau|^2\overline{\eta_1}}{(1-\eta\overline{\tau})^2}\cdot\frac{1}{\delta_q(\eta_1)}.$$

Since by Theorem 1.14 $\delta_q(\eta_1) \le 2\operatorname{Re} q(0) = \dfrac{2}{\operatorname{Re} p_1(0)} = \dfrac{2}{\operatorname{Re}\beta}$, we have

$$|\gamma| \ge \frac{1}{2}\operatorname{Re}\beta.$$

Note that equality is impossible since otherwise q (and hence p_1 and p) are constant. But $\angle \lim_{z\to\eta_1} p(z) = 0$, which means that $p(z) \equiv 0$.

This proves assertion (i).

(ii) Let now $\tau \in \partial\Delta$. In this case, we know already that

$$\beta = f'(\tau) = \angle \lim_{z\to\tau} f'(z) > 0.$$

Without loss of generality, let us assume that $\tau = 1$ and $\eta = -1$. In other words, we assume that $f \in \mathcal{G}^*[1,-1]$. We have to show that in that case $\gamma = \angle \lim_{z\to-1} f'(z) \le -\beta$, and equality holds if and only if f is a complete vector field.

Indeed, suppose to the contrary that $\gamma > -\beta$. Then the function $g \in \operatorname{Hol}(\Delta,\mathbb{C})$ defined by

$$g(z) = f(z) + \frac{\gamma}{2}(z^2-1)$$

belongs to the class $\mathcal{G}^*[1,-1]$, because this class is a real cone. In addition,

$$g'(1) = \beta + \gamma > 0,$$

while

$$g'(-1) = \gamma - \gamma = 0.$$

Then both points 1 and -1 are sink points of the semigroup generated by g, which is impossible. This contradiction shows that $\gamma \le -\beta$. If $\gamma = -\beta$, then $g'(1) = g'(-1) = 0$, and consequently g vanishes identically; hence

$$f(z) = -\frac{\gamma}{2}(z^2 - 1).$$

Thus f belongs to $\mathrm{aut}(\Delta)$, and the flow $\mathcal{S} = \{F_t\}_{t\in\mathbb{R}}$ consists of hyperbolic automorphisms of Δ. The lemma is proved. □

Now we are ready to prove our theorems. Since Theorem 8.2 is a compliment of Theorem 8.1, we give their proofs simultaneously.

Proofs of Theorems 8.1 *and* 8.2. We prove implications (i)⟹(ii)⟹(iii)⟹(i) of Theorem 8.1 successively, while assertions (a), (b) and (c) of Theorem 8.2 will be obtained in the process. Let $\mathcal{S} = \{F_t\}_{t\ge 0}$ be a semiflow on Δ generated by $f \in \mathcal{G}^*[\tau]$ with $\beta = f'(\tau)$, $\operatorname{Re}\beta > 0$. Let $h \in \mathrm{Hol}(\Delta, \mathbb{C})$ be the associated spirallike (starlike) function on Δ defined by equation (8.1.5) with $\mu = \beta$. Then h satisfies Schröder's equation (8.1.6)

$$h(F_t(z)) = e^{-t\beta}h(z) \tag{8.1.7}$$

for all $t \ge 0$ and $z \in \Delta$.

Step 1 ((i)⟹(ii)). If now $f \in \mathcal{G}^*[\tau, \eta]$ for some $\eta \in \partial\Delta$, that is $f(\eta)\big(= \angle\lim\limits_{z\to\eta} f(z)\big) = 0$ and $\gamma = f'(\eta)\big(= \angle\lim\limits_{z\to\eta} f'(z)\big)$ exists finitely, then by Lemma 8.1 the function h belongs to the class $\mathrm{Spiral}[\tau, \eta]$ with

$$Q_h(\eta) = \angle\lim_{z\to\eta} \frac{(z-\eta)h'(z)}{h(z)} = \frac{\beta}{\gamma}.$$

Since $\gamma \ne 0$ (actually, $\gamma < 0$), $Q_h(\eta)$ is finite.

In turn, Theorem 4.3 implies that there is a nonempty canonical $\dfrac{-\beta}{\gamma}$-spiral wedge $W \subset h(\Delta)$ with vertex at the origin such that for each $w \in W$ the spiral curve $\left\{e^{-t\beta}w,\ t \in \mathbb{R}\right\}$ belongs to W.

Define the simply connected domain $\Omega \subset \Delta$ by

$$\Omega = h^{-1}(W).$$

Then the family $\widetilde{F}_t : \Omega \mapsto \Omega$,

$$\widetilde{F}_t(z) = h^{-1}\left(e^{-t\beta}h(z)\right), \quad z \in \Omega,\ t \in \mathbb{R},$$

forms a flow (one-parameter group) of holomorphic self-mappings of Ω. Comparing the latter formula with (8.1.7), we see that for $t \ge 0$, $\widetilde{F}_t(z) = F_t(z)$ whenever $z \in \Omega$ and $\left(F_t|_{\Omega}\right)^{-1} = \widetilde{F_{-t}}$. Thus $\mathcal{S} \subset \mathrm{Aut}(\Omega)$.

Step 2 ((ii)$\Longrightarrow$(iii)). Let again $\mathcal{S} = \{F_t\}_{t\geq 0}$ be a semiflow generated by $f \in \mathcal{G}^*[\tau]$ so that

$$\lim_{t\to\infty} F_t(z) = \tau \in \partial\Delta \quad \text{and} \quad \operatorname{Re}\beta > 0, \quad \text{where } \beta = f'(\tau), \tag{8.1.8}$$

and let $\Omega \subset \Delta$ be a simply connected domain such that $\mathcal{S} \subset \operatorname{Aut}(\Omega)$. Let $\psi : \Delta \mapsto \Omega$ be any Riemann conformal mapping of Δ onto Ω. Consider the flow $\{G_t\}_{t\in\mathbb{R}} \subset \operatorname{Aut}(\Delta)$ defined by

$$G_t(z) = \psi^{-1}(F_t(\psi(z))), \; t \in \mathbb{R}. \tag{8.1.9}$$

In this case, ψ is a conjugation for G_t and F_t for each $t \in \mathbb{R}$, i.e.,

$$\psi(G_t(z)) = F_t(\psi(z)), \quad z \in \Delta, \; t \in \mathbb{R}. \tag{8.1.10}$$

Denote by $g \in \operatorname{aut}(\Delta)$ the generator of $\{G_t\}_{t\in\mathbb{R}}$:

$$g(z) = \lim_{t\to 0} \frac{z - G_t(z)}{t}.$$

Then by (8.1.10), ψ satisfies the differential equation

$$\psi'(z) \cdot g(z) = f(\psi(z)). \tag{8.1.11}$$

First we show that the family $\{G_t\}_{t\in\mathbb{R}} \subset \operatorname{Aut}(\Delta)$ consists of hyperbolic automorphisms or, which is the same, that it contains neither elliptic nor parabolic automorphisms.

Indeed, suppose $\{G_t\}_{t\in\mathbb{R}}$ contains an elliptic automorphism. Then there is a point $a \in \Delta$ such that $G_t(a) = a$ for all $t \in \mathbb{R}$; hence $g(a) = 0$ and $\operatorname{Re} g'(a) = 0$. By (8.1.11), $f(\psi(a)) = 0$; and thus $\psi(a) = \tau$. On the other hand, differentiating (8.1.11) with respect to z and setting $z = a$, we get $g'(a) = f'(\tau)$. Hence $\operatorname{Re} f'(\tau) = 0$, which contradicts (8.1.8).

Thus $\{G_t\}$ has no interior fixed point in Δ; hence there are boundary points ζ_1 and ζ_2 such that

$$\lim_{t\to\infty} G_t(z) = \zeta_1 \in \partial\Delta, \quad z \in \Delta, \tag{8.1.12}$$

and

$$\lim_{t\to-\infty} G_t(z) = \zeta_2 \in \partial\Delta, \quad z \in \Delta. \tag{8.1.13}$$

To show that the family $\{G_t\}_{t\in\mathbb{R}}$ does not contain a parabolic automorphism it is sufficient to prove that $\zeta_1 \neq \zeta_2$.

To this end, we again consider the associated spirallike (starlike) function h defined by equation (8.1.5) with $\mu = \beta$ and normalized by the conditions $h(\tau) = 0$, $h'(\tau) = 1$ if $\tau \in \Delta$ or by $h(\tau) = 0$ and $h(0) = 1$ if $\tau \in \partial\Delta$. Define $h_0 \in \operatorname{Hol}(\Delta, \mathbb{C})$ by

$$h_0(z) = h(\psi(z)). \tag{8.1.14}$$

Since h satisfies Schröder's equation (8.1.7), it follows from (8.1.10) that for all $t \geq 0$,

$$\begin{aligned} h_0(G_t(z)) &= h(\psi(G_t(z)) = h(F_t(\psi(z)) \\ &= e^{-t\beta} h(\psi(z)) = e^{-t\beta} h_0(z). \end{aligned}$$

Since the mapping $G_t \in \text{Hol}(\Delta)$ is an automorphism of Δ for each $t \in \mathbb{R}^+$, we have, in fact,

$$h_0(G_t(z)) = e^{-t\beta} h_0(z) \tag{8.1.15}$$

for all $t \in \mathbb{R}$.

From (8.1.15) we conclude that h_0 (hence, h) is a univalent spirallike (starlike) function on Δ. Moreover, (8.1.15) and Corollary 2.17 in [113] imply that

$$\angle \lim_{z \to \zeta_1} h_0(z) = 0,$$

while

$$\angle \lim_{z \to \zeta_2} h_0(z) = \infty.$$

Thus $\zeta_1 \neq \zeta_2$, and it follows that $\{G_t\}_{t \in \mathbb{R}}$ consists of hyperbolic automorphisms.

Now observe that $W = h_0(\Delta)$ is a spirallike (starlike) wedge with vertex at the origin belonging to $h(\Delta)$. Since all the points of $\partial h(\Delta)$ are admissible, $\psi = h^{-1} \circ h_0$ is a homeomorphism of $\overline{\Delta}$ onto $\overline{\Omega}$; hence $\partial\Omega$ is a Jordan curve. Now (8.1.10) implies that

$$\lim_{t \to \infty} \psi(G_t(z)) = \lim_{t \to \infty} F_t(\psi(z)) = \tau \tag{8.1.16}$$

and

$$\lim_{t \to -\infty} \psi(G_t(z)) = \lim_{t \to -\infty} F_t(\psi(z)) = \eta \tag{8.1.17}$$

for some $\eta \in \overline{\Delta}$. Applying again Corollary 2.17 in [113], we obtain

$$\psi(\zeta_1) := \lim_{z \to \zeta_1} \psi(z) = \tau \tag{8.1.18}$$

and

$$\psi(\zeta_2) := \angle \lim_{z \to \zeta_2} \psi(z) = \eta. \tag{8.1.19}$$

Thus $\eta \neq \tau$ and, moreover, $\eta \in \partial\Delta$. Indeed, if η is an interior point of Δ, then

$$\eta = \psi(\zeta_2) = \psi(G_t(\zeta_2)) = F_t(\psi(\zeta_2)) = F_t(\eta), \quad t \geq 0,$$

i.e., it must be an interior fixed point for all $F_t \in \mathcal{S}$, $t \geq 0$, which is impossible.

So $\tau \in \partial\Omega$ by (8.1.18), and $\eta \in \partial\Delta \cap \partial\Omega$ by (8.1.19).

To show that equation (8.1.4) has a locally univalent (even univalent) solution $\varphi \in \text{Hol}(\Delta)$ for some $\alpha > 0$, we use a Möbius transformation $m \in \text{Aut}(\Delta)$ such that $m(1) = \zeta_1$ and $m(-1) = \zeta_2$. Then $\psi_1 = \psi \circ m$ is a conformal mapping of Δ onto Ω with normalization

$$\psi_1(1) = \tau, \quad \psi_1(-1) = \eta.$$

For $s \in (-1, 1)$, define another conformal mapping φ_s of Δ onto Ω by

$$\varphi_s(z) := \psi_1\left(\frac{z-s}{1-zs}\right), \quad -1 < s < 1.$$

Clearly $\varphi_s(1) = \psi_1(1) = \tau$ and $\varphi_s(-1) = \psi_1(-1) = \eta$. Note also that $\ell = \{z = \varphi_s(0) \, (= \psi_1(-s)), \ s \in [-1, 1]\}$ is a continuous curve joining the points $z = 1$ and $z = -1$, and so $\ell_1 = \{z = h(\varphi_s(0))(= h(\psi_1(-s))), \ s \in [-1, 1]\}$ is a continuous curve joining $h(\tau) = 0$ and $h(\eta) = \infty$. Hence, there exists $s \in (-1, 1)$ such that $|h(\varphi_s(0))| = 1$.

Thus there exists a homeomorphism $\varphi(= \varphi_s)$ of $\overline{\Delta}$ onto $\overline{\Omega}$ holomorphic in Δ such that $\varphi(1) = \tau$, $\varphi(-1) = \eta$ and $h(\varphi(0)) = e^{i\theta}$ for some $\theta \in \mathbb{R}$.

Since the mapping ψ in our previous consideration was arbitrary, we can replace it by φ. In this case, the "new" flow $\{G_t\}_{t \in 0}$ defined by

$$G_t(z) = \varphi^{-1}(F_t(\varphi(z)))$$

is a one-parameter group of hyperbolic automorphisms of Δ having the fixed points $z = 1$ and $z = -1$ on $\partial\Delta$. In turn, its generator $g \in \mathrm{Hol}(\Delta, \mathbb{C})$ must have the form

$$g(z) = \frac{\alpha}{2}(z^2 - 1), \tag{8.1.20}$$

where $\alpha = g'(1) > 0$.

Hence, equation (8.1.11) (with φ in place of ψ) becomes (8.1.4)

$$\alpha\varphi'(z)(z^2 - 1) = 2f(\varphi(z)). \tag{8.1.21}$$

Combining this with (8.1.5), we show that α must be greater than or equal to $-\gamma > 0$. Namely, defining $h_0 \in \mathrm{Spiral}[1]$ as in (8.1.14) by

$$h_0(z) = h(\varphi(z)), \tag{8.1.22}$$

we have from (8.1.21) and (8.1.5) that

$$\beta h_0(z) = \frac{\alpha}{2}(z^2 - 1)h_0'(z) \tag{8.1.23}$$

with $h_0(0) = h(\varphi(0)) = e^{i\theta}$ for some $\theta \in [0, 2\pi)$. Solving this equation, we obtain

$$h_0(z) = e^{i\theta}\left(\frac{1-z}{1+z}\right)^{\beta/\alpha} \tag{8.1.24}$$

with $\alpha = g'(1)$. It is clear that the image $h_0(\Delta)$ is a canonical $\frac{\beta}{\alpha}$-spiral wedge.

On the other hand, by Theorem 4.3, the maximal spiral wedge contained in $h(\Delta)$ is $W_{-\nu,\theta}$, where

$$\nu = \angle \lim_{z\to\eta} \frac{\beta(z-\eta)h'(z)}{h(z)} = \angle \lim_{z\to\eta} \frac{z-\eta}{f(z)} = \frac{\beta}{\gamma}. \tag{8.1.25}$$

Thus γ is finite and $\varphi = h^{-1} \circ h_0$ is a well-defined self-mapping of Δ if and only if $\alpha \geq -\gamma$. This completes the proof of the implication ((ii)$\Longrightarrow$(iii)) of Theorem 8.1, as well as assertions (a) and (b) of Theorem 8.2. Note in passing that we have also proved the implication (ii)$\Longrightarrow$(i) of Theorem 8.1.

Step 3 ((iii)$\Longrightarrow$(i)). Suppose now that $\varphi \in \mathrm{Hol}(\Delta)$ is locally univalent and satisfies (8.1.4) for some $\alpha \in \mathbb{R}^+$. Solving this differential equation explicitly, we get

$$\alpha \int_{\varphi(0)}^{\varphi(z)} \frac{dw}{f(w)} = \int_0^z \frac{2dz}{z^2-1} = \log\left(\frac{1-z}{1+z}\right). \tag{8.1.26}$$

Since $\varphi'(z) \neq 0,\ z \in \Delta$, we have by (8.1.4) that there is no $z \in \Delta$ such that $\varphi(z) = \tau$. So if ℓ is a curve joining 0 and z, the curve $\varphi(\ell)$ joining $\varphi(0)$ and $\varphi(z)$ does not contain τ.

Consider now the differential equation (8.1.5) with initial data $h(\varphi(0)) = 1$. Separating variables in this equation, we see that

$$\beta \int_{\varphi(0)}^{\varphi(z)} \frac{dw}{f(w)} = \int_1^{h(\varphi(z))} \frac{dh}{h} = \log(h(\varphi(z)). \tag{8.1.27}$$

Comparing (8.1.26) with (8.1.27), we have

$$\log\left(h(\varphi(z)\right) = \frac{\beta}{\alpha} \log\left(\frac{1-z}{1+z}\right),$$

or

$$h(\varphi(z)) = \left(\frac{1-z}{1+z}\right)^{\frac{\beta}{\alpha}}.$$

This equality implies that the set $\left\{\left(\frac{1-z}{1+z}\right)^{\frac{\beta}{\alpha}} : z \in \Delta\right\}$ is a subset of $h(\Delta)$, so this set is different from $\mathbb{C}\setminus\{0\}$. It follows by Lemma 4.1 and Proposition 4.2 that in this case $\left|\frac{\beta}{\alpha} - 1\right| \leq 1$, the function $h_0 := \left(\frac{1-z}{1+z}\right)^{\beta/\alpha}$ is univalent on Δ. Its image $W = h_0(\Delta)$ is a spiral wedge with vertex at the origin. So, by Theorem 4.3, there is a point $\eta \in \partial\Delta$ such that $h(\eta) = \infty$ and $Q_h(\eta)$ exists finitely with $\arg Q_h(\eta) = \arg \beta$ and $\left|\frac{\beta}{\alpha}\right| \leq |Q_h(\eta)|$.

Finally, we note that $\varphi(z) = h^{-1}(h_0(z))$ is, in fact, a univalent function on Δ. Now, applying Lemma 8.1 with $\mu = \beta$, we complete the proof of the implication (iii)$\Longrightarrow$(i) of Theorem 8.1, as well as assertion (c) of Theorem 8.2.

Theorems 8.1 and 8.2 are proved. □

8.2 Maximal FIDs. Flower structures

Definition 8.2. *A (backward) flow invariant domain (FID) $\Omega \subset \Delta$ for $\mathcal{S}$ is said to be* **maximal** *if there is no $\Omega_1 \supset \Omega$, $\Omega_1 \neq \Omega$, such that $\mathcal{S} \subset \mathrm{Aut}(\Omega_1)$.*

Theorem 8.3. *Let $f \in \mathcal{G}^*[\tau, \eta]$ for some $\tau \in \overline{\Delta}$, $\eta \in \partial\Delta$ with $\gamma = f'(\eta)\,(< 0)$, and let φ be a (univalent) solution of (8.1.4) with some $\alpha \geq -\gamma$ normalized by $\varphi(1) = \tau$ and $\varphi(-1) = \eta$. The following assertions are equivalent:*

(i) *$\Omega = \varphi(\Delta)$ is a maximal FID;*

(ii) *$\alpha = -\gamma$;*

(iii) *φ is isogonal at the boundary point $z = -1$.*

Proof. We already know by (8.1.22) and (8.1.24) that $\varphi = h^{-1} \circ h_0$, where h is the spirallike (starlike) function associated to f and $h_0(z) = e^{i\theta}\left(\dfrac{1-z}{1+z}\right)^{\beta/\alpha}$ with $\beta = f'(\tau)$, $\operatorname{Re}\beta > 0$ and $\alpha \geq -\gamma$. We also know that $h_0(\Delta)$ is a canonical spiral wedge $W_{\frac{\beta}{\alpha},\theta_*}$ with some θ_*. So,

$$h_0(\Delta) = W_{\frac{\beta}{\alpha},\theta_*} \subset h(\Delta).$$

Thus $\Omega = \varphi(\Delta) = h^{-1}\left(W_{\frac{\beta}{\alpha},\theta_*}\right)$ is maximal FID if and only if the canonical spiral wedge $W_{\frac{\beta}{\alpha},\theta_*}$ is maximal. In turn, by Theorem 4.3, this wedge $W_{\frac{\beta}{\alpha},\theta_*}$ is maximal if and only if $\frac{\beta}{\alpha} = -\nu$. Comparing this fact with (8.1.25), we obtain the equivalence of assertions (i) and (ii) of the theorem.

We prove the equivalence of assertions (ii) and (iii) for the case where $\tau = 1$. Namely, let $f_1 \in \mathcal{G}^*[1, \eta]$ with $f_1'(1) = \beta_1 > 0$ and $f_1'(\eta) = \gamma_1 < 0$. Let ψ be a univalent solution of equation (8.1.4), i.e.,

$$\alpha\psi'(z)(z^2 - 1) = 2f_1(\psi(z)) \tag{8.2.1}$$

for some $\alpha \geq -\gamma_1$, normalized by $\psi(1) = 1$, $\psi(-1) = \eta$.

Substituting in formula (8.1.22) the explicit form of h_0 (see (8.1.24)) and the integral representation (3.4.7) with $\tau = 1$ for the spirallike function h and taking into account that $Q_h(\eta) = \nu = \frac{\beta_1}{\gamma_1}$ (cf. (8.1.25)), we get

$$\begin{aligned} &(\psi(z) - 1)(1 - \psi(z)\bar{\eta})^{\beta_1/\gamma_1} \\ &\cdot \exp\left[-(2 + \beta_1/\gamma_1)\int_{\partial\Delta} \log(1 - \psi(z)\bar{\zeta})d\sigma(\zeta)\right] = C_1\left(\frac{1-z}{1+z}\right)^{\beta_1/\alpha} \end{aligned}$$

or

$$\frac{\psi(z)-\eta}{z+1} = (z+1)^{-1-\gamma_1/\alpha} \cdot \frac{C_2(1-z)^{\gamma_1/\alpha}}{(1-\psi(z))^{\gamma_1/\beta_1}}$$
$$\cdot \exp\left[\frac{2\gamma_1+\beta_1}{\beta_1}\int_{\partial\Delta}\log(1-\psi(z)\bar{\zeta})d\sigma(\zeta)\right].$$

Note that one can choose an analytic branch of the multivalued function $C_2\dfrac{(1-z)^{\gamma_1/\alpha}}{(1-\psi(z))^{\gamma_1/\beta_1}}$. We denote this branch by $\chi(z)$. It is a continuous function which does not vanish at the point $z=-1$. Hence its argument is a well-defined continuous function at this point. Thus

$$\arg\frac{\psi(z)-\eta}{z+1}$$
$$= \arg\left((z+1)^{-1-\gamma_1/\alpha}\right) + \arg\chi(z) + \frac{2\gamma_1+\beta_1}{\beta_1}\int_{\partial\Delta}\arg(1-\psi(z)\bar{\zeta})d\sigma(\zeta).$$

Applying the Lebesgue Bounded Convergence Theorem, we conclude that the limit of the last summand as $z\to -1$ exists finitely. Therefore, the function ψ is isogonal at the point $z=-1$ if and only if the limit

$$\lim_{z\to-1}\arg\left((z+1)^{-1-\gamma_1/\alpha}\right)$$

exists. Obviously, this happens if and only if the exponent vanishes, i.e., $\alpha=-\gamma_1$.

Now let $\tau\in\overline{\Delta}$ be arbitrary, and let $f\in\mathcal{G}^*[\tau,\eta]$ with $f'(\tau)=\beta$, $\operatorname{Re}\beta>0$, and $f'(\eta)=\gamma<0$. Let φ be a univalent solution of equation (8.1.4) for some $\alpha\geq-\gamma$, normalized by $\varphi(1)=\tau$, $\varphi(-1)=\eta$. Denote by h the spirallike function associated to f, that is, h satisfies equation (8.1.5) with $\mu=\beta$. As above, let h_0 be the function which maps the disk Δ onto a spiral wedge, namely, $h_0(z)=e^{i\theta}\left(\frac{1-z}{1+z}\right)^{\beta/\alpha}$, such that $\varphi=h^{-1}\circ h_0$.

Find a conformal mapping Φ of Δ such that $\Phi(1)=\tau$, $\Phi(-1)=\eta$, and $h_1=h\circ\Phi$ is a spirallike function with respect to a boundary point. Note here that the domain $D=\Phi(\Delta)$ has a corner of opening π at the point η because Φ maps a circular arc containing $z=-1$ onto a circular arc which contains $z=\eta$. By Theorem 3.7 of [113], the limit $\lim\limits_{z\to-1}\arg\frac{\Phi(z)-\eta}{z+1}$ exists. Hence Φ is isogonal at the point -1. Moreover, by Theorem 1.9, the function Φ satisfies the Visser–Ostrowski condition

$$\angle\lim_{z\to-1}\frac{\Phi(z)-\eta}{z+1}=1. \tag{8.2.2}$$

Now write

$$\varphi=h^{-1}\circ h_0=\Phi\circ\left(h_1^{-1}\circ h_0\right)=\Phi\circ\psi, \tag{8.2.3}$$

where $\psi = h_1^{-1} \circ h_0$. One sees that

$$\psi(-1) := \lim_{s \to -1^+} \psi(s) = \lim_{s \to -1^+} h_1^{-1}(h_0(s)) = -1$$

and

$$\psi(1) := \lim_{s \to 1^-} \psi(s) = \lim_{s \to 1^-} h_1^{-1}(h_0(s)) = 1.$$

Using this notation, we have

$$\arg \frac{\Phi(z) - \eta}{z+1} = \arg \frac{\Phi(\psi(z)) - \eta}{\psi(z)+1} + \arg \frac{\psi(z)+1}{z+1}. \tag{8.2.4}$$

Thus (8.2.2) and (8.2.4) imply that φ is isogonal at the point η if and only if ψ is isogonal at the point $z = -1$.

Now we check that function ψ satisfies equation (8.2.1). We have seen already that $\beta h_1(z) = h_1'(z) f_1(z)$, where $f_1 \in \mathcal{G}^*[1, -1]$ is defined by $f_1(z) = \frac{f(\Phi(z))}{\Phi'(z)}$. Using (8.2.2), we get

$$f_1'(-1) = \angle \lim_{z \to -1} \frac{f_1(z)}{z+1} = \angle \lim_{z \to -1} \frac{f(\Phi(z))}{\Phi(z) - \eta} \cdot \frac{\Phi(z) - \eta}{(z+1)\Phi'(z)} = \gamma.$$

Furthermore,

$$\begin{aligned} h_0(z) = h_1(\psi(z)) &= \frac{1}{\beta} h_1'(\psi(z)) f_1(\psi(z)) \\ &= \frac{1}{\beta} \frac{(h_1(\psi(z)))'}{\psi'(z)} f_1(\psi(z)) = \frac{h_0'(z)}{\beta \psi'(z)} f_1(\psi(z)). \end{aligned}$$

Substituting $h_0(z) = e^{i\theta} \left(\frac{1-z}{1+z}\right)^{\beta/\alpha}$ in the last equality and differentiating, we see that equation (8.2.1) holds. But we have already shown that in that case ψ (hence, φ) is isogonal if and only if $\alpha = -f_1'(-1) = -\gamma$. This completes the proof. □

Remark 8.1. In general, a maximal FID for $\mathcal{S}$ need not be unique. Theorem 8.3 states that if $\mathcal{S} = \{F_t\}_{t \geq 0}$ is generated by $f \in \mathcal{G}^*[\tau]$, then its *FID is not empty if and only if there is a point* $\eta \in \partial\Delta$, *such that* $f(\eta) = \angle \lim_{z \to \eta} f(z) = 0$ *and* $f'(\eta) = \angle \lim_{z \to \eta} f'(z)$ *exists finitely with* $f'(\eta) < 0$. This point η is a repelling fixed point for $\mathcal{S} = \{F_t\}_{t \geq 0}$, namely, $F_t(\eta) = \eta$ and $\left.\frac{\partial F_t(z)}{\partial z}\right|_{z=\eta} = e^{-tf'(\eta)} > 1$ (see Theorem 2.12). Moreover, there is *a one-to-one correspondence between maximal flow invariant domains for* $\mathcal{S}$ *and such repelling fixed points.*

Theorem 8.4. *Let* $f \in \mathcal{G}^*[\tau, \eta_k]$ *for some sequence* $\{\eta_k\} \in \partial\Delta$, *i.e.,* $f(\eta_k) = 0$ *and* $\gamma_k = f'(\eta_k) > -\infty$.

The following assertions hold.

(i) *There is* $\delta > 0$ *such that* $\gamma_k < -\delta < 0$ *for all* $k = 1, 2, \ldots$.

(ii) *For each* $a < -\delta < 0$ *there is at most a finite number of the points* η_k *such that* $a \leq \gamma_k < -\delta$.

Consequently equation (8.1.4) *has a (univalent) solution* $\varphi \in \mathrm{Hol}(\Delta)$ *for each* $\alpha \geq -\max\{\gamma_k\} > -\delta$.

(iii) *If* φ_k *is a solution of* (8.1.4) *normalized by* $\varphi_k(1) = \tau$, $\varphi_k(-1) = \eta_k$ *with* $\alpha = \gamma_k$ *and* $\Omega_k = \varphi_k(\Delta)$ *(i.e.,* Ω_k *are maximal), then for each pair* Ω_{k_1} *and* Ω_{k_2} *such that* $\eta_{k_1} \neq \eta_{k_2}$ *either* $\overline{\Omega_{k_1}} \cap \overline{\Omega_{k_2}} = \{\tau\}$ *or* $\overline{\Omega_{k_1}} \cap \overline{\Omega_{k_2}} = l$, *where* l *is a continuous curve joining* τ *with a point on* $\partial\Delta$.

Proof. Assertions (i) and (ii) of the theorem are direct consequences of Lemma 8.2. To prove assertion (iii), we first note that the inclusion $\tau \in \cap_k \partial\Omega_k$ follows by assertion (a) of Theorem 8.2.

Also observe that for each pair k_1 and k_2 such that $\eta_{k_1} \neq \eta_{k_2}$, the set $\Omega_{k_1,k_2} = \Omega_{k_1} \cap \Omega_{k_2}$ is empty. Indeed, otherwise Ω_{k_1,k_2} is a FID for $\mathcal{S}$. Hence, it must contain a point $\eta \in \partial\Omega_{k_1,k_2} \cap \partial\Delta$ such that

$$\eta = \angle \lim_{t \to -\infty} F_t(z)$$

whenever $z \in \Omega_{k_1,k_2}$. Hence we should have a contradiction $\eta = \eta_{k_1} = \eta_{k_2}$.

Let us suppose now that for a pair k_1 and k_2 there is a point $z_0 \neq \tau$, $z_0 \in \Delta$, such that $z_0 \in \partial\Omega_{k_1} \cap \partial\Omega_{k_2}$. Then the whole curve

$$\ell = \{z \in \Delta : \ z = F_t(z_0), \ t \geq 0\}$$

ending at τ must belong to both $\overline{\Omega}_{k_1}$ and $\overline{\Omega}_{k_2}$, hence to $l \subset \partial\Omega_{k_1} \cap \partial\Omega_{k_2}$, since $\Omega_{k_1} \cap \Omega_{k_2} = \emptyset$.

Finally, we have that $f \in \mathrm{Hol}(\Delta, \mathbb{C})$ is locally Lipschitzian. Therefore, if $\zeta \in \Delta$ is an interior end point of ℓ, $\zeta \neq \tau$, then there is $\delta > 0$ such that the Cauchy problem

$$\begin{cases} \dfrac{\partial u(t,\zeta)}{\partial t} + f(u(t,\zeta)) = 0, \quad t \geq 0, \\ u(0,\zeta) = \zeta, \end{cases}$$

has a solution $u(t,\zeta)\,(= F_t(\zeta))$ for all $t \in [-\delta, \infty)$; and the curve $\ell_1 = \{z \in \Delta : z = u(t,\zeta), \ t \in [-\delta, \infty)\}$ also belongs to $\partial\Omega_{k_1} \cap \partial\Omega_{k_2}$. But ℓ_1 properly contains ℓ, which is impossible. So ζ must belong to $\partial\Delta$. The proof is complete. □

8.3 Examples

We illustrate the content of our theorems in the following examples.

Example 8.1. Consider a generator $f \in \mathcal{G}^*[0]$ defined by

$$f(z) = z(1 - z^n), \quad n \in \mathbb{N}.$$

Solving the Cauchy problem, we find

$$F_t(z) = \frac{ze^{-t}}{\sqrt[n]{1 - z^n + z^n e^{-nt}}}.$$

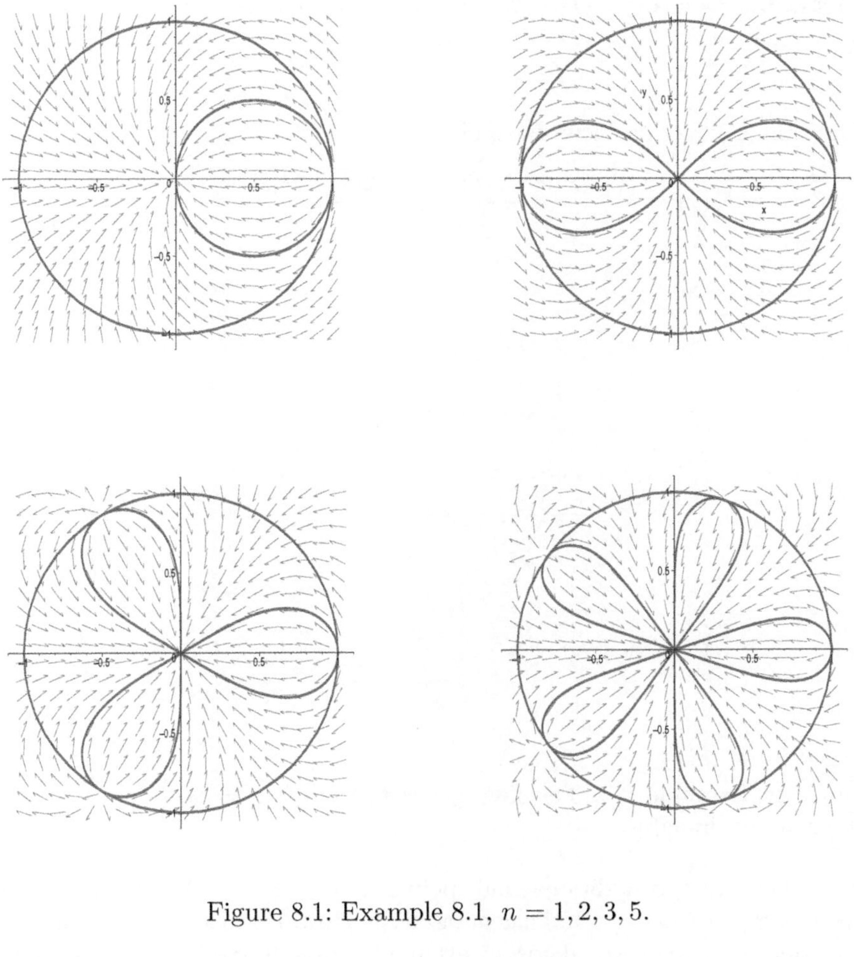

Figure 8.1: Example 8.1, $n = 1, 2, 3, 5$.

In this case, f has n additional null points $\eta_k = e^{\frac{2\pi i k}{n}}$, $k = 1, 2, \ldots, n$, on the unit circle with finite angular derivative $\gamma = f'(\eta_k) = -n$. So the generated semiflow has n repelling fixed points, and there are n maximal flow invariant domains. One can show that the functions

$$\varphi_k(z) = e^{\frac{2\pi i k}{n}} \sqrt[n]{\frac{1-z}{2}}$$

are the solutions of (8.1.4) with $\alpha = n$ satisfying $\varphi_k(1) = 0$ and $\varphi_k(-1) = \eta_k$ which map Δ onto n FID's Ω_k (for $n = 2$, these domains form a lemniscate) with $\overline{\Omega_i} \bigcap \overline{\Omega_j} = \{0\}$ when $i \neq j$. The family $\{F_t\}_{t\in\mathbb{R}}$ forms a group of automorphisms of each one of these domains. See Figure 8.1 for $n = 1, 2, 3$ and 5. For $n = 1$, for instance, it can be seen explicitly that $F_t(\varphi(z))$ is well-defined for all $t \in \mathbb{R}$ and tends to $\eta = 1$ when $t \to -\infty$.

Example 8.2. Consider a generator $f \in \mathcal{G}^*[1]$ defined by

$$f(z) = -\frac{(1-z)(1+z^2)}{1+z}.$$

Solving the Cauchy problem, we find

$$F_t(z) = \frac{(1+z^2)e^{2t} - (1-z)\sqrt{2(1+z^2)e^{2t} - (1-z)^2}}{(1+z^2)e^{2t} - (1-z)^2}.$$

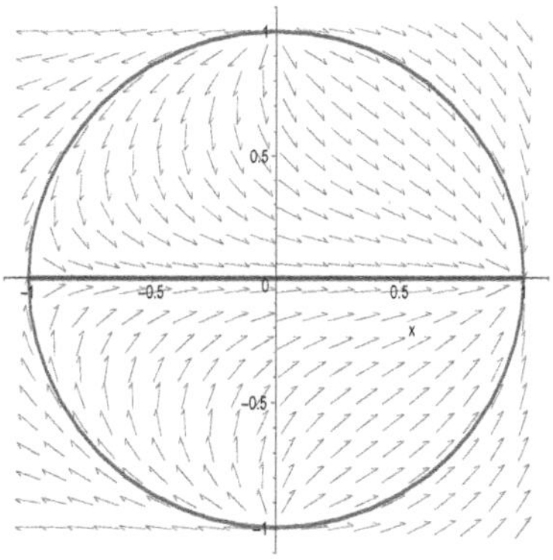

Figure 8.2: Example 8.2. The flow generated by $f(z) = -\frac{(1-z)(1+z^2)}{1+z}$ and two flow-invariant domains.

Since f has the two additional null points $\eta_{1,2} = \pm i \in \partial\Delta$ with finite angular derivative $\gamma = f'(\pm i) = -2$, the generated semiflow has two repelling fixed points. Thus, there are two maximal flow invariant domains Ω_1 and Ω_2. One can show that these domains Ω_j coincide with the upper and the lower half-disks (see Figure 8.2). So we have $\overline{\Omega_1} \bigcap \overline{\Omega_2} = \{-1 < x < 1\}$. In each of these two domains, the family $\{F_t\}_{t\in\mathbb{R}}$ is well defined and forms a group of automorphisms.

The following example shows that a maximal flow invariant domain may be even dense in the open unit disk.

Example 8.3. Let $f \in \mathcal{G}^*[0]$ be given by

$$f(z) = z\,\frac{1-z}{1+z}\,.$$

In this case, $\tau = 0$ and $\eta = 1$. Also, we have $f'(0) = 1$ and $f'(1) = -\frac{1}{2}$. Solving equation (8.1.4) with $\alpha = \frac{1}{2}$, one can write its solution in the form $\varphi(z) = h^{-1}(h_0(z))$, where h is the Koebe function $h(z) = \dfrac{z}{(1-z)^2}$ and $h_0(z) = \left(\dfrac{1-z}{1+z}\right)^2$. We shall see below that each solution of (8.1.4) has a similar representation.

Thus φ maps Δ onto the maximal flow invariant domain $\Omega = \varphi(\Delta) = \Delta \setminus \{-1 \le x \le 0\}$; see Figure 8.3. (All the pictures were obtained by using the vector field drawing tool in Maple 9.)

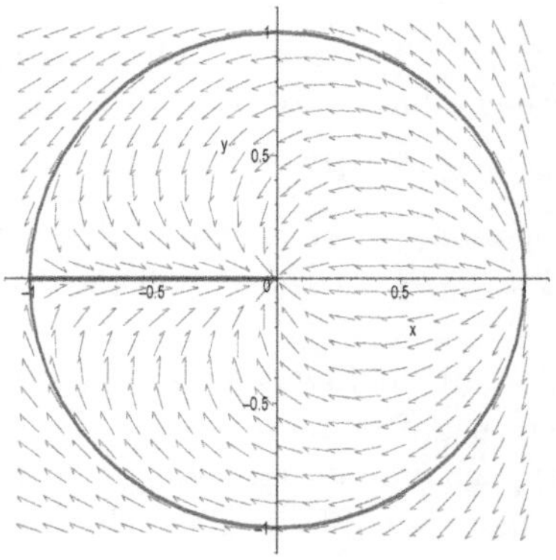

Figure 8.3: Example 8.3. The flow generated by $f(z) = z\frac{1-z}{1+z}$ and the dense flow-invariant domain.

8.4 Angular characteristics of flow invariant domains

In Theorem 8.3 above we have proved that for a univalent solution φ of (8.1.4) with some $\alpha \ge -\gamma$, the image $\Omega = \varphi(\Delta)$ is a maximal FID if and only if $\alpha = -\gamma$ and if and only if φ is isogonal at the boundary point $z = -1$. Repeating its proof for arbitrary $\alpha \ge -\gamma$, one can complete this result and obtain the following angular description of flow invariant domains (not necessary maximal).

Theorem 8.5 (see [60]). *Let $f \in \mathcal{G}$ and let $\eta \in \partial\Delta$ be a boundary regular null point of f, different from the Denjoy–Wolff point $\tau \in \overline{\Delta}$ of the generated semigroup $\mathcal{S} = \{F_t\}_{t\geq 0}$, i.e.,*

$$\gamma = f'(\eta) := \angle \lim_{z\to\eta} \frac{f(z)}{z-\eta} < 0.$$

Let $\varphi \in \mathrm{Hol}(\Delta)$ satisfy the differential equation (8.1.4) with some $\alpha \geq -\gamma$ such that $\varphi(1) = \tau$ and $\varphi(-1) = \eta$, and let $\Omega = \varphi(\Delta)$ be a (backward) flow-invariant domain for the semigroup $\mathcal{S}$. Assume that $\beta = \mathrm{Re}\, f'(\tau) > 0$. Then

(i) *$\partial\Omega$ has a corner of opening $\dfrac{\pi|\gamma|}{\alpha}$ at the point $\eta = \varphi(-1) \in \partial\Delta \cap \partial\Omega$;*

(ii) *If $\tau \in \partial\Delta$, then the opening of the corner of Ω at the point $\tau = \varphi(1) \in \partial\Omega\cap\partial\Delta$ is $\dfrac{\pi\beta}{\alpha}$.*

Corollary 8.1. *Let $\mathcal{S} = \{F_t\}_{t\geq 0} \subset \mathrm{Hol}(\Delta)$ be a semigroup of hyperbolic type. Suppose that $\mathcal{S}$ has a backward flow-invariant domain $\Omega \subset \Delta$. If $\partial\Omega$ has a corner of opening π at the Denjoy–Wolff point $\tau \in \partial\Delta$, then $\mathcal{S}$ consists of hyperbolic automorphisms of Δ.*

Proof. Let $\mathcal{S} = \{F_t\}_{t\geq 0} \subset \mathrm{Hol}(\Delta)$ be a semigroup of hyperbolic type, and let f be its generator. Since $\mathcal{S}$ has a backward flow invariant domain $\Omega \subset \Delta$, there is a boundary regular null point $\eta \in \partial\Omega\cap\partial\Delta$ of f such that $\gamma = f'(\eta) := \angle \lim_{z\to\eta} \frac{f(z)}{z-\eta} < 0$. By Theorem 8.5, the opening of the corner of $\partial\Omega$ at the point $\tau \in \partial\Omega \cap \partial\Delta$ is $\frac{\pi\beta}{\alpha}$, where $\beta = f'(1)$ and $\alpha \geq -\gamma$. At the same time, by our assumption, $\frac{\pi\beta}{\alpha} = \pi$. Hence, $\beta = \alpha \geq -\gamma$.

We claim that the last inequality is possible if and only if $\beta = -\gamma$ and the maximal backward flow invariant domain is all of Δ. In other words, in this case f generates a group of hyperbolic automorphisms of Δ. Indeed, let us set $\tau = 1$ and $\eta = -1$ and assume that $-\gamma \in (0,\beta)$. Then the function g defined by

$$g(z) = f(z) + \frac{\gamma}{2}\,(z^2-1)$$

belongs to the class $\mathcal{G}^+[1]$, because this class is a real cone. In addition, we have

$$g'(1) = \beta + \gamma > 0$$

while

$$g'(-1) = \gamma - \gamma = 0.$$

Then $g \neq 0$ and both the points 1 and -1 are the Denjoy–Wolff points of the semigroup generated by g, which is impossible. This contradiction implies that $\beta \leq -\gamma$. Moreover, the same considerations show that $\beta = -\gamma$ if and only if $g(z) = 0$. Hence $f(z)$ has the form

$$f(z) = \frac{\gamma}{2}(1-z^2) = \frac{\beta}{2}(z^2-1).$$

Thus f generates a group of hyperbolic automorphisms of Δ. This completes our proof. $\square$

Theorem 8.6. *Let $f \in \mathcal{G}(\Delta)$ have a boundary null point $\eta \in \partial\Delta$ such that the angular limit*

$$\gamma := f'(\eta) = \angle \lim_{z\to\eta} \frac{f(z)}{z-\eta}$$

exists and is a negative real number. Let $\mathcal{S} = \{F_t\}_{t\geq 0}$ be the semigroup generated by f. Suppose that Ω is the maximal flow-invariant domain such that $\eta \in \partial\Omega$, and φ is a Riemann mapping of Δ onto Ω which can be defined by the differential equation (8.1.4). *Then for each $w \in \Omega$,*

$$\lim_{t\to-\infty} \arg\left(1 - \bar{\eta}F_t(w)\right) = \arg\frac{1+\psi(w)}{1-\psi(w)} = \operatorname{Im}\left(\int_{\varphi(0)}^{w} \frac{\gamma d\zeta}{f(\zeta)}\right), \tag{8.4.5}$$

where $\psi = \varphi^{-1} : \Omega \mapsto \Delta$ and the integration path lies in Ω. The right-hand side of (8.4.5) *does not depend on the choice of a solution φ of* (8.1.4).

Proof. Consider the family of functions $\{h_t\}_{t\geq 0}$ defined by

$$h_t(z) = \frac{1-\bar{\eta}F_t(\varphi(z))}{|1-\bar{\eta}F_t(\varphi(0))|}, \quad z \in \Delta. \tag{8.4.6}$$

This is a normal family since $\operatorname{Re} h_t(z) > 0$.

Therefore by Montel's Theorem, each sequence in this family contains a subsequence $\{h_{t_k}\}_{k=0}^{\infty}$, $t_k \to \infty$, which converges locally uniformly on Δ either to a holomorphic function h or to infinity. However, the last case is impossible because of the obvious equality $|h_t(0)| = 1$.

Now it follows by the Hurwitz Theorem that h is either a univalent function on Δ, or a constant. To see that the last case is again impossible, we differentiate $h_t(z)$ to obtain

$$h'(z) = \lim_{k\to\infty} \frac{-\bar{\eta}F'_{t_k}(\varphi(z))\varphi'(z)}{|1-\bar{\eta}F_{t_k}(\varphi(0))|}$$

and

$$\frac{h'(z)}{h(z)} = \lim_{k\to\infty} \frac{-\bar{\eta}F'_{t_k}(\varphi(z))\varphi'(z)}{1-\bar{\eta}F_t(\varphi(z))}. \tag{8.4.7}$$

Since $\dfrac{f(F_t(z))}{f(z)} = F'_t(z)$ for all $t > 0$, we obtain by (8.1.4) and (8.4.7) that

$$\begin{aligned}
\frac{h'(z)}{h(z)} &= \lim_{k\to\infty} \frac{-\bar{\eta}f\left(F_{t_k}(\varphi(z))\right)\varphi'(z)}{f(\varphi(z))\left(1-\bar{\eta}F_{t_k}(\varphi(z))\right)} \\
&= \lim_{k\to\infty} \frac{2f(F_{t_k}(\varphi(z))}{\gamma(z^2-1)\left(\eta - F_{t_k}(\varphi(z))\right)} \\
&= \frac{-2}{\gamma(z^2-1)} \cdot \lim_{k\to\infty} \frac{f(F_{t_k}(\varphi(z))}{F_{t_k}(\varphi(z)) - \eta}.
\end{aligned}$$

Now we claim that $F_{t_k}(\varphi(z))$ converges non-tangentially. Indeed, for each point $w \in \Omega$, consider a flow-invariant domain Ω_1 defined in the following way. Let $\Omega_{\max}$ be the maximal flow-invariant domain which contains w. The trajectory $\{F_t(w) : t \in \mathbb{R}\}$ divides $\Omega_{\max}$ into two subdomains. Let Ω_1 be one of them. By Theorem 8.1, the Riemann mapping φ onto Ω_1 is a solution of the differential equation (8.1.4) with some $\alpha \geq -\gamma$. Since this domain is not maximal, $\alpha > -\gamma$. Now Proposition 8.5 implies that the opening of the corner of $\partial\Omega_1$ at the point η is less than π. Hence, $\{F_t(w) : t \in \mathbb{R}\} \left(\supset F_{t_k}(\varphi(z))\right)$ converges non-tangentially to η as $t \to -\infty$.

Because of the equality $\angle \lim\limits_{z\to\eta} \dfrac{f(z)}{z-\eta} = \gamma$, we conclude that the limit function h must satisfy the differential equation

$$\frac{h'(z)}{h(z)} = \frac{2}{1-z^2}.$$

Solving this equation we get $h(z) = h(0) \cdot \dfrac{1+z}{1-z}$. Since $\operatorname{Re} h_t(z) > 0$ for all $z \in \Delta$ and $|h_t(0)| = 1$, we necessarily have $h(0) = 1$, so $h(z) = \frac{1+z}{1-z}$. Therefore, for each convergent sequence $\{h_{t_k}\}_{k=1}^{\infty}$,

$$\lim_{k\to\infty} h_{t_k}(z) = \lim_{k\to\infty} \frac{1-\bar{\eta}F_{t_k}(\varphi(z))}{|1-\bar{\eta}F_{t_k}(\varphi(0))|} = \frac{1+z}{1-z}.$$

Since the limit does not depend on the choice of $\{h_{t_k}\}_{k=1}^{\infty}$, we obtain that

$$\lim_{t\to-\infty} \frac{1-\bar{\eta}F_t(\varphi(z))}{|1-\bar{\eta}F_t(\varphi(0))|} = \frac{1+z}{1-z}.$$

Hence

$$\lim_{t\to-\infty} \arg\left(1-\bar{\eta}F_t(\varphi(z))\right) = \arg\frac{1+z}{1-z},$$

or, denoting by $\psi : \Omega \mapsto \Delta$ the inverse mapping to φ,

$$\lim_{t\to-\infty} \arg\left(1-\bar{\eta}F_t(w)\right) = \arg\frac{1+\psi(w)}{1-\psi(w)}, \quad w(=\varphi(z)) \in \Omega. \tag{8.4.8}$$

Note that one can rewrite (8.1.4) in the form

$$\frac{\gamma}{f(w)} = \frac{2\psi'(w)}{1-\psi^2(w)}.$$

Hence, integrating this equation, we see that

$$\log\left(\frac{1+\psi(w}{1-\psi(w)}\right) = \int_{w_0}^{w} \frac{\gamma d\zeta}{f(\zeta)},$$

where $\psi(w_0) = 0$, or, which is the same, $w_0 = \varphi(0)$. Extracting the imaginary part, we prove (8.4.5).

To complete the proof, we note that the mapping φ is well-defined by (8.1.4) up to composition with an automorphism of the unit disk preserving the points $z = 1$ and $z = -1$. Denoting by φ_1 another Riemann mapping of Δ onto Ω, we have

$$\varphi_1(z) = \varphi\left(\frac{z+s}{1+zs}\right) \quad \text{for some } s \in (-1, 1).$$

In this case the mapping ψ_1, inverse to φ_1, has the form

$$\psi_1(w) = \frac{\psi(w) - s}{1 - s\psi(w)}.$$

Consequently,

$$\frac{1+\psi_1(w)}{1-\psi_1(w)} = \frac{(1-s)(1+\psi(w))}{(1+s)(1-\psi(w))},$$

and the argument of this quotient does not depend on s. □

Corollary 8.2. *Let $\mathcal{S}_f = \{F_t\}_{t\geq 0}$ and $\mathcal{S}_g = \{G_t\}_{t\geq 0}$ be two semigroups of holomorphic self-mappings on Δ with the same Denjoy–Wolff point $\tau \in \overline{\Delta}$ and a common boundary regular repelling fixed point $\eta \in \partial\Delta$. The following assertions are equivalent.*

(i) *The semigroups coincide up to rescaling, i.e., there is a positive real number a such that $F_t = G_{at}$ for all $t \geq 0$.*

(ii) *There is a nonempty open set $D \subset \Delta$, where $\mathcal{S}_f$ and $\mathcal{S}_g$ have a similar angular asymptotic behavior, i.e.,*

$$\lim_{t\to-\infty} \arg \frac{1-\bar{\eta}F_t(w)}{1-\bar{\eta}G_t(w)} = 0 \quad \textit{for all } w \in D.$$

(iii) *There is a common backward flow-invariant domain Ω for both $\mathcal{S}_f$ and $\mathcal{S}_g$ with $\eta \in \partial\Omega$.*

Proof. As above, denote the semigroup generators by f and g, respectively.

Step 1. If assertion (i) holds, then $f = ag$. If it is the case, then $\mathcal{S}_f$ and $\mathcal{S}_g$ have the same flow-invariant domains; and Theorem 8.6 implies that

$$\lim_{t\to-\infty} \arg\left(1 - \bar{\eta}F_t(w)\right) = \lim_{t\to-\infty} \arg\left(1 - \bar{\eta}G_t(w)\right)$$

for all w in any of their common flow-invariant domain which contains η. So, (i)$\Rightarrow$(ii).

Step 2. Assume now that assertion (ii) holds. Denote by Ω_1 and Ω_2 the maximal flow-invariant domains for $\mathcal{S}_f$ and $\mathcal{S}_g$, respectively, the boundary of which contains η. Let φ_1 and φ_2 be univalent mappings of Δ onto Ω_1 and Ω_2, respectively, normalized by $\varphi_i(1) = \tau$ and $\varphi_i(-1) = \eta$, $i = 1, 2$; and let ψ_1 and ψ_2 be their inverse mappings φ_1 and φ_2, respectively. By formula (8.4.5), for all $w \in D$,

$$\arg \frac{1 + \psi_1(w)}{1 - \psi_1(w)} = \arg \frac{1 + \psi_2(w)}{1 - \psi_2(w)}.$$

By the Maximum Principle for harmonic functions, we conclude that there exists a positive constant K such that

$$\frac{1 + \psi_1(w)}{1 - \psi_1(w)} = K \frac{1 + \psi_2(w)}{1 - \psi_2(w)}$$

on D. Therefore,

$$\psi_1(w) = \frac{\psi_2(w) - \frac{1-K}{1+K}}{1 - \psi_2(w)\frac{1-K}{1+K}} \qquad \text{for all} \quad w \in D.$$

Note here that the number $s = \frac{1-K}{1+K}$ is real and of modulus less than 1. Hence, using the inverse functions, we have for all $z \in D_1 := \psi_2(D)$,

$$\varphi_1(z) = \varphi_2\left(\frac{z + s}{1 + zs}\right).$$

By uniqueness this equality holds for all $z \in \Delta$. Noting that the mapping $\frac{z+s}{1+zs}$ is an automorphism of the unit disk, we obtain that $\Omega_1 = \varphi_1(\Delta) = \varphi_2(\Delta) = \Omega_2$, that is (iii) holds.

Step 3. Let now $\mathcal{S}_f$ and $\mathcal{S}_g$ have a common backward flow-invariant domain Ω. Denote by φ a univalent mapping of Δ onto Ω such that $\varphi(1) = \tau$ and $\varphi(-1) = \eta$. Then by equation (8.1.4), we have for all $z \in \Delta$,

$$\varphi'(z)(z^2 - 1) = \frac{2f(\varphi(z))}{\alpha_f} = \frac{2g(\varphi(z))}{\alpha_g},$$

where $\alpha_f \geq -f'(\eta)$ and $\alpha_g \geq -g'(\eta)$. In other words, we have $f(w) = \frac{\alpha_f}{\alpha_g} g(w)$ on Ω. Again by the uniqueness, this equality holds for all $z \in \Delta$. Hence, the generated semigroups coincide up to rescaling.

The proof is complete. □

8.5 Additional remarks

Let $F \in \mathrm{Hol}(\Delta)$ be a single self-mapping of Δ which can be embedded into a continuous semigroup, i.e., there is a semiflow $\mathcal{S} = \{F_t\}_{t \geq 0}$ such that $F = F_1$. In this case, all the fractional iterations F_t of F have the same collection of boundary fixed points for all $t \geq 0$ (see [33]). In turn, Theorem 8.1 asserts the existence of backward fractional iterations of F defined on a FID Ω whenever F has a repelling boundary fixed point η, i.e.,

$$A = F'(\eta) = \lim_{z \to \eta} F'(z) > 1. \tag{8.5.9}$$

As a matter of fact, for a single mapping which is not necessarily embedded into a semiflow (not even necessarily univalent on Δ), the existence of backward integer iterations under condition (8.5.9) was proved in [107]. This fact has provided the existence of conjugations near repelling points. More precisely, the main result in [107] asserts that

- *if* $\eta = 1$, $a = \dfrac{A-1}{A+1}$ *and* $G(z) = \dfrac{z-a}{1-az}$, *then there is* $\varphi \in \mathrm{Hol}(\Delta)$ *with* $\varphi(1) = 1$ *which is a conjugation for* F *and* G, *i.e.,*

$$\varphi(G(z)) = F(\varphi(z)).$$

However, for the case in which F can be embedded into a continuous semigroup $\mathcal{S} = \{F_t\}$, it is not clear whether φ is a conjugation for the whole semiflow $\mathcal{S}$ and the flow produced by G.

It is natural to expect a more precise result under stronger requirements. A direct consequence of the proof of our Theorem 8.1 is the following assertion for conjugations.

Proposition 8.1. *Let* $F \subset \mathrm{Hol}(\Delta)$ *be embedded into a semiflow* $\mathcal{S} = \{F_t\}_{t \geq 0}$ *of hyperbolic type and let* $\eta \in \partial\Delta$ *be a repelling fixed point of* F *with* $A = F'(\eta) > 1$. *Then for each* $B \geq A$ *and the automorphism* $G(= G_B) \in \mathrm{Aut}(\Delta)$ *defined by*

$$G(z) = \frac{z+b}{1+zb},$$

where $b = \frac{B-1}{B+1}$, *there is a homeomorphism* $\varphi(= \varphi_B)$ *of* $\overline{\Delta}$, $\varphi \in \mathrm{Hol}(\Delta)$, *such that* $\varphi(\eta) = -1$ *and*

$$\varphi(G(z)) = F(\varphi(z)), \quad z \in \Delta.$$

Moreover, for all $t \in \mathbb{R}$ *and* $w \in \varphi(\Delta)$, *the flow* $\{F_t(w)\}_{t \in \mathbb{R}}$ *is well-defined with* $F_1 = F$ *and*

$$F_t(\varphi(z)) = \varphi(G_t(z)), \quad \textit{for all } t \in \mathbb{R},$$

where

$$G_t(z) = \frac{z+1+e^{-\alpha t}(z-1)}{z+1-e^{-\alpha t}(z-1)}, \quad t \in \mathbb{R},$$

with $\alpha = \log B$.

In addition, $\varphi_B(\Delta) \subseteq \varphi_A(\Delta)$, *with* $\varphi_A(\Delta) = \varphi_B(\Delta)$ *if and only if* $A = B$.

Our approach to construct conjugations is different from that used in [107].

Further, following the work of Baker [11], Karlin and McGregor [90] considered the local embedding problem of holomorphic functions with two fixed points into a continuous group. In particular, they studied a class $\mathcal{L}$ of functions holomorphic in the extended complex plane $\overline{\mathbb{C}}$ except for an at most countable closed set in $\overline{\mathbb{C}}$ and proved the following result.

• *Let F be a function of class $\mathcal{L}$ with two fixed points z_0 and z_1, such that the segment $[z_0, z_1]$ is in the domain of regularity of F and is mapped onto itself. Assume that $0 < |F'(z_0)| < 1 < |F'(z_1)|$ and that for z in the open segment (z_0, z_1), $F(z) \neq z$, $F'(z) \neq 0$. Then there is a continuous one-parameter group $\{F_t\}_{t\in\mathbb{R}}$ of functions with common fixed points z_0 and z_1 and invariant segment $[z_0, z_1]$ such that $F_1(z) = F(z)$ if and only if $F(z)$ is a linear fractional transformation on $\mathbb{C}$.*

First we note that the condition that F map $[z_0, z_1]$ into itself implies that $F'(z)$ is real on this segment.

Suppose now that F is linear fractional, $F(z) \not\equiv z$, and let z_0 and z_1 be its finite fixed points, $z_0 \neq z_1$. The following simple assertion can be obtained by using the linear model of mappings having two fixed points 0 and ∞ and applying the Julia–Carathéodory theorem.

Lemma 8.3. *The following are equivalent.*

(i) *There is an open disk D such that either $z_0 \in \partial D$ and $z_1 \notin \overline{D}$, or $z_0 \in D$ and $z_1 \in \partial D$, which is F-invariant.*

(ii) *Each open disk D such that $z_0 \in \overline{D}$ and $z_1 \notin D$ is F-invariant.*

(iii) *The segment $[z_0, z_1]$ is F-invariant and $|F'(z_0)| \leq 1$.*

(iv) *If $a = F'(z_0)$ then $0 < a < 1$.*

Since Schröder's equation

$$h(F(z)) = F'(z_0)h(z), \tag{8.5.10}$$

with linear-fractional F has a linear-fractional solution h, we have that h is starlike; hence F can be embedded into a one-parameter semigroup $\{F_t\}_{t\in\mathbb{R}}$ on each disk D containing z_0 and such that $z_1 \notin D$. This disk is F_t-invariant for all $t \geq 0$.

In turn, for the embedding property into a continuous group, we obtain the following assertion by using our Theorems 8.1 and 8.2 and Theorem 1 in [90].

Proposition 8.2. *Let F be a function of class $\mathcal{L}$ with two different fixed points z_0 and z_1. Assume that $0 < |F'(z_0)| < 1 < |F'(z_1)|$, and that for all z in the open segment (z_0, z_1), $F(z) \neq z$, $F'(z) \neq 0$. The following assertions are equivalent.*

(i) *For each open disk D such that $z_0 \in \overline{D}$ and $z_1 \notin D$, there is a semiflow $\mathcal{S} = \{F_t\}_{t\geq 0}$ with $F_1 = F$ such that $\mathcal{S} \subset \mathrm{Hol}(D)$.*

(ii) *For each domain* Ω *bounded by two circles passing through* z_0 *and* z_1*, there is a one-parameter flow* $\mathcal{S} = \{F_t\}_{t\in\mathbb{R}}$ *such that* $\mathcal{S} \subset \mathrm{Aut}(\Omega)$ *and* $F = F_1$.

(iii) *The function* F *is linear fractional with* $0 < F'(z_0) < 1$.

Consequently, in this case, for any disk D *such that* $z_0 \in \overline{D}$ *and* $z_1 \in \partial D$*, the maximal (backward) flow-invariant domain is the disk* $\Omega \subset D$ *whose boundary passes through* z_0 *and is internally tangent to* ∂D *at* z_1.

Chapter 9

Appendices

9.1 Controlled Approximation Problems

In this section, we use autonomic dynamical systems to study approximation problems for starlike and spirallike functions with respect to a boundary point and for some related classes of functions. Since such systems are time-independent, their solutions form one-parameter semigroups of holomorphic self-mappings of the open unit disk.

9.1.1 Setting of approximation problems

1. A perturbation problem for spirallike function. The first approximation problem we discuss here can be described as follows. Let $f \in \mathcal{G}^+[1]$, and let $h \in \text{Spiral}[1]$ be a spirallike function which, by Theeorem 2.14, satisfies the equation

$$\mu h(z) = h'(z)f(z). \tag{9.1.1}$$

Consider the perturbed equation

$$\mu_\tau h_\tau(z) = h'_\tau(z)f_\tau(z),$$

where $f_\tau \in \mathcal{G}[\tau]$, i.e., $f_\tau \in \mathcal{G}$, $f_\tau(\tau) = 0$ for some $\tau \in \Delta$, and $\mu_\tau = f'_\tau(\tau)$, are such that f_τ converges to f locally uniformly on Δ when τ goes to 1 unrestrictedly. (Note that such a perturbation is always possible, see, for example, the Berkson–Porta representation (2.2.7).) We ask:

◊ *Does the net $\{h_\tau\}$ converge to h as $\tau \to 1$?*

The following simple example shows that, in general, the answer is negative.

Example 9.1. Let $f(z) = (z-1) \in \mathcal{G}^+[1]$. Then the function $h(z) = 1 - z$ satisfies equation (9.1.1) with $\mu = 1$:

$$h(z) = h'(z)f(z).$$

Define now $f_\tau(z) = \dfrac{(z-\tau)(1-z\tau)}{1-z}$, $\tau \in (0,1)$. Obviously, $f_\tau \in \mathcal{G}[\tau]$ (by Theorem 2.7), and f_τ converges to f as $\tau \to 1^-$. Consider the perturbed problem

$$\mu_\tau h_\tau(z) = h'_\tau(z) f_\tau(z), \tag{9.1.2}$$

where $\mu_\tau = f'_\tau(\tau) = 1+\tau$, $f_\tau \to f$ as $\tau \to 1^-$. Then the function

$$h_\tau(z) = \frac{(\tau - z)(1-z\tau)^{\frac{1}{\tau}}}{\tau}$$

is a solution of equation (9.1.2) satisfying $h_\tau(0) = h(0) = 1$. Letting τ tend to the boundary point 1, we obtain that the limit function

$$\lim_{\tau \to 1^-} h_\tau(z) = (1-z)^2,$$

which is different from $h(z)$.

At the same time if we choose $f_\tau \in \mathcal{G}[\tau]$ in a different way, say $f_\tau(z) = z - \tau$, we see that $h_\tau = 1 - \frac{z}{\tau}$ defined as a solution of equation (9.1.2) converges to h.

Thus one can consider the following perturbation problem.

$\Diamond$ *For any $\tau \in \Delta$, find a perturbed function $f_\tau \in \mathcal{G}[\tau]$ converging to f as τ goes to 1 unrestrictedly and such that the solution of (9.1.2) converges to the original solution of equation (9.1.1) uniformly on compact subsets of Δ.*

Geometrically, an affirmative answer to this question would give us a constructive method for approximation of spirallike (starlike) functions with respect to a boundary point by spirallike (starlike) functions with respect to interior points.

We solve this problem as follows. Given $\tau \in \Delta$, we find a transformation $\Phi_\tau : \mathrm{Hol}(\Delta, \mathbb{C}) \mapsto \mathrm{Hol}(\Delta, \mathbb{C})$ which takes $h \in \mathrm{Spiral}[1]$ to $h_\tau \in \mathrm{Spiral}[\tau]$, $\tau \in \Delta$ (i.e., $h_\tau = \Phi_\tau(h)$), and such that $\Phi_\tau(h)$ tends to h when τ tends to 1 (see Theorem 9.4).

To do this, we need first to consider some approximation and interpolation problems for the classes $\mathcal{G}$ and $\mathcal{P}$ which are of independent interest.

2. An approximation problem for generators. As we already saw in Section 5.4 a solution of the differential equation

$$\beta h(z) = h'(z) f(z), \quad \beta = f'(1) > 0,$$

belongs to the class $\mathrm{Star}_1[1]$. Therefore, by Proposition 4.2, if a solution h of equation (9.1.1) is univalent, then μ must lie in the region $\Omega = \big\{w \neq 0 : \ |w-\beta| \le \beta, \text{ or } |w+\beta| \le \beta\big\}$.

Actually, the instability phenomenon we have seen in Example 9.1 above follows from the fact that $\mu_\tau = f'_\tau(\tau)$ does not necessarily converge to μ as $\tau \to 1$ even if $\mu = f'(1)$. Therefore, one can pose the following problem on controlled approximation for functions in the class $\mathcal{G}_h[1] \subset \mathcal{G}[1]$.

◊ *Let* $f \in \mathcal{G}_h[1]$ *with* $f'(1) = \beta > 0$. *For* $\tau \in \Delta$ *and given* $\mu \in \Omega_+ := \{w \neq 0 : |w - \beta| \leq \beta\}$, *find a net* $\{f_\tau\}$, $f_\tau \in \mathcal{G}[\tau]$, *converging locally uniformly to* f *as* $\tau \to 1$ *unrestrictedly and such that* $\mu_\tau = f'_\tau(\tau)$ *converges to* μ.

3. An approximation problem for semigroups. For each $f \in \mathcal{G}^*[\tau]$ with $\operatorname{Re} f'(\tau) > 0$, $\tau \in \overline{\Delta}$, the generated semigroup $\mathcal{S} = \{F_t\}_{t \geq 0}$ converges to the point τ as $t \to \infty$. Moreover, the rate of convergence of the semigroup can be estimated by the derivative $f'(\tau)$ of the generator f at the point τ (for the boundary case $\tau \in \partial\Delta$, this fact follows by Theorem 2.10; for the general case $\tau \in \overline{\Delta}$, we refer to [57], see also [130]).

Note that the number μ in equation (9.1.1) is not necessarily real, while the angular derivative $\beta = f'(1)$ is a nonnegative real number. So, in light of the instability phenomenon mentioned above, the following question seems to be natural.

◊ *Let* $f \in \mathcal{G}_h[1]$ *generate the semigroup* $\mathcal{S} = \{F_t\}_{t \geq 0}$. *Suppose that a net* $\{f_\tau\}$, $f_\tau \in \mathcal{G}[\tau]$, $\tau \in \Delta$, *converges to* f *locally uniformly as* $\tau \to 1$. *Do the corresponding elements* $F_{t,\tau}$, $t \geq 0$, *of the semigroups* $\mathcal{S}_\tau$ *generated by* f_τ *converge to* F_t *for all* $t \geq 0$ *as* τ *tends to* 1?

Here the difficulty is that the value $F_{t,\tau}(z)$ must lie in a neighborhood of $\tau \in \Delta$ for any $z \in \Delta$ and for t sufficiently large, while $F_t(z)$ is close to 1.

Nevertheless, we shall show that for each $r \in (0, 1)$ and $T > 0$, the net $\{F_{t,\tau}(z)\}$ converges to $F_t(z)$ uniformly on the set $[0, T] \times (r\Delta)$.

4. An approximation problem for functions with positive real part. By the Berkson–Porta representation (see Theorem 2.7), each cone $\mathcal{G}^+[\tau]$, $\tau \in \partial\Delta$, or $\mathcal{G}[\tau]$, $\tau \in \Delta$, can be parameterized by elements of the cone $\mathcal{P}$ of functions with positive real part. It turns out that the approximation problems above lead us to the following interpolation question about the approximation of functions of class $\mathcal{P}$. Recall that a holomorphic function $f \in \operatorname{Hol}(\Delta, \mathbb{C})$ is said to be conformal at a boundary point $\tau \in \partial\Delta$ if its angular derivative $f'(\tau)$ exists finitely and $f'(\tau) \neq 0$.

◊ *Let* $q \in \mathcal{P}$ *be conformal at the boundary point* 1. *For* $\tau \in \Delta$ *and given* $\varphi \in \left(-\frac{\pi}{2}, \frac{\pi}{2}\right)$, *does there exist a net* $\{q_\tau\} \subset \mathcal{P}$ *converging locally uniformly to* q *as* $\tau \to 1$ *unrestrictedly and such that*

$$\arg q_\tau(\tau) = \varphi \quad \textit{for all } \tau \in \Delta.$$

9.1.2 Solutions of approximation problems

Recall that by the Riesz–Herglotz formula (1.4.2)

$$p(z) = \oint_{\partial\Delta} \frac{1 + z\overline{\zeta}}{1 - z\overline{\zeta}}\, d\sigma_p(\zeta) + i \operatorname{Im} p(0),$$

the angular limit

$$\delta_p(\tau) = \angle \lim_{z\to 1}(1 - z\overline{\tau})p(z) \tag{9.1.3}$$

exists and is a nonnegative real number. We call this number the charge of the function $p \in \mathcal{P}$ at the boundary point $\tau \in \partial\Delta$ (see Section 1.4 for details). It follows by Theorem 1.14 that

$$\delta_p(\tau) = 2 \inf_{z\in\Delta} \frac{1 - |z|^2}{|1 - z|^2} \operatorname{Re} p(z).$$

As above, we denote by $\mathcal{P}^+[\tau]$ the subclass of $\mathcal{P}$ consisting of functions with strictly positive charges at $\tau \in \partial\Delta$, so, $p \in \mathcal{P}^+[1]$ if and only if $\delta_p(1) > 0$. Thus, $p \in \mathcal{P}^+[1]$ if and only if the function $q \in \mathcal{P}$ defined by $q = \frac{1}{p}$ is conformal at $\tau = 1$ with $q(1) = 0$. Moreover, in this case, $q'(1)$ is a negative real number.

The following assertion, which gives a solution to the problem in Section 9.1.1 (4), is the key for our further considerations.

Theorem 9.1. *Let $q \in \mathcal{P}$ be conformal at $\tau = 1$ with $q(1) = 0$. Then for all $\tau \in \Delta$ and each $\gamma \in \mathbb{C}$ such that*

$$\operatorname{Re}\gamma \geq \frac{\alpha}{2}, \quad \textit{where} \quad \alpha = -q'(1)\,(>0)\,, \tag{9.1.4}$$

there exist functions $\{q_\tau\}_{\tau\in\Delta} \subset \mathcal{P}$ converging to q uniformly on compact subsets of Δ when τ tends to 1 unrestrictedly and such that

$$q_\tau(\tau) = \gamma(1 - |\tau|^2) \to 0 \quad \textit{as} \quad \tau \to 1.$$

In particular, if γ is real and $\gamma \geq \frac{\alpha}{2}$, the values $q_\tau(\tau)$ are real numbers.

Proof. For $\gamma \in \mathbb{C}$, consider the function

$$r(z) = \frac{zq(z) + \gamma z^2 - \overline{\gamma} - 2iz \operatorname{Im}\gamma}{-(1 - z)^2}. \tag{9.1.5}$$

Clearly, $r \in \operatorname{Hol}(\Delta, \mathbb{C})$ and $r(0) = \overline{\gamma}$.

We claim that $r \in \mathcal{P}$ if and only if γ satisfies inequality (9.1.4).

Indeed, consider the function

$$f(z) = zq(z) + \gamma z^2 - \overline{\gamma} - 2iz \operatorname{Im}\gamma, \tag{9.1.6}$$

which is the numerator of (9.1.5).

It follows by Theorem 2.6 that $f \in \mathcal{G}$. In addition,

$$f(1) \left(= \angle \lim_{z\to 1} f(z)\right) = \gamma - \overline{\gamma} - 2i \operatorname{Im}\gamma = 0$$

and

$$f'(1) \left(= \angle \lim_{z\to 1} \frac{f(z)}{z-1}\right)$$
$$= q'(1) + \angle \lim_{z\to 1} \frac{\gamma z^2 - \bar{\gamma} - 2iz \operatorname{Im}\gamma}{z-1} = q'(1) + 2\operatorname{Re}\gamma.$$

Thus,

$$f'(1) = 2\operatorname{Re}\gamma - \alpha \geq 0 \tag{9.1.7}$$

if and only if condition (9.1.4) holds. If it is the case, the Julia–Wolff–Carathéodory Theorem 2.10 implies that the point $\tau = 1$ is the Denjoy–Wolff point of the semigroup generated by f. On the other hand, by the uniqueness of the Berkson–Porta representation (see Theorem 2.7) of the class $\mathcal{G}$, the inequality in (9.1.7) is equivalent to

$$\operatorname{Re} \frac{f(z)}{-(1-z)^2} \geq 0.$$

Comparing (9.1.7) with (9.1.5) and (9.1.6) proves our claim.

Now for $\tau \in \Delta$, define the function q_τ on Δ by

$$q_\tau(z) = \frac{1}{z}\left[(z-\tau)(1-z\bar{\tau})\, r(z) + \bar{\gamma}\tau - \gamma\bar{\tau}z^2 + 2iz\operatorname{Im}\gamma\right] \tag{9.1.8}$$

or

$$q_\tau(z) = \frac{1}{z} g_\tau(z), \tag{9.1.9}$$

where

$$g_\tau(z) = (z-\tau)(1-z\bar{\tau})\, r(z) + \bar{\gamma}\tau - \gamma\bar{\tau}z^2 + 2iz\operatorname{Im}\gamma. \tag{9.1.10}$$

Since $r(0) = \bar{\gamma}$, we have $g_\tau(0) = 0$. Hence, q_τ is holomorphic in Δ.

Moreover, since $\mathcal{G}$ is a real cone, we have by Theorems 2.6 and 2.7 that $g_\tau \in \mathcal{G}$ for all $\tau \in \Delta$. Consequently, $g_\tau \in \mathcal{G}[0]$.

Again by Theorem 2.7, $\operatorname{Re} q_\tau(z) \left(= \operatorname{Re} \dfrac{g_\tau(z)}{z}\right) \geq 0$ for all $z \in \Delta$.

In addition, since

$$q(z) = \frac{1}{z}\, g(z), \tag{9.1.11}$$

where

$$g(z) = -\left(1-z\right)^2 r(z) - \gamma z^2 + \bar{\gamma} + 2iz\operatorname{Im}\gamma, \tag{9.1.12}$$

we have by (9.1.9) and (9.1.10) that $q_\tau(z)$ converges to $q(z)$ as τ tends to 1 unrestrictedly.

Finally, direct calculation shows that $q_\tau(\tau) = \gamma\left(1 - |\tau|^2\right)$. □

Example 9.2. Consider the function $q(z) = 1 - z$, which obviously has positive real part. Then the function r defined by (9.1.5) with $\gamma = 1$ has the form

$$r(z) = \frac{1}{1-z}.$$

Substituting this function into formula (9.1.8), we find the approximating functions

$$q_\tau(z) = \bar{\tau}(1-z) + \frac{|1-\tau|^2}{1-z}.$$

We see that $q_\tau \to q$ as τ tends to 1 unrestrictedly.

(i) Choosing the sequence of real numbers $\tau_n^{(1)} = 1 - \frac{1}{n}$, we get the approximating sequence $q_n^{(1)}$ of functions with positive real part. In Figure 9.1, we see the images of the unit circle $\partial\Delta$ under the approximating functions $q_1^{(1)}$, $q_2^{(1)}$ and $q_4^{(1)}$ as well as the image of q. It is worth noting that $q_1^{(1)}(\Delta) = \{w : \operatorname{Re} w > \frac{1}{2}\}$ and that the images $q_n^{(1)}(\Delta)$ increase in the sense that $q_n^{(1)}(\Delta) \subset q_{n+1}^{(1)}(\Delta)$. As n tends to infinity, these images tend to the right half-plane, while the original function q is bounded. (Note that the Carathéodory Kernel Theorem is not applicable here since the functions q_n are not univalent.)

(ii) Choosing another sequence $\tau_n^{(2)}$ converging to 1, say, $\tau_n^{(2)} = 1 - \frac{3(1-i)}{n}$, we get a different approximating sequence $q_n^{(2)}$ of positive real part functions. In Figure 9.2, we see the images of the unit circle $\partial\Delta$ under the functions $q_4^{(2)}$, $q_6^{(2)}$ and $q_{12}^{(2)}$. Once again, in spite of the boundedness of q, for n large enough the image $q_n^{(2)}(\Delta)$ almost covers the right half-plane.

Corollary 9.1. *Let $p \in \mathcal{P}^+[1]$ with the charge $\delta_p(1) = \beta > 0$. Then for all $\tau \in \Delta$ and each $\mu \in \mathbb{C}$ such that*

$$|\mu - \beta| \le \beta, \ \mu \ne 0, \tag{9.1.13}$$

there exist functions $\{p_\tau\}_{\tau\in\Delta} \subset \mathcal{P}$ which converge to p uniformly on compact subsets of Δ as τ tends to 1 and such that

$$p_\tau(\tau) = \frac{\mu}{1-|\tau|^2} \to \infty \quad \text{as } \tau \to 1. \tag{9.1.14}$$

In particular, if μ is real and

$$0 < \mu \le 2\beta,$$

the values $p_\tau(\tau)$ are real numbers.

Proof. Setting

$$q(z) = \frac{1}{p(z)},$$

we have $q \in \mathcal{P}$ with $q(1) = 0$ and

$$q'(1) = \angle\lim_{z\to1} \frac{q(z)}{z-1} = \angle\lim_{z\to1} \frac{1}{(z-1)p(z)} = -\frac{1}{\beta}\,(:= -\alpha).$$

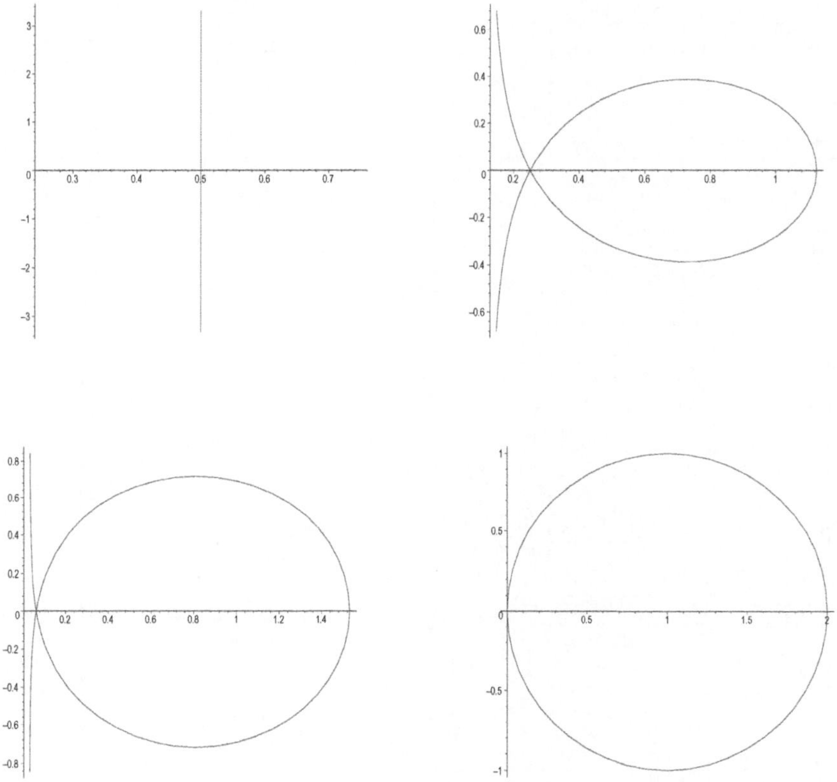

Figure 9.1: Example 9.2 (i), the images of $q_1^{(1)}$, $q_2^{(1)}$, $q_4^{(1)}$ and of q.

Using Theorem 9.1 for each $\gamma \in \mathbb{C}$ such that

$$2\,\mathrm{Re}\,\gamma \geq \frac{1}{\beta} \tag{9.1.15}$$

and all $\tau \in \Delta$, one can find $q_\tau \in \mathcal{P}$ converging to q as $\tau \to 1$ and such that

$$q_\tau(\tau) = \gamma\left(1 - |\tau|^2\right).$$

Setting now

$$p_\tau(z) = \frac{1}{q_\tau(z)}$$

and $\gamma = \frac{1}{\mu}$, we obtain that (9.1.13) is equivalent to (9.1.15) and p_τ converges locally uniformly to p.

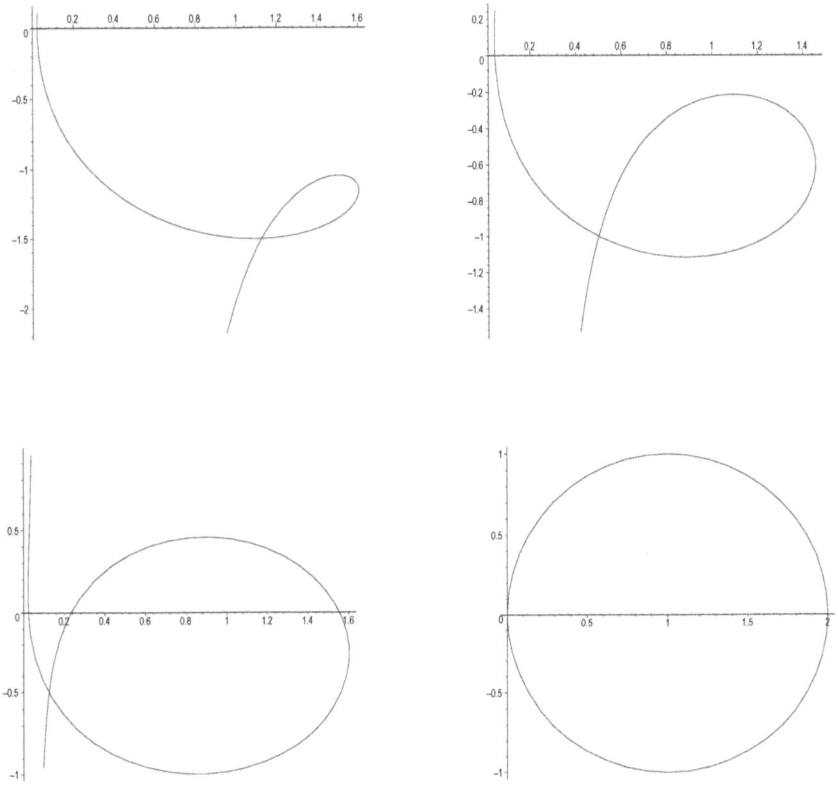

Figure 9.2: Example 9.2 (ii), the images of $q_4^{(2)}$, $q_6^{(2)}$, $q_{12}^{(2)}$.

Finally,

$$p_\tau(\tau) = \frac{1}{q_\tau(\tau)} = \frac{1}{\gamma\,(1-|\tau|^2)} = \frac{\mu}{1-|\tau|^2},$$

and we are done. □

It is now easy to prove the following theorem, which solves the problem in Section 9.1.1 (2).

Theorem 9.2. *Let $f \in \mathcal{G}$ satisfy the conditions $f(1) = 0$ and $f'(1) = \beta > 0$, i.e., $f \in \mathcal{G}_h[1]$. Then for each $\mu \in \Omega_+ = \{w \neq 0 : \; |w - \beta| \leq \beta\}$ and for all $\tau \in \Delta$, there exist generators $\{f_\tau\}$ with $f_\tau \in \mathcal{G}[\tau]$ such that $f_\tau'(\tau) = \mu$ and f_τ converges to f uniformly on compact subsets of Δ as τ tends to 1 unrestrictedly.*

In particular, setting $\mu = \beta$ we have $f_\tau'(\tau) = \beta \in (0, \infty)$ for all $\tau \in \Delta$.

Proof. By the Berkson–Porta formula (2.2.7), f must be of the form

$$f(z) = -(1-z)^2\, p(z), \tag{9.1.16}$$

where $p \in \mathcal{P}^+[1]$ with $\delta_p(1) = \beta > 0$.

By Corollary 9.1, for each μ satisfying (9.1.13) and $\tau \in \Delta$, one can find $p_\tau \in \mathcal{P}$ with $p_\tau(\tau) = \dfrac{\mu}{1-|\tau|^2}$ and such that p_τ converges to p as $\tau \to 1$.

Now we construct $f_\tau \in \mathcal{G}$ by the Berkson–Porta representation (2.2.7)

$$f_\tau(z) = (z-\tau)\,(1-z\bar{\tau})\,p_\tau(z).$$

Since $f_\tau'(\tau) = \left(1-|\tau|^2\right) p_\tau(\tau)$, we obtain our assertion. □

In some sense a converse assertion is also true.

Theorem 9.3. *Let $f \in \mathcal{G}_h[1]$ with $f(1) = 0$ and $f'(1) = \beta > 0$, and let $f_n \in \mathcal{G}[\tau_n]$, $\tau_n \in \Delta$, $n = 1, 2, \ldots$. Suppose that f_n converges to f uniformly on compact subsets of Δ. Then*

(i) *the sequence $\{\tau_n\}$ converges to 1;*

(ii) *if the sequence $\mu_n := f_n'(\tau_n)$ converges to $\mu \neq 0$, then μ must satisfy condition (9.1.13), i.e., $|\mu - \beta| \le \beta$.*

Proof. (i) For each $n = 1, 2, \ldots$, the function $f_n \in \mathcal{G}[\tau_n]$ has the form $f_n(z) = (z-\tau_n)\,(1-z\overline{\tau_n})\,p_n(z)$ with $\operatorname{Re} p_n(z) \ge 0$ everywhere.

Since $\overline{\Delta}$ is a compact subset of $\mathbb{C}$ and $\{p_n\}$ is a normal family on Δ, one can choose a subsequence $\{n_k\} \subset \mathbb{N}$ such that τ_{n_k} converges to $\tau \in \overline{\Delta}$ and p_{n_k} converges either to $\widetilde{p} \in \mathcal{P}$ or $\widetilde{p} = \infty$. In any case, the sequence $\{f_{n_k}\} \subset \mathcal{G}$ converges to either $f(z) = (z-\tau)\,(1-z\bar{\tau})\,\widetilde{p}(z)$ with $\operatorname{Re}\widetilde{p}(z) \ge 0$ and $\tau \in \overline{\Delta}$ or to infinity. Since the latter case is impossible, the uniqueness of the Berkson–Porta representation and (9.1.16) imply that $\tau = 1$ and $\widetilde{p} = p$.

(ii) Assume now that $\mu_n := f_n'(\tau_n)$ converges to $\mu \in \mathbb{C}$, $\mu \neq 0$, and consider the differential equations

$$\mu_n h_n(z) = h_n'(z) f_n(z),$$

normalized by the conditions $h_n(0) = 1$. This initial value problem has a unique solution which is a univalent function on Δ spirallike with respect to an interior point and $h_n(\tau_n) = 0 \in h_n(\Delta)$.

Explicitly, h_n can be written as

$$h_n(z) = \exp\left[\mu_n \int_0^z \frac{dz}{f_n(z)}\right]. \tag{9.1.17}$$

Now for each $r \in (0,1)$, one can find $n_0 \in \mathbb{N}$ such that for all $n > n_0$, the points τ_n do not belong to the closed disk $\overline{\Delta_r} = \{|z| \le r < 1\}$. In other words, for all

$n > n_0$ the functions f_n do not vanish in this disk. Then the functions h_n defined by (9.1.17) converge to a function $h \in \mathrm{Hol}(\Delta_r, \mathbb{C})$ uniformly on this disk. Since r is arbitrary, we have that actually $h \in \mathrm{Hol}(\Delta, \mathbb{C})$ and has the form

$$h(z) = \exp\left[\mu \int_0^z \frac{dz}{f(z)}\right]. \tag{9.1.18}$$

In addition, it follows by Hurwitz's theorem that h is either a univalent function on Δ or a constant. The latter case is impossible because of the equality $\dfrac{h'(z)}{h(z)} = \dfrac{\mu}{f(z)}$, which follows by (9.1.18) and $\mu \neq 0$.

On the other hand, we already know that the initial value problem

$$\beta\widetilde{h}(z) = \widetilde{h}'(z), f(z), \quad h(0) = 1$$

has also the unique solution

$$\widetilde{h}(z) = \exp\left[\beta \int_0^z \frac{dz}{f(z)}\right], \tag{9.1.19}$$

which is a starlike function with respect to the boundary point $\left(\widetilde{h}(1) = 0\right)$.

Comparing (9.1.18) and (9.1.19), we obtain

$$h(z) = \left[\widetilde{h}(z)\right]^{\frac{\mu}{\beta}}.$$

In addition, $\widetilde{h}$ satisfies the Visser–Ostrowski condition:

$$\angle \lim_{z \to 1} \frac{(z-1)\widetilde{h}'(z)}{\widetilde{h}(z)} = 1.$$

Now Theorem 3.3 implies that the smallest wedge which contains $h(\Delta)$ is exactly of angle π. Since h is univalent, we get that $\mathrm{Re}\,\dfrac{\beta}{\mu} \geq \dfrac{1}{2}$, which is equivalent to (9.1.13). $\square$

As a consequence of Theorems 3.2 and 3.3, we obtain the following result.

Theorem 9.4. *Let $f \in \mathcal{G}_h[1]$ with $f'(1) = \beta > 0$. Then the initial value problem*

$$\lambda h(z) = h'(z)f(z), \quad h(0) = 1 \tag{9.1.20}$$

has a univalent solution $h \in \mathrm{Hol}(\Delta, \mathbb{C})$ if and only if the complex number λ belongs to the set $\Omega = \Omega_+ \cup \Omega_-$, where

$$\Omega_\pm = \{\omega \in \mathbb{C} : \ |\omega \mp \beta| \leq \beta, \ \omega \neq 0\}.$$

Moreover,

- *for each* $\mu \in \Omega_+$ *and each* $\tau \in \Delta$, *there are* $h_\tau \in \mathrm{Spiral}[\tau]$ *which converge to the function*
$$h_1(z) = h^{\frac{\mu}{\lambda}}(z)$$
as τ *tends to* 1 *unrestrictedly. In particular, if* $\lambda \in \Omega_+$, *then choosing* $\mu = \lambda$, *we find* $h_\tau \in \mathrm{Spiral}[\tau]$ *which converge to the original function* h *as* $\tau \to 1$.

- *for each* $\mu \in \Omega_-$ *and for each* $\tau \in \Delta$, *there are meromorphic functions* $\widetilde{h}_\tau$ *with a unique simple pole at* τ *and such that* $\widetilde{h}_\tau$ *converges to the holomorphic function*
$$\widetilde{h}(z) = h^{\frac{\mu}{\lambda}}(z) \tag{9.1.21}$$
as τ *tends to* 1 *unrestrictedly.*

Proof. Take any $\mu \in \Omega_+$ and any $\tau \in \Delta$. By Theorem 9.2, one can choose generators $f_\tau \in \mathcal{G}[\tau]$ such that the net $\{f_\tau\}$ converges uniformly on compact subsets to f and satisfies the conditions $f_\tau(\tau) = 0$ and $f'_\tau(\tau) = \mu$. Then, as in the proof of the second part of Theorem 9.3, one shows that functions $h_\tau \in \mathrm{Spiral}[\tau]$ defined by
$$h_\tau = \exp\left[\mu \int_0^z \frac{dz}{f_\tau(z)}\right]$$
converge to a univalent function
$$h_1(z) = \exp\left[\mu \int_0^z \frac{dz}{f(z)}\right] \tag{9.1.22}$$
which satisfies the equation
$$\mu h_1(z) = h_1'(z) f(z). \tag{9.1.23}$$
If now $\lambda \in \Omega_+$, then setting $\mu = \lambda$, we see that $h = h_1$ must be univalent. If $\lambda \in \Omega_-$, then setting $\mu = -\lambda$, and comparing differential equations (9.1.20) and (9.1.23), we see that $h = h_1^{-1}$. Since $h_1(z) \neq 0$, $z \in \Delta$, h is a well-defined univalent function on Δ. In addition, it is clear that h_1^{-1} is a locally uniform limit of meromorphic functions h_τ^{-1} with poles at τ.

Conversely, assume that for some $\lambda \in \mathbb{C}$, $\lambda \neq 0$, equation (9.1.20) has a univalent solution in Δ. If $\mathrm{Re}\,\lambda > 0$, then setting $\mu = \beta$ in (9.1.22), we see as in the proof of Theorem 9.3 that the image of the function $h_1(z) = h^{\frac{\beta}{\lambda}}(z)$ must lie in the wedge of the angle π, which is the smallest one containing $h_1(\Delta)$. Then, by Proposition 4.2, we have $\lambda \in \Omega_+$. If $\mathrm{Re}\,\lambda < 0$, then the same considerations show that $-\lambda \in \Omega_+$, and we are done. □

9.1.3 Perturbation formulas

Theorem 9.5. *Let* $f \in \mathcal{G}_h[1]$ *with* $f'(1) = \beta > 0$ *and let* $\lambda \in \Omega_+ = \{w \in \mathbb{C} : |w - \beta| \leq \beta,\ w \neq 0\}$. *Assume that* $h \in \mathrm{Hol}(\Delta, \mathbb{C})$ *is the solution of equation (9.1.20)*
$$\lambda h(z) = h'(z) f(z)$$

normalized by the conditions $h(0) = 1$ and $h(1) = 0$. Then for each $\mu \in \Omega_+$ and for each $\tau \in \Delta$, the function $h_\tau \in \mathrm{Hol}(\Delta, \mathbb{C})$ defined by

$$h_\tau(z) = [h(z)]^{\frac{\mu}{\lambda}} \frac{(z-\tau)(1-z\bar{\tau})^{\mu/\bar{\mu}}}{-(1-z)^{1+\mu/\bar{\mu}}}$$

is univalent on Δ and belongs to the class Spiral$[\tau]$ *with $h_\tau(\tau) = 0$ and $h_\tau(0) = \tau$.*

If, in particular, $\mu = \lambda$, then h_τ converges to h whenever τ tends to 1. Thus h is a univalent function on Δ spirallike (starlike) with respect to a boundary point with $h(1) = 0$.

Proof. Let $h \in \mathrm{Hol}(\Delta, \mathbb{C})$ be a solution of the differential equation (9.1.20) with $f \in \mathcal{G}_h[1]$ defined by $f(z) = \dfrac{-(1-z)^2}{q(z)}$, where

$$\operatorname{Re} q(z) > 0 \quad \text{and} \quad \angle \lim_{z\to 1} \frac{1-z}{q(z)} = \beta.$$

Then

$$\frac{h'(z)}{\lambda h(z)} = \frac{q(z)}{-(1-z)^2}\,. \tag{9.1.24}$$

As in the proof of Theorem 9.1, for γ complex with $2\operatorname{Re}\gamma \geq \frac{1}{\beta}$ we define $r \in \mathcal{P}^+[1]$ by

$$r(z) = \frac{zq(z) + (\gamma z + \bar{\gamma})(z-1)}{-(1-z)^2}\,. \tag{9.1.25}$$

Comparing (9.1.24) and (9.1.25), we have

$$r(z) = \frac{zh'(z)}{\lambda h(z)} + \frac{\gamma z + \bar{\gamma}}{1-z}\,. \tag{9.1.26}$$

Once again, for each point $\tau \in \Delta$, we define

$$q_\tau(z) = \frac{1}{z}\left[(z-\tau)(1-z\bar{\tau})r(z) + \bar{\gamma}\tau - \gamma\bar{\tau}z^2 + (\gamma - \bar{\gamma})z\right].$$

Then the differential equation

$$\mu h_\tau(z) = h_\tau'(z)(z-\tau)(1-z\bar{\tau})\frac{1}{q_\tau(z)} \tag{9.1.27}$$

has a holomorphic solution h_τ if and only if $\mu = \dfrac{1}{\gamma} \in \Omega_+$. Moreover, $h_\tau \in$ Spiral$[\tau]$ by Theorems 2.7 and 2.14. Since $h_\tau(\tau) = 0$, the function $g_\tau(z) := \dfrac{h_\tau(z)}{z-\tau}$ is well-defined. Now, using (9.1.26) and (9.1.27), we calculate

$$\begin{aligned}\frac{g_\tau'(z)}{g_\tau(z)} &= \mu\left[\frac{h'(z)}{\lambda h(z)} + \frac{\gamma z+\bar{\gamma}}{z(1-z)} - \frac{\bar{\gamma}}{z(1-z\bar{\tau})}\right] \\ &= \mu\left[\frac{h'(z)}{\lambda h(z)} + \frac{\gamma}{1-z} + \frac{\bar{\gamma}}{1-z} + \frac{\bar{\gamma}(-\bar{\tau})}{1-z\bar{\tau}}\right] \\ &= \frac{\mu h'(z)}{\lambda h(z)} + \frac{1}{1-z} + \frac{\mu\bar{\gamma}}{1-z} + \frac{\mu\bar{\gamma}(-\bar{\tau})}{1-z\bar{\tau}}.\end{aligned}$$

Then

$$h_\tau(z) = (z-\tau)g_\tau(z) = C\left(h(z)\right)^{\frac{\mu}{\lambda}} \cdot \frac{(z-\tau)(1-z\bar{\tau})^{\frac{\mu}{\bar{\mu}}}}{(1-z)^{1+\frac{\mu}{\bar{\mu}}}}$$

with some constant C. So, to satisfy the normalization $h_\tau(0) = \tau$, we must set $C = -1$. The proof is complete. □

Example 9.3. Consider the starlike function with respect to a boundary point

$$h(z) = (1-z)^{0.8}.$$

It satisfies equation (9.1.20) with $f(z) = z-1$ and $\lambda = 0.8$:

$$0.8h(z) = h'(z)\cdot(z-1).$$

Setting $\mu = \lambda = 0.8$, we see that by Theorem 9.5, h can be approximated by the functions

$$h_\tau(z) = \frac{(\tau-z)(1-z\bar{\tau})}{(1-z)^{1.2}}.$$

(i) Choosing, in particular, the sequence of real numbers $\tau_n^{(1)} = 1-\frac{3}{n}$, we get the approximating sequence $h_n^{(1)}$ of starlike functions with respect to (different) interior points. In Figure 9.3, we see the images of h as well as the images of the approximating functions $h_6^{(1)}$, $h_{10}^{(1)}$ and $h_{30}^{(1)}$. Note that in this case, the intersection $\bigcap_{n=0}^{\infty} h_n^{(1)}(\Delta)$ contains the left half-plane, while the image $h(\Delta)$ lies in the right half-plane.

(ii) Choosing another sequence $\tau_n^{(2)}$ converging to 1, say, $\tau_n^{(2)} = 1-(1-i)\frac{3}{n}$, we get a different approximating sequence $h_n^{(2)}$ of starlike functions with respect to interior points. In Figure 9.4, we see the images of the approximating functions $h_6^{(2)}$, $h_{10}^{(2)}$ and $h_{30}^{(2)}$. Once again, all of the images $h_n^{(2)}(\Delta)$ contain the left half-plane.

(iii) On the other hand, setting $\mu = 1+i$ and choosing $\tau_n^{(3)} = 1-\frac{3}{n}$, we find the sequence $h_n^{(3)}$ of spirallike functions with respect to interior points, which converges to the function

$$\widetilde{h}(z) = \left(h(z)\right)^{\frac{1+i}{0.8}} = (1-z)^{1+i},$$

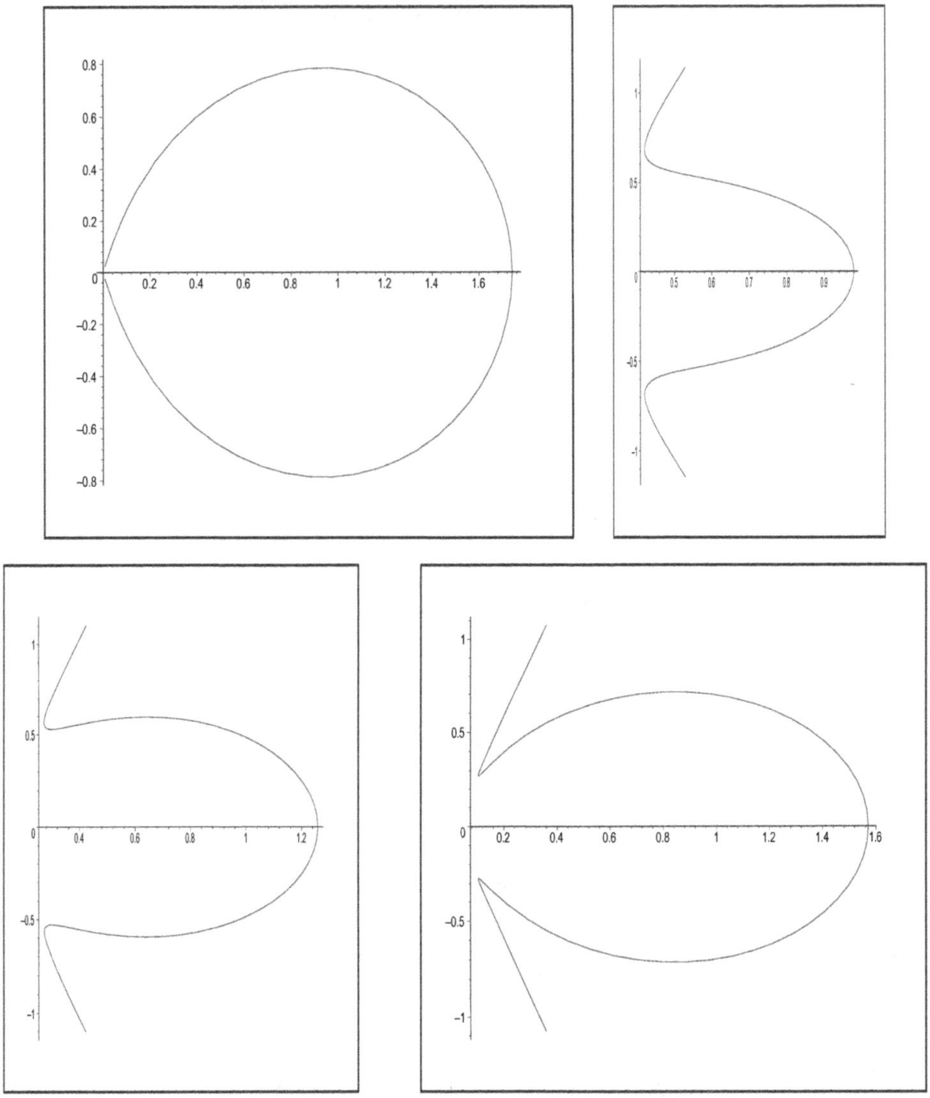

Figure 9.3: Example 9.3 (i), the images of $h, h_6^{(1)}, h_{10}^{(1)}$ and $h_{30}^{(1)}$.

which is spirallike with respect to a boundary point. In Figure 9.5, one can see images of a number of approximating functions and the image of $\widetilde{h}$.

Using Theorem 9.4 above, one can prove the uniform stability of a semigroup generated by $f \in \mathcal{G}_h[1]$ under any perturbation of its generator (see Problem in Section 9.1.1 (3)).

Let $\{F_t\}_{t\geq 0}$ be a semigroup generated by $f \in \mathcal{G}_h[1]$ and let $f_n \in \mathcal{G}[\tau_n]$, $\tau_n \in$

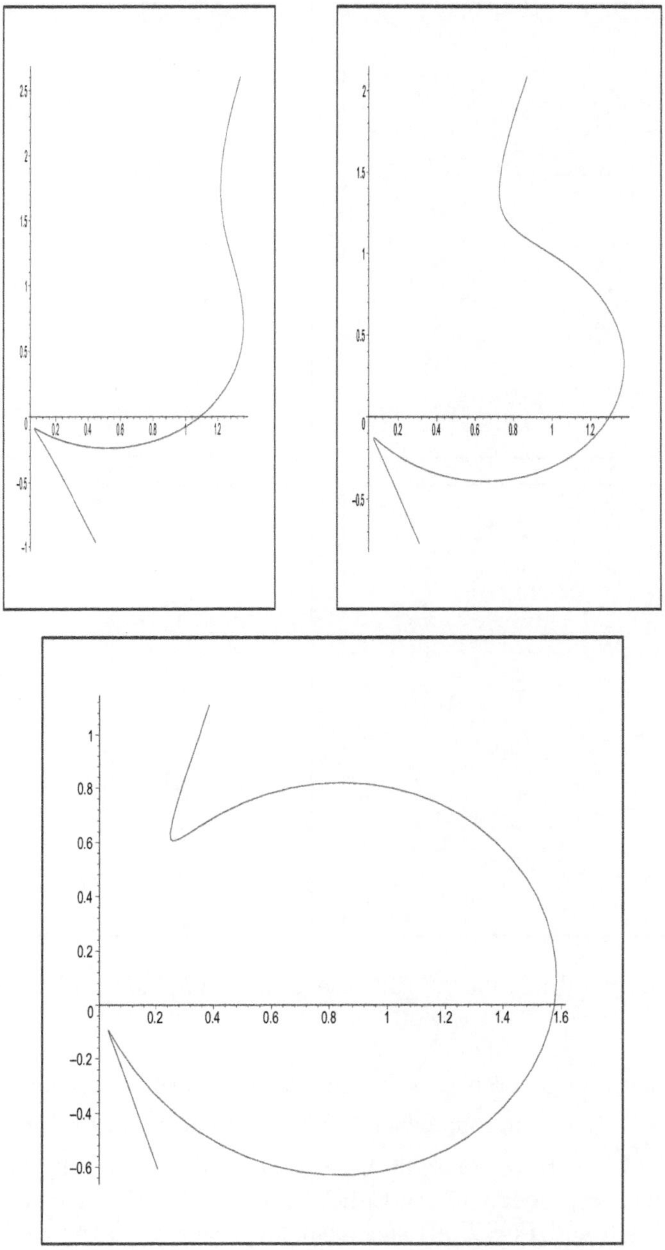

Figure 9.4: Example 9.3 (ii), the images of $h_6^{(2)}, h_{10}^{(2)}$ and $h_{30}^{(2)}$.

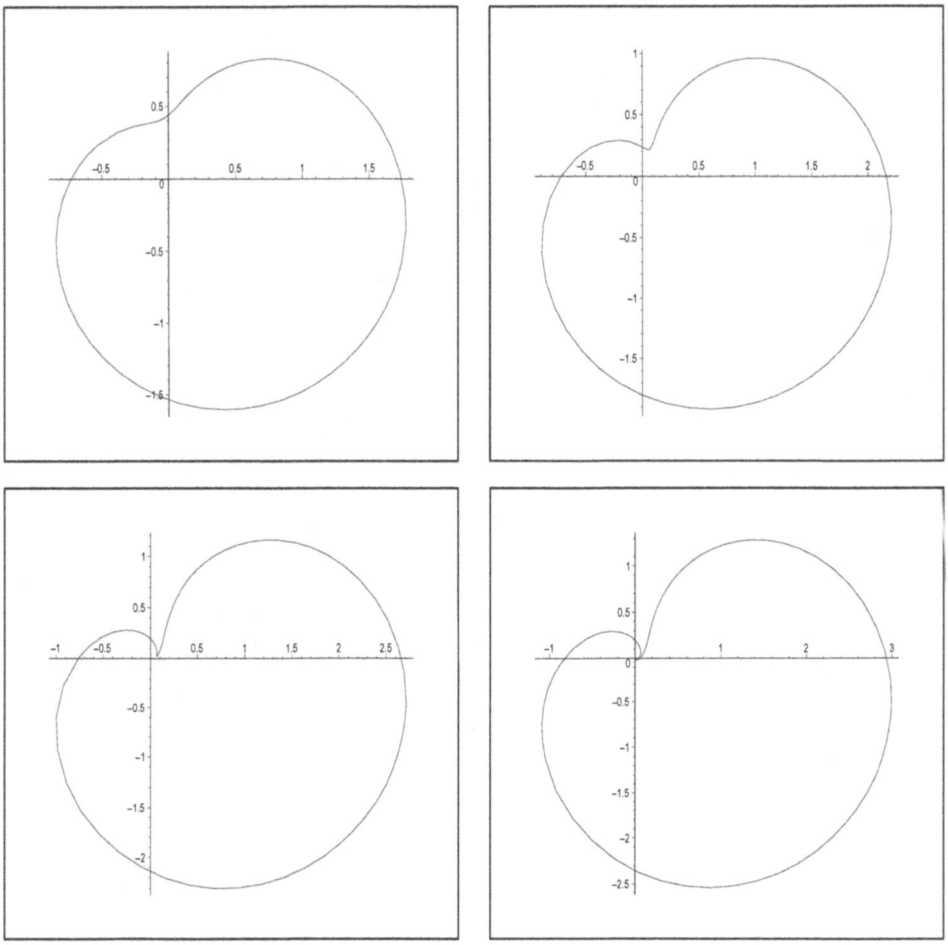

Figure 9.5: Example 9.3 (iii), the images of $h_6^{(3)}, h_{10}^{(3)}, h_{30}^{(3)}$ and $\widetilde{h}$.

Δ, be a sequence of generators converging to f uniformly on compact subsets of Δ. If $\left\{F_t^{(n)}\right\}_{t\geq 0}$ is a semigroup generated by f_n, then it follows that $F_t^{(n)}(z) \to \tau_n$ for each z as t tends to ∞. In addition, since f_n and f are locally Lipschitzian on Δ, the uniqueness property of the Cauchy problem implies that for each $a \in \Delta$, there exist $r > 0$ and $T(= T(r))$ such that for each z from the circle $|z - a| < r$ and each $t \in [0, T(r)]$, the sequence of values $F_t^{(n)}(z)$ converges to $F_t(z)$ as n tends to infinity. The question is:

$\Diamond$ *Whether this convergence is uniform on each compact subset of* $\Delta \times \mathbb{R}^+$*, in other words, whether for each* $0 < r < 1$ *and each* $0 < T < \infty$ *which does not* *depend on r, the sequence $F_t^{(n)}$ converges to F_t uniformly on the set $\overline{\Delta_r} \times [0,T]$, where $\overline{\Delta_r} = \{z \in \mathbb{C} : |z| \le r\}$?*

We now show that the answer is affirmative.

Theorem 9.6. *Let $\{F_t\}_{t\ge 0}$ be a semigroup generated by $f \in \mathcal{G}_h[1]$, and let $f_n \subset \mathcal{G}$ be a sequence of generators converging to f uniformly on compact subsets of Δ such that $f_n(\tau_n) = 0$ with $\tau_n \in \Delta$. Assume also that the closure of the set $\{\mu_n = f_n'(\tau_n)\}$ does not contain the origin. Then the sequence $\left\{F_t^{(n)}(z)\right\}_{t\ge 0}$ of semigroups generated by f_n converges to $F_t(z)$ uniformly on compact subsets of $\Delta \times \mathbb{R}^+$.*

Proof. By passing to a subsequence, if necessary, we may suppose that the sequence $\mu_n = f_n'(\tau_n)$ is convergent, say, to $\mu \in \mathbb{C}$. It follows from our assumption that $\mu \neq 0$ and from Theorem 9.3 that μ must satisfy the condition $|\mu - \beta| \le \beta$. Then the sequence $\{h_n\}$ defined by

$$h_n(z) = \exp\left[\mu_n \int_0^z \frac{dz}{f_{\tau_n}(z)}\right]$$

converges to the function h_1 defined by (9.1.22).

In addition, we have

$$F_t^{(n)}(z) = h_n^{-1}\left(e^{-t\mu_n} h_n(z)\right).$$

Thus $F_t^{(n)}(z)$ converges uniformly on compact subsets of $\mathbb{R}^+ \times \Delta$ to a function $G : \mathbb{R}^+ \times \Delta \mapsto \Delta$ defined by

$$G(t,z) = h_1^{-1}\left(e^{-t\mu} h_1(z)\right). \tag{9.1.28}$$

On the other hand, we already know that if h is a solution of the equation $\beta h(z) = h'(z) f(z)$, then

$$F_t(z) = h^{-1}\left(e^{-t\beta} h(z)\right) \tag{9.1.29}$$

and

$$h_1 = h^{\frac{\mu}{\beta}}. \tag{9.1.30}$$

Then we get from (9.1.28) and (9.1.30) that

$$h^{\frac{\mu}{\beta}}\left(G(t,z)\right) = e^{-t\mu} h^{\frac{\mu}{\beta}}(z)$$

or

$$h\left(G(t,z)\right) = e^{-t\beta} h(z).$$

The latter equation and (9.1.29) imply that $G(t,z) = F_t(z)$ for all $z \in \Delta$ and $t \in \mathbb{R}^+$. □

Since $\mathcal{G}_h[1]$ is a real cone, we can always consider equation (9.1.20) with f normalized by $f'(1) = 1$. Then Definition 4.3 can be reformulated as follows:

♦ *A function* $h \in \mathrm{Hol}(\Delta, \mathbb{C})$ *belongs to the class* $\mathrm{Spiral}_\lambda[1]$ *if it satisfies the equation*

$$\lambda h(z) = h'(z) f(z), \tag{9.1.31}$$

where $f \in \mathcal{G}_h[1]$ *with* $f'(1) = 1$ *and* $\lambda \in \Omega_+ = \{w : |w-1| \le 1, w \neq 0\}$.

Theorem 9.7. *Let* $\lambda \in \Omega_+ = \{w : |w-1| \le 1\}$, $\lambda \neq 0$. *Then a function* $h \in \mathrm{Hol}(\Delta, \mathbb{C})$ *belongs to the class* $\mathrm{Spiral}_\lambda[1]$ *if and only if it admits the representation*

$$h(z) = (1-z)^{2\lambda} \left[\frac{h_*(z)}{z} \right]^\lambda, \tag{9.1.32}$$

where h_* *is a starlike function of class* S^* *which satisfies the condition*

$$2 \inf \left(\frac{1-|z|^2}{|1-z|^2} \operatorname{Re} \frac{z h_*'(z)}{h_*(z)} \right) = 1.$$

Proof. Let $h \in \mathrm{Spiral}[1]$ be λ-spirallike. Then it satisfies the equation

$$\lambda h'(z) = h(z) f(z) \left(= -h(z)(1-z)^2 \cdot \frac{1}{q(z)} \right),$$

where $f(z) \in \mathcal{G}_h[1]$ with $f'(1) = 1$ (or, which is the same, $q'(1) = -1$ and $q \in \mathcal{P}$).

By Theorem 9.5, the functions $h_\tau \in \mathrm{Hol}(\Delta, \mathbb{C})$ defined by

$$h_\tau(z) = h^{\frac{\mu}{\lambda}}(z) \frac{(z-\tau)(1-z\bar{\tau})^{\frac{\mu}{\bar{\mu}}}}{-(1-z)^{1+\frac{\mu}{\bar{\mu}}}}$$

are spirallike with respect to an interior point ($h(\tau) = 0$) and satisfy the equations

$$\mu h_\tau(z) = h_\tau'(z)(z-\tau)(1-z\bar{\tau}) \frac{1}{q_\tau(z)},$$

where $q_\tau \in \mathcal{P}$ is defined by the formula

$$q_\tau(z) = \frac{1}{z} \left[(z-\tau)(1-z\bar{\tau}) r(z) + \bar{\gamma}\tau - \gamma\bar{\tau} z^2 + 2iz \operatorname{Im} \gamma \right],$$

where

$$r(z) = \frac{z q(z) + \gamma z^2 - \bar{\gamma} - 2iz \operatorname{Im} \gamma}{-(1-z)^2}, \quad \gamma = \frac{1}{\mu}.$$

If, in particular, we set $\mu = 1$ and $\tau = 0$, then we get $\gamma = 1$; and $h_0(z) = h^{\frac{1}{\lambda}} \cdot \frac{z}{-(1-z)^2}$, hence $h_* = -h_0$, is of the class S^* and satisfies the equation

$$h_*(z) = h_*'(z) f_0(z) = h_*'(z) \cdot \frac{z}{q_0(z)},$$

where $q_0(z) = r(z)$.

Since

$$\delta_r(1) = \angle \lim_{z\to 1}(1-z)r(z) = \angle \lim_{z\to 1} \frac{zq(z)}{z-1} + \angle \lim_{z\to 1} \frac{z^2-1}{z-1} = 1,$$

we obtain

$$\operatorname{Re} q_0(z) = \operatorname{Re} \frac{zh_*'(z)}{h_*(z)} \geq \frac{1}{2} \frac{1-|z|^2}{|1-z|^2}.$$

Conversely, let $h \in \operatorname{Hol}(\Delta, \mathbb{C})$ admit the representation

$$h(z) = (1-z)^{2\lambda} \left[\frac{h_*(z)}{z}\right]^{\lambda},$$

where h_* satisfies the required condition. Differentiating logarithmically, we obtain

$$\frac{1}{\gamma} \frac{h'(z)}{h(z)} = -\frac{2}{1-z} + \frac{1}{z}\left[\frac{zh_*'(z)}{h_*(z)} - 1\right] = -\frac{2}{1-z} + \frac{1}{z}\left[q_0(z) - 1\right].$$

Consider the function $q \in \operatorname{Hol}(\Delta, \mathbb{C})$ defined by

$$q(z) = -\frac{(1-z)^2}{\lambda} \frac{h'(z)}{h(z)}.$$

We have to show that $q \in \mathcal{P}$ and $q'(1) = -1$. This will imply that h satisfies equation (9.1.31) with $f(z) = -(1-z)^2 \frac{1}{q(z)}$.

Indeed, if $g(z) = zq(z)$, we have

$$g(z) = 1 - z^2 - (1-z)^2 q_0(z).$$

Both terms in the right-hand side of this equality are in $\mathcal{G}$. Hence g also belongs to $\mathcal{G}$, since $\mathcal{G}$ is a real cone.

Thus q must belong to $\mathcal{P}$ by the Berkson–Porta formula (2.2.7) with $\tau = 0$.

Finally, a direct calculation shows that

$$q'(1) = \angle \lim_{z\to 1} \frac{q(z)}{z-1} = \angle \lim_{z\to 1} \frac{1-z^2}{z-1} + \angle \lim_{z\to 1}(1-z)q_0(z) = -1,$$

and we are done. □

Define $\widetilde{S}^*$ as the subclass of S^* of all starlike functions $h \in \operatorname{Hol}(\Delta, \mathbb{C})$, $h(0) = 0$, which satisfy the condition

$$2 \inf \frac{1-|z|^2}{|1-z|^2} \operatorname{Re} \frac{zh'(z)}{h(z)} = 1.$$

Corollary 9.2. *For each* $\lambda \in \Omega_+ = \{w : |w-1| \leq 1,\ w \neq 0\}$, *the class* $\widetilde{S}^*$ *is homeomorphic to the class* $\mathrm{Spiral}_\lambda[1]$.

Using representation (9.1.32), one checks easily the following characterization of spirallike functions with respect to a boundary point (cf., Theorem 4.1).

Corollary 9.3 (Generalized Robertson condition). *A function* $h \in \mathrm{Hol}(\Delta, \mathbb{C})$ *normalized by the condition* $h(0) = 1$ *is of class* Spiral[1] *if and only if, for some* $\lambda \in \Omega_+ = \{w : |w-1| \le 1,\ w \ne 0\}$,

$$\mathrm{Re}\left[\frac{2z}{\lambda}\frac{h'(z)}{h(z)} + \frac{1+z}{1-z}\right] \ge 0.$$

9.2 Weighted semigroups of composition operators

Another useful application of the results given above is the following description of the spectrum of the infinitesimal generator of a one-parameter semigroup of composition operators in the infinite dimensional Frechét space $E = \mathrm{Hol}(\Delta, \mathbb{C})$.

For $f \in \mathcal{G}$, define a linear operator Γ_f on E by the formula

$$(\Gamma_f h)(z) := h'(z) f(z). \tag{9.2.33}$$

Recall that the spectrum $\sigma(\Gamma_f)$ of the operator Γ_f is the set of all complex numbers $\lambda \in \mathbb{C}$ for which $\lambda I - \Gamma_f$ is not continuously invertible, where I is the identity operator on $\mathrm{Hol}(\Delta, \mathbb{C})$. The point spectrum $\sigma_p(\Gamma_f)$ of Γ_f is the subset of $\sigma(\Gamma_f)$ which consists of its eigenvalues, i.e.,

$$\sigma_p(\Gamma_f) = \{\lambda \in \mathbb{C} : (\lambda I - \Gamma_f) h = 0 \text{ for some } h \ne 0\}.$$

In this case, a nontrivial solution of the equation $(\lambda I - \Gamma_f) h = 0$ is called an eigenvector of Γ_f corresponding to eigenvalue λ.

Generally speaking, in infinite dimensional spaces, the spectrum of a linear operator does not coincide with its point spectrum. However, since $f \in \mathcal{G}^+[1]$ does not vanish in Δ, we have $\sigma_p(\Gamma_f) = \sigma(\Gamma_f) = \mathbb{C}$.

This, for example, implies immediately that the linear semigroup generated by Γ_f (and hence, the nonlinear semigroup $\mathcal{S} = \{F_t\}_{t\ge 0}$ generated by f) cannot be holomorphically extended into a sector containing the positive real axis (see, for example, [148, Chapter IX.10]).

Over the past few decades a lot of works address a wide range of topics concerning properties of composition operators on classical Banach spaces of analytic functions. In particular, composition operators have been studied extensively in the setting of Hardy or Bergman spaces on Δ (see [40] and [127]).

In the case of an interior null point of a generator, the eigenvalues of Γ_f form a discrete set, while in case of the boundary Denjoy–Wolff point, the spectrum of Γ_f may be essentially larger.

Indeed, as we have already seen in Theorem 9.4, the eigenfunction h corresponding to the eigenvalue 2β, where $\beta = f'(1)$, is univalent and starlike with respect to a boundary point. Then $h \in H^p$ for each $p < \frac{1}{2}$ (see, for example, [72]).

Hence, for each positive λ the function $h_1(z) := (h(z))^{\frac{2\beta}{\lambda}} \in H^q$ when $q\lambda < \beta$. Obviously, h_1 is an eigenfunction corresponding to λ. Therefore, for any Hardy space H^q we have $\sigma(\Gamma_f) \supset \left(0, \frac{\beta}{q}\right)$.

This, inter alia, motivates consideration of the univalence of eigenfunctions of composition operators on the (locally convex) Fréchet space $\mathrm{Hol}(\Delta, \mathbb{C})$. In a slightly more general setting, we describe the structure of the spectrum of weighted composition operators in the context of the k-valency of the corresponding eigenfunctions. Namely, we use the following definition.

Definition 9.1 (see [75], p. 89). *A function f meromorphic in a domain D is said to be k-valent in D if for each w_0 (infinity included) the equation $f(z) = w_0$ has at most k roots in D (where the roots are counted in accordance with their multiplicity) and if there is some w_1 such that the equation $f(z) = w_1$ has exactly k roots in D.*

Let E be a space of meromorphic functions in Δ, and let $\mathcal{S} = \{F_t\}_{t\geq 0}$ be a semigroup of holomorphic self-mappings of Δ generated by $f \in \mathcal{G}_h[1]$, $f'(1) = \beta$. For a suitable $w \in E$, one can define a (weighted) composition semigroup of linear operators $T_t : E \mapsto E$, $t \geq 0$, by the formula

$$T_t(h)(z) = \frac{w(F_t(z))}{w(z)} h(F_t(z)).$$

This semigroup is generated by the operator $\widetilde{\Gamma}_f$ defined by

$$\widetilde{\Gamma}_f h = h'f + hf\frac{w'}{w}, \quad h \in \mathrm{Hol}(\Delta, \mathbb{C}). \tag{9.2.34}$$

When $w \equiv 1$, this reduces to the unweighted semigroup of composition operators $\{C_t\}_{t\geq 0}$ generated by Γ_f, where

$$C_t(h)(z) := h(F_t(z)).$$

It is clear that for each λ in the point spectrum $\sigma_p(\widetilde{\Gamma}_f)$, the eigenspace E_λ corresponding to λ is one-dimensional. For $k \in \mathbb{N}\bigcup\{\infty\}$, denote by $\sigma^{(k)}$ the subset of $\sigma_p(\widetilde{\Gamma}_f)$ such that for each $\lambda \in \sigma^{(k)}$ the function wh is k-valent whenever $h \in E_\lambda$.

Theorem 9.8. *Let $f \in \mathcal{G}_h[1]$, $f'(1) = \beta$, and let the operator $\widetilde{\Gamma}_f : E \mapsto E$ be defined by (9.2.34). Then the spectrum $\sigma_p\left(\widetilde{\Gamma}_f\right)$ is the whole complex plane $\mathbb{C}$. Moreover,*

$$\sigma_p\left(\widetilde{\Gamma}_f\right) = \bigcup_{k\in\mathbb{N}\cup\{\infty\}} \sigma^{(k)},$$

where for $k \in \mathbb{N}$,

$$\sigma^{(k)} = (k\Omega_+ \cup k\Omega_-) \setminus ((k-1)\Omega_+ \cup (k-1)\Omega_-)$$

with $\Omega_{\pm} = \{\omega \neq 0 : \ |\omega \mp \beta| \leq \beta\}$, *and*

$$\sigma^{(\infty)} = \{\lambda \in \mathbb{C} : \ \mathrm{Re}\,\lambda = 0\}.$$

In addition, for each $\lambda \neq 0$ *and for each element* h *of the eigenspace corresponding to* λ, *the following modified Robertson inequality holds:*

$$\mathrm{Re}\left[\frac{2\beta}{\lambda}\frac{zh'(z)}{h(z)} + \frac{1+z}{1-z} + \frac{2\beta}{\lambda}\frac{zw'(z)}{w(z)}\right] > 0. \tag{9.2.35}$$

Figure 9.6 illustrates the sets $\sigma^{(k)}$ described in Theorem 9.8.

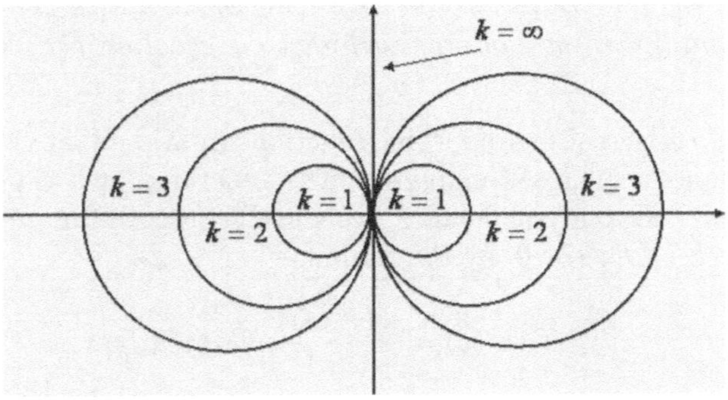

Figure 9.6: The sets $\sigma^{(k)}$ of k-valence of eigenfunctions.

Proof. The eigenvalue problem for the operator $\widetilde{\Gamma}_f$ is the differential equation

$$h'f + hf\frac{w'}{w} = \lambda h. \tag{9.2.36}$$

To solve it, we denote by h^* the starlike function with respect to a boundary point which satisfies

$$\beta h^* = h^{*\prime} f \tag{9.2.37}$$

and is normalized by $h^*(0) = 1, \ h^*(1) = 0$ (h^* is the eigenfunction of Γ_f corresponding to the eigenvalue β).

Then (9.2.36) and (9.2.37) imply that

$$\frac{(wh)'}{wh} = \frac{\lambda}{\beta}\frac{h^{*\prime}}{h^*},$$

and hence

$$w(z)h(z) = a\left(h^*(z)\right)^{\lambda/\beta}, \quad a \in \mathbb{C}. \tag{9.2.38}$$

For a given eigenvalue λ, (9.2.38) describes the eigenspace corresponding to λ. So it is enough to prove our assertion for the case $w \equiv 1$ or, which is the same, $\widetilde{\Gamma}_f = \Gamma_f$.

To this end, let $\lambda \in \mathbb{C}$. First suppose that $\operatorname{Re}\lambda > 0$ and $\lambda \in k\Omega_+ \setminus (k-1)\Omega_+$. Then $\lambda_1 = \frac{\lambda}{k}$ belongs to Ω_+. Thus, by Theorem 9.4, the solution h_1 of the initial value problem

$$h_1'(z)f(z) = \lambda_1 h_1(z), \quad h_1(0) = 1,$$

is univalent. Set $h(z) = h_1(z)^k$. Evidently, this function satisfies the equation

$$\Gamma_f h(z)\, (= h'(z)f(z)) = \lambda h(z),$$

and is at most k-valent. To show that h is k-valent, consider the univalent solution h_2 of the initial value problem

$$h_2'(z)f(z) = 2\beta h_2(z), \quad h_2(0) = 1.$$

This function is starlike with respect to a boundary point, and the smallest wedge which contains the image $h_2(\Delta)$ is exactly of angle 2π. Therefore, for each $\epsilon > 0$, there is real ψ and $r > 0$ small enough such that the image $h_2(\Delta)$ contains the curve

$$\left\{ r\exp\left(\phi\frac{\operatorname{Im}\lambda}{\operatorname{Re}\lambda}\right)\exp\left(i\phi - i\log r\frac{\operatorname{Im}\lambda}{\operatorname{Re}\lambda}\right) : \psi \le \phi \le \psi + 2\pi(1-\epsilon)\right\}.$$

Since $\lambda \notin (k-1)\Omega_+$, one can choose $\epsilon < 1 - \frac{2\beta(k-1)\operatorname{Re}\lambda}{|\lambda|^2}$. Then there are k different points $z_0, z_1, \dots, z_{k-1}$ in Δ for which

$$h_2(z_m) = r\exp\left(\phi_m\frac{\operatorname{Im}\lambda}{\operatorname{Re}\lambda}\right)\exp\left(i\phi_m - i\log r\frac{\operatorname{Im}\lambda}{\operatorname{Re}\lambda}\right),$$
$$\phi_m = \psi + 2\pi m\frac{2\beta\operatorname{Re}\lambda}{|\lambda|^2}, \qquad m = 0, 1, \dots, k-1.$$

Note that $h(z) = h_2(z)^{\lambda/2\beta}$. Now a simple calculation shows that $h(z_m)$ does not depend on $m = 0, 1, \dots, k-1$, i.e., h is k-valent and $\lambda \in \sigma^{(k)}$.

The case $\operatorname{Re}\lambda < 0$ is reduced to the previous one by replacing h by $1/h$.

Finally, suppose $\operatorname{Re}\lambda = 0$. As usual, denote by $\{F_t\}_{t\ge 0}$ the semigroup generated by f. Thus, the function $u(t,z) := F_t(z)$ satisfies the Cauchy problem

$$\begin{cases} \dfrac{\partial u(t,z)}{\partial t} + f(u(t,z)) = 0, \\ u(0,z) = z \in \Delta. \end{cases}$$

Solving this problem for $z = 0$, we have

$$\int_0^{u(t,0)} \frac{du}{f(u)} = -\int_0^t dt = -t.$$

Since the function $1/f$ is holomorphic on the open unit disk Δ, we conclude that, for each point z which belongs to the curve $\Lambda := \{u(t,0) : t \geq 0\}$ (joining the origin with the boundary point $z = 1$), the integral

$$\int_0^z \frac{du}{f(u)}$$

takes real values which tend to $-\infty$ as $z \in \Lambda$ tends to 1.

Returning to the eigenvalue problem

$$\Gamma_f(h)(z)\,(:= h'(z)f(z)) = \lambda h(z)$$

and separating variables in this differential equation, we have

$$\int_0^z \frac{dh(z)}{h(z)} = \int_0^z \frac{\lambda dz}{f(z)},$$

so that

$$\log\left(\frac{h(z)}{h(0)}\right) = \lambda \int_0^z \frac{dz}{f(z)}.$$

Thus, for each point $z \in \Lambda$,

$$h(z) = h(0)\exp\left(-\lambda t\right),$$

where $z = u(t,0)$. We claim that this function is infinite-valent. When $\lambda = 0$, this is evident. Suppose $\lambda \neq 0$; then $\lambda = ia$, $a \neq 0$. Choosing the sequence $z_n = u\left(t_0 + \frac{2\pi n}{a}, 0\right) \in \Lambda \subset \Delta$ for any fixed $t_0 \geq 0$, we see immediately that the value $h(z_n)$ does not depend on n.

Now, let h^* be the solution of the differential equation (9.2.37) normalized by $h^*(0) = 1$, $h^*(1) = 0$. As in the proof of Theorem 9.3, we conclude that the image of the function $h^*(\Delta)$ must lie in a wedge of angle π. Since h^* is starlike with respect to a boundary point, it follows from a result of Lyzzaik [102] (see also Theorem 3.2 above) that

$$\operatorname{Re}\left[\frac{2zh^{*\prime}(z)}{h^*} + \frac{1+z}{1-z}\right] > 0.$$

Substituting (9.2.38) into the latter inequality, we get (9.2.35). The proof is complete. □

Combining this assertion with previous results, one can obtain additional geometrical information on (characteristics of) eigenfunctions.

Example 9.4. For each $a \in (0,1)$, we rewrite (9.2.35) in the form

$$\operatorname{Re}\left[\frac{2a\beta}{\lambda}\,\frac{zh'(z)}{h(z)} + \frac{1+z}{1-z}\right] > \operatorname{Re}\left[(1-a)\,\frac{1+z}{1-z} - \frac{2a\beta}{\lambda}\,\frac{zw'(z)}{w(z)}\right]. \tag{9.2.39}$$

Suppose that $w(z) = (1-z)^\delta$ for some $\delta \in \mathbb{C}$.

Let c be a positive number less than 1. In this case, for each λ such that $\frac{\delta}{\lambda}$ is real with $\frac{\delta}{\lambda} < -\frac{c}{a\beta}$, a simple calculation shows that the right-hand side in (9.2.39) is greater than c. Indeed, for each $z \in \partial\Delta$,

$$\begin{aligned}\operatorname{Re}\left[(1-a)\frac{1+z}{1-z} - \frac{2a\beta}{\lambda}\frac{zw'(z)}{w(z)}\right] &= \frac{1}{|1-z|^2}\operatorname{Re}\left[-\frac{2a\beta\delta}{\lambda} + \frac{2a\beta\delta z}{\lambda}\right]\\ &= -\frac{2a\beta\delta}{\lambda}\operatorname{Re}\frac{1-z}{|1-z|^2} \geq -\frac{a\beta\delta}{\lambda} > c.\end{aligned}$$

Letting $a \to 1^-$, we conclude that

$$\operatorname{Re}\left[\frac{2\beta}{\lambda}\frac{zh'(z)}{h(z)} + \frac{1+z}{1-z}\right] > c.$$

Therefore, by Theorem 4.10, the image $h(\Delta)$ of the eigenfunction h covers the image of the function $(1-z)^{\frac{c\lambda}{\beta}}$.

Example 9.5. Consider now the (nonanalytic) weight $w(z) = \frac{(1-z)^2}{z}$. By formula (9.2.38), each eigenfunction has the form

$$h(z) = a\,\frac{z}{(1-z)^2}\left(h^*(z)\right)^{\frac{\lambda}{\beta}}.$$

In this situation, for all positive λ less than 2β, inequality (9.2.35) becomes

$$\operatorname{Re}\frac{zh'(z)}{h(z)} > \left(1 - \frac{\lambda}{2\beta}\right)\frac{1-|z|^2}{|1-z|^2} > 0,$$

i.e., all eigenfunctions are starlike with respect to an interior point ($h(0) = 0$).

In particular, the function

$$h_0(z) = \frac{z}{(1-z)^2}\,h^*(z)$$

is the eigenfunction corresponding to the eigenvalue $\lambda = \beta$. Since $h^* \in \mathrm{Spiral}_1[1]$, Theorem 9.7 implies that h_0 is a starlike function of class $S^*[0]$ which satisfies the condition

$$2\inf\left(\frac{1-|z|^2}{|1-z|^2}\operatorname{Re}\frac{zh_0'(z)}{h_0(z)}\right) = 1.$$

Finally, we observe that the above theorem implies the following nice characterization of multivalent starlike functions with respect to a boundary point suggested by Bshouty and Lyzzaik (see [23] and [24]; cf. also [97] and [99]).

Theorem 9.9. *Let* $h \in \mathrm{Hol}(\Delta, \mathbb{C})$ *satisfy the equation*

$$-(1-z)^2\,\frac{h'(z)}{h(z)} = 4\,\frac{1-\omega(z)}{1+\omega(z)}, \tag{9.2.40}$$

where $\omega \in \mathrm{Hol}(\Delta)$ *is a holomorphic self-mapping of* Δ *with the boundary fixed point* $\tau = 1$ *and positive multiplier*

$$\alpha = \angle \lim_{z \to 1} \frac{1 - \omega(z)}{1 - z}.$$

Then

(i) $h(\Delta)$ *is a starlike domain;*

(ii) h *is a* k*-valent function if and only if* $k - 1 < \alpha \leq k$.

In particular, if $\tau = 1$ *is the Denjoy–Wolff point for* ω *(whence* $0 < \alpha \leq 1$*), then* h *is univalent and the smallest wedge which contains* $h(\Delta)$ *is of angle* $2\alpha\pi$*, i.e.,* $h \in \mathrm{Star}_{2\alpha}[1]$.

Proof. Let

$$p(z) = \frac{\alpha}{2} \frac{1 + \omega(z)}{1 - \omega(z)}.$$

This function has positive real part. Moreover,

$$\angle \lim_{z \to 1} (1 - z) p(z) = \angle \lim_{z \to 1} \frac{\alpha}{2} \frac{1 - z}{1 - \omega(z)} (1 + \omega(z)) = 1.$$

Define a generator $f \in \mathcal{G}(\Delta)$ by the Berkson–Porta formula

$$f(z) = -(1 - z)^2 p(z).$$

It is clear that $f \in \mathcal{G}_h[1]$ with $\beta = f'(1) = 1$.

Since h satisfies the equation

$$2\alpha h(z) = h'(z) f(z),$$

and α is positive, applying Theorem 9.8 with $\lambda = 2\alpha$ and $\beta = 1$, we obtain the result.

If $\tau = 1$ is the Denjoy–Wolff point for ω, i.e., $0 < \alpha \leq 1$, then

$$\angle \lim_{z \to 1} \frac{(z - 1) h'(z)}{h(z)} = 2\alpha \leq 2;$$

and the assertion follows by Theorem 3.3. The proof is complete. □

Bibliography

[1] M. Abate, Converging semigroups of holomorphic maps, *Atti Accad. Naz. Lincei Rend. Cl. Sci. Fis. Mat. Natur.* **82** (1988), 223–227.

[2] M. Abate, *Iteration theory of holomorphic maps on taut manifolds*, Mediterranean Press, Cosenza, 1989.

[3] M. Abate, The infinitesimal generators of semigroups of holomorphic maps, *Ann. Mat. Pura Appl.* **161** (1992), 167–180.

[4] D. Aharonov, M. Elin, S. Reich and D. Shoikhet, Parametric representations of semi-complete vector fields on the unit balls in $\mathbb{C}^n$ and Hilbert space, *Atti Accad. Naz. Lincei Cl. Sci. Fis. Mat. Natur. Rend. Lincei (9) Mat. Appl.* **10** (1999), 229–253.

[5] D. Aharonov, M. Elin and D. Shoikhet, Spirallike functions with respect to a boundary point, *J. Math. Anal. Appl.* **280** (2003), 17–29.

[6] D. Aharonov, S. Reich and D. Shoikhet, Flow invariance conditions for holomorphic mappings in Banach spaces, *Math. Proc. of the Royal Irish Acad.* **99A** (1999), 93–104.

[7] J. W. Alexander, Functions which map the interior of the unit circle upon simple regions, *Annals of Math.* **17** (1915), 12–22.

[8] I. A. Aleksandrov,*Parametric continuations in the theory of univalent functions*, Nauka, Moscow, 1976.

[9] M. Arslan, Rigidity of analytic functions at the boundary, Preprint, *arXiv:math.CV/0605026.*

[10] K. B. Athreya and P. E. Ney, *Branching Processes*, Springer-Verlag, Berlin, 1972.

[11] I. N. Baker, Fractional iteration near a fixpoint of multiplier 1, *J. Aust. Math. Soc.* **4** (1964), 143–148.

[12] I. N. Baker and Ch. Pommerenke, On the iteration of analytic functions in a halfplane. II. *J. London Math. Soc. (2)* **20** (1979), 255–258.

[13] L. Baracco, D. Zaitsev and G. Zampieri, A Burns-Krantz type theorem for domains with corners, *Math. Ann.* **336** (2006), 491–504.

[14] D. Behan, Commuting analytic functions without fixed points, *Proc. Amer. Math. Soc.* **37** (1973), 114–120.

[15] G. Belitskii, V. Tkachenko, *One-dimensional Functional Equations*, Oper. Theory. Adv. Appl. vol.144, Birkhäuser, 2003.

[16] E. Berkson, R. Kaufman and H. Porta, Möbius transformations of the disc and one-parameter groups of isometries of H^p, *Trans. Amer. Math. Soc.* **199** (1974), 223–239.

[17] E. Berkson and H. Porta, Semigroups of analytic functions and composition operators, *Michigan Math. J.* **25** (1978), 101–115.

[18] P. S. Bourdon and J. H. Shapiro, Cyclic phenomena for composition operators, *Mem. Amer. Math. Soc.* **125** (1997), no. 596.

[19] F. Bracci, Fixed points of commuting holomorphic mappings other than the Wolff point, *Trans. Amer. Math. Soc.* **355** (2003), 2569–2584.

[20] F. Bracci, P. Poggi-Corradini, On Valiron's theorem, *Future Trends in Geometric Function Theory. RNC Workshop Jyvaskyla 2003*, Rep. Univ. Jyvaskyla Dept. Math. Stat. **92** (2003), 39–55.

[21] F. Bracci, R. Tauraso and F. Vlacci, Identity principles for commuting holomorphic self-maps of the unit disc, *J. Math. Anal. Appl.* **270** (2002), 451–473.

[22] L. Brickman, Φ-like analytic functions, I, *Bull. Amer. Math. Soc.* **79** (1973), 555–558.

[23] D. Bshouty and A. Lyzzaik, Univalent functions starlike with respect to a boundary point, *Complex Analysis and Dynamical Systems II*, 83–87, *Contemp. Math.* **382** *Amer. Math. Soc., Providence, RI* 2005.

[24] D. Bshouty and A. Lyzzaik, Personal communication.

[25] D. M. Burns and S. G. Krantz, Rigidity of holomorphic mappings and a new Schwarz lemma at the boundary, *J. Amer. Math. Soc.* **7** (1994), 661–676.

[26] C. Carathéodory, Uber die Winkelderivierten von beschränkten Analytischen Funktionen, *Sitzungsber. Preuss. Acad. Viss. Berlin*, Phys.-Math. Kl. (1929), 39–54.

[27] L. Carleson and T. W. Gamelin, *Complex dynamics*, Universitext: Tracts in Mathematics, Springer-Verlag, New York, 1993.

[28] D. Chelst, Generalized Schwarz lemma at the boundary, *Proc. Amer. Math. Soc.* **129** (2001), 3275–3278.

[29] M. Chen and S. Owa, Generalization of Robertson functions, in: *Complex analysis and its applications (Hong Kong, 1993)*, 159–165, Pitman Res. Notes Math. Ser., 305, Longman Sci. Tech., Harlow, 1994.

[30] K. Ciozda, Sur quelques problémes extrémaux dans les classes des fonctions convexes vers l'axe réel négatif, *Ann. Polon. Math.* **38** (1980), 311–317.

[31] K. Ciozda, Sur la classe des fonctions convexes vers l'axe réel négatif, *Bull. Acad. Polon. Sci. Ser. Sci. Math.* **27** (1979), 255–261.

[32] M. D. Contreras and S. Díaz-Madrigal, Analytic flows on the unit disk: angular derivatives and boundary fixed points, *Pacific J. Math.* **222** (2005), 253–286.

[33] M. D. Contreras, S. Díaz-Madrigal and Ch. Pommerenke, Fixed points and boundary behaviour of the Kœnigs function, *Ann. Acad. Sci. Fenn. Math.* **29** (2004), 471–488.

[34] M. D. Contreras, S. Díaz-Madrigal and Ch. Pommerenke, On boundary critical points for semigroups of analytic functions, *Math. Scand.* **98** (2006), 125–142.

[35] M. D. Contreras, S. Díaz-Madrigal and Ch. Pommerenke, Some remarks on the Abel equation in the unit disk, *J. Lond. Math. Soc.* (2) **75** (2007), 623–634.

[36] M. D. Contreras, S. Díaz-Madrigal and Ch. Pommerenke, Second angular derivatives and parabolic iteration in the unit disk, Preprint, 2005.

[37] M. D. Contreras, S. Díaz-Madrigal and Ch. Pommerenke, Iteration in the unit disk: the parabolic zoo, in: *Complex and Harmonic Analysis, Proceedings of the International Conference, Thessaloniki, 2006*, 63–92, DEStech Publications, Inc., 2007.

[38] C. C. Cowen, Iteration and the solution of functional equations for functions analytic in the unit disk, *Trans. Amer. Math. Soc.* **265** (1981), 69–95.

[39] C. C. Cowen, Commuting analytic functions, *Trans. Amer. Math. Soc.* **283** (1984), 685–695.

[40] C. C. Cowen and B. D. MacCluer, *Composition Operators on Spaces of Analytic Functions*, CRC Press, Boca Raton, FL, 1995.

[41] A. Denjoy, Sur l'itération des fonctions analytiques, *C. R. Acad. Scie.* **182** (1926), 255–257.

[42] J. Dong and L. Liu,Distortion properties of a class of functions starlike with respect of a boundary point, *Heilongjiang Daxue Ziran Kexue Xuebao* **15** (1998), 1–6.

[43] P. Duren, *Univalent Functions*, Springer–Verlag, New York, 1983.

[44] P. Duren, *Harmonic Mappings in the Plane*, Cambridge University Press, 2004.

[45] C. J. Earle and R. S. Hamilton,A fixed point theorem for holomorphic mappings, *Proc. Symp. Pure Math.*, Vol. 16, Amer. Math. Soc., Providence, R.I. (1970), 61–65.

[46] E. Egervàry, Abbildungseigenschaften der arithmetischen Mittel der geometrischen Reihe, *Math. Z.* **42** (1937), 221–230.

[47] M. Elin, Covering and distortion theorems for spirallike functions with respect to a boundary point, *Intern. J. Pure Appl. Math.* **28** (2006), 387–400.

[48] M. Elin, A. Goldvard,, S. Reich and D. Shoikhet, Dynamics of spirallike functions, *Complex analysis and Dynamical Dystems*, 41–57, *Contemp. Math.* **364** *Amer. Math. Soc., Providence, RI*, 2004.

[49] M. Elin, V. Goryainov, S. Reich and D. Shoikhet, Fractional iteration and functional equations for functions analytic on the unit disk, *Comput. Methods Funct. Theory* **2** (2002), 353–366.

[50] M. Elin, M. Levenshtein, S. Reich and D. Shoikhet, Rigidity results for holomorphic mappings on the unit disk, in: *Complex and Harmonic Analysis, Proceedings of the International Conference, Thessaloniki, 2006, 93–110*, DEStech Publications, Inc., 2007.

[51] M. Elin, M. Levenshtein, S. Reich, and D. Shoikhet, Commuting semigroups of holomorphic mappings, *Math. Scand.*, **103** (2008), 295–319.

[52] M. Elin, M. Levenshtein, S. Reich and D. Shoikhet, Two rigidity theorems for holomorphic generators of continuous semigroups, *J. Nonlinear Convex Anal.* **9** (2008), 59–64.

[53] M. Elin, M. Levenshtein, D. Shoikhet, and R. Tauraso, Rigidity of holomorphic generators and one-parameter semigroups, *Dynam. Systems Appl.* **16** (2007), 251–266.

[54] M. Elin, S. Reich and D. Shoikhet, Holomorphically accretive mappings and spiral-shaped functions of proper contractions, *Nonlinear Analysis Forum* **5** (2000), 149–161.

[55] M. Elin, S. Reich and D. Shoikhet, A semigroup approach to the geometry of domains in complex Banach spaces, *Nonlinear Analysis* **47** (2001), 3271–3280.

[56] M. Elin, S. Reich and D. Shoikhet, Dynamics of inequalities in geometric function theory, *Journ. of Inequal. & Appl.* **6** (2001), 651–664.

[57] M. Elin, S. Reich and D. Shoikhet, Asymptotic behavior of semigroups of ρ-nonexpansive and holomorphic mappings on the Hilbert ball, *Ann. Mat. Pura Appl.* (4) **181** (2002), 501–526.

[58] M. Elin, S. Reich and D. Shoikhet, Complex Dynamical Systems and the Geometry of Domains in Banach Spaces, *Dissertationes Math. (Rozprawy Mat.)* **427** (2004), 62 pp.

[59] M. Elin, S. Reich and D. Shoikhet, A Julia–Carathéodory theorem for hyperbolically monotone mappings in the Hilbert ball, *Israel J. Math.* **164** (2008), 397–411.

[60] M. Elin, S. Reich, D. Shoikhet and F. Yacobzon, Asymptotic behavior of one-parameter semigroups and rigidity of holomorphic generators, *Compl. Anal. Oper. Theory* **2** (2008), 55–86.

[61] M. Elin, S. Reich, D. Shoikhet and F. Yacobzon, Rates of convergence of one-parameter semigroups with boundary Denjoy–Wolff fixed points, in: *Fixed Point Theory and its Applications, 43–58, Yokohama Publishers*, 2008.

[62] M. Elin and D. Shoikhet, Dynamic extension of the Julia–Wolff–Carathéodory theorem, *Dynam. Systems Appl.* **10** (2001), 421–438.

[63] M. Elin and D. Shoikhet, Univalent functions of proper contractions spirallike with respect to a boundary point, in: *Multidimensional Complex Analysis* Krasnoyarsk, (2002), 28–36.

[64] M. Elin and D. Shoikhet, Semigroups with boundary fixed points on the unit Hilbert ball and spirallike mappings, in: *Geometric Function Theory in Several Complex Variables, 82–117, World Sci. Publishing, River Edge, NJ*, 2004.

[65] M. Elin and D. Shoikhet, Angle distortion theorems for starlike and spirallike functions with respect to a boundary point, *Intern. Journ. Math. Math. Sci.*, 2006, Article ID 81615, 1–13.

[66] M. Elin, D. Shoikhet and V. Volkovich,Semigroups of holomorphic mappings on the unit disk with a boundary fixed point, *Int. J. Pure Appl. Math.* **12** (2004), 427–453.

[67] M. Elin, D. Shoikhet, F. Yacobzon, Linearization models for parabolic type semigroups, *J. Nonlinear Convex Anal.*, *J. Nonlinear Convex Anal.* **9** (2008), 205–214.

[68] M. Elin, D. Shoikhet and L. Zalcman, A flower structure of backward flow invariant domains for semigroups, *C. R. Acad. Sci. Paris, Ser. 1, Math.* **346** (2008), 293–296.

[69] M. Elin, D. Shoikhet and L. Zalcman, A flower structure of backward flow invariant domains for semigroups, *Ann. Acad. Sci. Fenn. Math.*, **33** (2008), 3–34.

[70] M. Elin, D. Shoikhet and L. Zalcman, Controlled approximation for some classes of holomorphic functions, *Complex analysis and Dynamical Systems III*, 63–92, *Contemp. Math.* **455**, *Amer. Math. Soc., Providence, RI*, 2008.

[71] T. W. Gamelin, *Conjugation theorems via Neumann series*, MRS Report 99-1 (1999), Math Department, UCLA.

[72] J. B. Garnett, *Bounded Analytic functions*, Academic Press, 1981.

[73] K. Goebel and S. Reich, *Uniform Convexity, Hyperbolic Geometry and Nonexpansive Mappings*, Marcel Dekker, New York and Basel, 1984.

[74] G. M. Golusin, *Geometric Theory of Functions of a Complex Variable*, Amer. Math. Soc., Providence, RI, 1969.

[75] A. W. Goodman, *Univalent Functions*, Vols. I, II, Mariner Publ. Co., Tampa, FL, 1983.

[76] V. V. Goryainov, Fractional iterates of functions analytic in the unit disk with given fixed points, *Matem. Sb.* **182** (1991), 29–46.

[77] K. R. Gurganus, Φ-like holomorphic functions in $\mathbb{C}^n$ and Banach space, *Trans. Amer. Math. Soc.* **205** (1975), 389–406.

[78] J. Hadamard, Two works on iteration and related questions, *Bull. Amer. Math. Soc.* **50** (1944), 67–75.

[79] T. E. Harris, *The Theory of Branching Processes*, Springer-Verlag, Berlin–Göttingen–Heidelberg, 1963.

[80] M. H. Heins, A generalization of the Aumann–Carathéodory "Starrheitssatz", *Duke Math. J.* **8** (1941), 312–316.

[81] U. Helmke and J. B. Moore, *Optimization and Dynamical Systems*, Springer, London, 1994.

[82] W. Hengartner and G. Schober, On schlicht mappings to domains convex in one direction, *Comm. Math. Helv.* **45** (1970), 303–314.

[83] G. Herglotz, Über Potenzrechen mit positiven, reelen Teil im Einheitskreis, *S.-B. Sächs. Akad. Wiss. Leipzig Math.-Natur. Kl.* **63** (1911), 501–511.

[84] J. A. Hummel, Multivalent starlike functions, *J. Analyse Math.* **18** (1967), 133–160.

[85] J. A. Hummel, Extremal properties of weakly starlike p-valent functions, *Trans. Amer. Math. Soc.* **130** (1968), 544–551.

[86] J. A. Hummel, The coefficients of starlike functions, *Proc. Amer. Math. Soc.* **22** (1969), 311–315.

[87] J. M. Isidro and L. L. Stacho, *Holomorphic Automorphism Groups in Banach Spaces: An Elementary Introduction*, North Holland, Amsterdam, 1984.

[88] G. Julia, *Principes géométriques d'analyse*, Gauthier–Villars, Paris, Part I (1930), Part II (1932).

[89] W. Kaplan, Close-to-convex schlicht functions, *Michigan Math. J.* **1** (1952), 169–185.

[90] S. Karlin and J. McGregor, Embedding iterates of analytic functions with two fixed points into continuous groups, *Trans. Amer. Math. Soc.* **132** (1968), 137–145.

[91] V. Khatskevich, S. Reich and D. Shoikhet, Complex dynamical systems on bounded symmetric domains, *Electronic J. Differential Equations* **19** (1997), 1–9.

[92] L. Kindermann, References on Iterative Roots and Fractional Iteration, *http://reglos.de/lars/ffx.html.*

[93] G. Kœnigs, Recherches sur les intégrales de certaines équations fonctionnelles, *Ann. Sci. École Norm.* **Sup. 1** (1884), 2–41.

[94] T. L. Kriete and B. D. MacCluer, A rigidity theorem for composition operators on certain Bergman spaces, *Michigan Math. J.* **42** (1995), 379–386.

[95] M. Kuczma, *On the Schröder equation*, Rozprawy Mat. **34** 1963.

[96] M. Kuczma, B. Choczewski and R. Ger, *Iterative functional equations. Encyclopedia of Mathematics and its Applications* **32** Cambridge University Press, Cambridge, 1990.

[97] A. Lecko, On the class of functions starlike with respect to a boundary point, *J. Math. Anal. Appl.* **261** (2001), 649–664.

[98] A. Lecko, On the class of functions convex in the negative direction of the imaginary axis, *J. Aust. Math. Soc.* **73** (2002), 1–10.

[99] A. Lecko and A. Lyzzaik, A note on univalent functions starlike with respect to a boundary point, *J. Math. Anal. Appl.* **282** (2003), 846–851.

[100] M. Levenshtein and S. Reich, A rigidity theorem for commuting holomorphic functions, *J. Nonlinear Convex Anal.*, accepted for publication.

[101] E. Lindelöf, Mémoire sur sertaines inégalités dans la théorie des fonctions monogénes et sur quelques properiétés nouvelles de ces fonctions dans le voisinage d'un point singulier essentiel, *Acta Soc. Sci. Fennicae* **35** *7* (1909).

[102] A. Lyzzaik, On a conjecture of M.S. Robertson, *Proc. Amer. Math. Soc.* **91** (1984), 108–110.

[103] S. Migliorini and F. Vlacci, A new rigidity result for holomorphic maps, *Indag. Math.* **13** (2002), 537–549.

[104] J. Milnor, *Dynamics in one complex variable*, 3-d edition, Annals of Mathematics Studies, 160. Princeton University Press, Princeton, NJ, 2006.

[105] P. Montel, *Leçons sur les Fonctions Univalentes ou Multivalentes*, Gauthier-Villars, Paris, 1933.

[106] R. Nevanlinna, Über die konforme Abbildung Sterngebieten, *Oeversikt av Finska-Vetenskaps Societeten Ferhandlinger* **63(A)**, No. 6, 1921.

[107] P. Poggi-Corradini, Canonical conjugation at fixed points other than Denjoy–Wolff point, *Ann. Acad. Sci. Fenn. Math.* **25** (2000), 487–499.

[108] P. Poggi-Corradini, On the uniqueness of classical semiconjugations for self-maps of the disk, *Comput. Methods Funct. Theory* **6** (2006), 403–421.

[109] P. Poggi-Corradini, Pointwise convergence on the boundary in the Denjoy–Wolff theorem, Preprint.

[110] Ch. Pommerenke, On the iteration of analytic functions in a halfplane, *J. London Math. Soc.* (2) **19** (1979), 439–447.

[111] Ch. Pommerenke, On the stationary measures of critical branching processes, *Z. Wahrscheinlichkeitstheorie* **55** (1981), 305–312.

[112] Ch. Pommerenke, On asymptotic iteration of analytic functions in the disk, *Analysis* **1** (1981), 45–61.

[113] Ch. Pommerenke, *Boundary Behavior of Conformal Maps*, Springer–Verlag, New York, Berlin, Heidelberg, 1992.

[114] T. Poreda, On generalized differential equations in Banach space, *Dissertationes Math. (Rozprawy Mat.)* **310**, 1991.

[115] S. Reich and D. Shoikhet, Generation theory for semigroups of holomorphic mappings in Banach spaces, *Abstr. Appl. Anal.* **1** (1996), 1–44.

[116] S. Reich and D. Shoikhet, Semigroups and generators on convex domains with the hyperbolic metric, *Atti Accad. Naz. Lincei Cl. Sci. Fis. Mat. Natur. Rend. Lincei (9) Mat. Appl.* **8** (1997), 231–250.

[117] S. Reich and D. Shoikhet, The Denjoy–Wolff theorem, *Ann. Univ. Mariae Curie-Skłodowska, Sect. A* **51** (1997), 219–240.

[118] S. Reich and D. Shoikhet, Metric domains, holomorphic mappings and nonlinear semigroups, *Abstr. Appl. Anal.* **3** (1998), 203–228.

[119] S. Reich and D.Shoikhet, The Denjoy–Wolff theorem, *Math. Encyclopaedia, Supplement 3*, Kluwer Academic Publishers, (2002), 121–123.

[120] S. Reich and D.Shoikhet, , *Fixed Points, Nonlinear Semigroups and the Geometry of Domains in Banach Spaces*, World Scientific Publisher, Imperial College Press, London, 2005.

[121] M. S. Robertson, On the theory of univalent functions, *Ann. of Math.* **37** (1936), 374–408.

[122] M. S. Robertson, Analytic functions starlike in one direction, *Amer. J. Math.* **58** (1936), 465–472.

[123] M. S. Robertson, Univalent functions starlike with respect to a boundary point, *J. Math. Anal. Appl.* **81** (1981), 327–345.

[124] S. Ruscheweyh, A subordination theorem for Φ-like functions, *J. London Math. Soc.* **13** *2* (1976), 275–280.

[125] E. Schröder, Über iterierte Funktionen, *Math. Ann.* **3** (1871), 296–322.

[126] B. A. Sevastyanov, *Branching Processes*, Nauka, Moscow, 1971.

[127] J. H. Shapiro, *Composition Operators and Classical Function Theory*, Springer, Berlin, 1993.

[128] T. Sheil-Small, Starlike univalent functions, *Proc. London Math. Soc.* (3) **21** (1970), 577–613.

[129] A. L. Shields, On fixed points of commuting analytic functions, *Proc. Amer. Math. Soc.* **15** (1964) 703–706.

[130] D. Shoikhet, *Semigroups in Geometrical Function Theory*, Kluwer, Dordrecht, 2001.

[131] D. Shoikhet, Note on the Kœnigs function of a one-parameter semigroup and a distortion theorem for starlike functions with respect to a boundary point, Preprint, 2002.

[132] D. Shoikhet, Representations of holomorphic generators and distortion theorems for spirallike functions with respect to a boundary point, *Int. J. Pure Appl. Math.* **5** (2003), 335–361.

[133] D. Shoikhet, Kœnigs type linearization models and asymptotic behavior of one-parameter semigroups, *Contemporary Mathematics. Fundamental Directions*, **21** (2007), 149–166

[134] D. Shoikhet, Another look at the Burns–Krantz theorem, *J. Anal. Math.* **105** (2008), 19–42.

[135] H. Silverman and E. M. Silvia, Subclasses of univalent functions starlike with respect to a boundary point, *Houston J. Math.* **16** (1990), 289–299.

[136] A. Siskakis, Semigroups of composition operators on spaces of analytic functions, a review. *Studies on composition operators (Laramie, WY, 1996)*, 229–252, *Contemp. Math.* **213** *Amer. Math. Soc., Providence, RI*, 1998.

[137] L. Špaček, Přispěvek k teorii funkci prostych, *Časopis Pěst. Mat.* **62** (1933), 12–19.

[138] G. Srekeres, Regular iteration of real and complex functions, *Acta Math.* **100** (1958), 203–258.

[139] R. Tauraso, Commuting holomorphic maps of the unit disc, *Ergodic Theory Dynam. Systems* **24** (2004), 945–953.

[140] R. Tauraso and F. Vlacci, Rigidity at the boundary for holomorphic self-maps of the unit disk, *Complex Variables Theory Appl.* **45** (2001), 151–165.

[141] P. Todorov, On the univalent functions starlike with respect to a boundary point, *Proc. Amer. Math. Soc.* **97** (1986), 602–604.

[142] H. Upmeier, *Jordan Algebras in Analysis, Operator Theory and Quantum Mechanics*, CBMS Reg. Conf. Ser. in Math., **67**, Amer. Math. Soc., Providence, RI, 1987.

[143] G. Valiron, *Fonctions Analytiques*, Presses Univ. France, Paris, 1954.

[144] J. K. Wald, On starlike functions, Ph.D. Thesis, Univ. of Delaware, Newark, Delaware, 1978.

[145] J. Wolff, Sur l'iteration des fonctions holomorphes dans une region, et dont les valeurs appartiennent a cette region, *C. R. Acad. Sci.* **182** (1926), 42–43.

[146] J. Wolff, Sur l'iteration des fonctions bornees, *C. R. Acad. Sci.* **182** (1926), 200–201.

[147] J. Wolff, Sur une generalisation d'un theoreme de Schwarz, *C. R. Acad. Sci.* **182** (1926), 918–920.

[148] K. Yosida, *Functional Analysis*, Springer, Berlin, 1968.

Subject Index

Author Index

Symbols

List of Figures

Operator Theory: Advances and Applications (OT)

Edited by
Joseph A. Ball (Blacksburg, VA, USA), **Harry Dym** (Rehovot, Israel), **Marinus A. Kaashoek** (Amsterdam, The Netherlands), **Heinz Langer** (Vienna, Austria), **Christiane Tretter** (Bern, Switzerland)

This series is devoted to the publication of current research in operator theory, with particular emphasis on applications to classical analysis and the theory of integral equations, as well as to numerical analysis, mathematical physics and mathematical methods in electrical engineering.

OT 208: Elin, M. / Shoikhet, D., Linearization Models for Complex Dynamical Systems (2010). Subseries **L**inear **O**perators and **L**inear **S**ystems. ISBN 978-3-0346-0508-3

OT 207: Axler, S. / Rosenthal, P. / Sarason, D. (eds.), A Glimpse at Hilbert Space Operators. Paul R. Halmos in Memoriam (2010). ISBN 978-3-0346-0346-1

OT 206: Lerer, L. / Olshevsky, V. / Spitkovsky, I.M. (eds.), Convolution Equations and Singular Integral Operators (2010). ISBN 978-3-7643-8955-0

OT 205: Schulze, B.-W. / Wong, M.W. (eds.), Pseudo-Differential Operators: Complex Analysis and Partial Differential Equations (2009). ISBN 978-3-0346-0197-9

OT 204: Frazho, A. / Bhosri, W. An Operator Perspective on Signals and Systems (2009). Subseries **L**inear **O**perators and **L**inear **S**ystems. ISBN 978-3-0346-0291-4

OT 202/203: Ball, J.A. / Bolotnikov, V. / Helton, J.W. / Rodman, L. / Spitkovsky, I.M. (eds.), Topics in Operator Theory. A Tribute to Israel Gohberg on the Occasion of his 80th Birthday (2010)
Volume 1: Operators, Matrices and Analytic Functions. ISBN 978-3-0346-0157-3
Volume 2: Systems and Mathematical Physics. ISBN 978-3-0346-0160-3
Set Volumes 1/2: ISBN 978-3-0346-0163-4

OT 201: Curbera, G.P. / Mockenhaupt, G. / Ricker, W.J. (eds.), Vector Measures, Integration and Related Topics (2009). ISBN 978-3-0346-0210-5

OT 200: Bart, H. / Gohberg, I. / Kaashoek, M.A., Ran, A.C.M., A State Space Approach to Canonical Factorization with Applications (2010). Subseries **L**inear **O**perators and **L**inear **S**ystems. ISBN 978-3-7643-8752-5

OT 199: Bini, D.A. / Mehrmann, V. / Olshevsky, V. / Tyrtsyhnikov, E. / Van Barel, M. (eds.), Numerical Methods for Structured Matrices and Applications. The Georg Heinig Memorial Volume (2010). ISBN 978-3-7643-8995-6

OT 198: Behrndt, J. / Förster, K.-H. / Trunk, C. (eds.), Recent Advances in Operator Theory in Hilbert and Krein Spaces (2009). ISBN 978-3-0346-0179-5

OT 197: Alpay, D. / Vinnikov, V. (eds.), Characteristic Functions, Scattering Functions and Transfer Functions. The Moshe Livsic Memorial Volume (2009). ISBN 978-3-0346-0182-5

OT 196: Maz'ya, V. / Soloviev, A., Boundary Integral Equations on Contours with Peaks (2009). ISBN 978-3-0346-0170-2

OT 195: Grobler, J.J. / Labuschagne, L.E. / Möller, M. (eds.), Operator Algebras, Operator Theory and Applications (2009). ISBN 978-3-0346-0173-3

OT 194: González, M. / Martínez-Abejón, A,, Tauberian Operators (2009). ISBN 978-3-7643-8997-0

OT 193: Cialdea, A. / Lanzara, F. / P.E. Ricci (eds.), Analysis, Partial Differential Equations and Applications. The Vladimir Maz'ya Anniversary Volume (2009). ISBN 978-3-7643-9897-2

Operator Theory: Advances and Applications (OT)

Edited by
Joseph A. Ball (Blacksburg, VA, USA), **Harry Dym** (Rehovot, Israel), **Marinus A. Kaashoek** (Amsterdam, The Netherlands), **Heinz Langer** (Vienna, Austria), **Christiane Tretter** (Bern, Switzerland)

This series is devoted to the publication of current research in operator theory, with particular emphasis on applications to classical analysis and the theory of integral equations, as well as to numerical analysis, mathematical physics and mathematical methods in electrical engineering.

BIRKHÄUSER

OT 192: Gohberg, I. / Leiterer, J., Holomorphic Operator Functions of One Variable and Applications. Methods from Complex Analysis in Several Variables (2009).
ISBN 978-3-0346-0125-2

OT 190/191: Adamyan, V. / Berezansky, Yu.M. / Gohberg, I. / Gorbachuk, M.L. / Gorbachuk, V. / Kochubei, A.N. / Langer, H. / Popov, G. (eds.), Modern Analysis and Applications. (2009)
Volume 1 (OT 190): Operator Theory and Related Topics. ISBN 978-3-7643-9918-4
Volume 2 (OT 191): Differential Operators and Mechanics. ISBN 978-3-7643-9920-7
Set Volumes 1/2: ISBN 978-3-7643-9924-5

OT 189: Rodino, L. / Wong, M.W. (eds.), New Developments in Pseudo-Differential Operators (2008). ISBN 978-3-7643-8968-0

OT 188: Behrndt, J. / Förster, K.-H. / Langer, H. / Trunk, C. (eds.), Spectral Theory in Inner Product Spaces and Applications (2008).
ISBN 978-3-7643-8910-9

OT 187: Ando, T. / Curto, R.E. / Jung, I.B. / Lee, W.Y. (eds.), Recent Advances in Operator Theory and Applications (2008).
ISBN 978-3-7643-8892-8

OT 186: Janas, J. / Kurasov, P. / Laptev, A. / Naboko, S. / Stolz, G. (eds.), Methods of Spectral Analysis in Mathematical Physics (2008).
ISBN 978-3-7643-8754-9

OT 185: Vasilevski, N., Commutative Algebras of Toeplitz Operators on the Bergman Space (2008).
ISBN 978-3-7643-8725-9

OT 184: Qin, Y., Nonlinear Parabolic-Hyperbolic Coupled Systems and Their Attractors (2008). Subseries **A**dvances in **P**artial **D**ifferential **E**quations. ISBN 978-3-7643-8813-3

OT 183: Nazaikinskii, V.E.. / Savin, A.Yu. / Sternin, B.Yu., Elliptic Theory and Noncommutative Geometry (2008). Subseries **A**dvances in **P**artial **D**ifferential **E**quations.
ISBN 978-3-7643-8774-7

OT 182: van der Mee, C., Exponentially Dichotomous Operators and Applications (2008). Subseries **L**inear **O**perators and **L**inear **S**ystems. ISBN 978-3-7643-8732-7

OT 181: Bastos, A. / Gohberg, I. / Lebre, A.B. / Speck, F.-O. (eds.), Operator Algebras, Operator Theory and Applications (2008).

OT 180: Okada, S. / Ricker, W.J. / Sánchez Pérez, E.A., Optimal Domain and Integral Extension of Operators Acting in Function Spaces (2008). ISBN 978-3-7643-8647-4

OT 179: Ball, J.A. / Eidelman, Y. / Helton, J.W. / Olshevsky, V. / Rovnyak, J. (eds.), Recent Advances in Matrix and Operator Theory (2007). ISBN 978-3-7643-8538-5

OT 178: Bart, H. / Gohberg, I. / Kaashoek, M.A. / Ran, A.C.M., Factorization of Matrix and Operator Functions: The State Space Method (2007).
ISBN 978-3-7643-8267-4

OT 177: López-Gómez, J. / Mora-Corral, C., Algebraic Multiplicity of Eigenvalues of Linear Operators (2007). ISBN 978-3-7643-8400-5

OT 176: Alpay, D. / Vinnikov, V. (Eds.), System Theory, the Schur Algorithm and Multidimensional Analysis (2007).
ISBN 978-3-7643-8136-3

OT 175: Förster, K.-H. / Jonas, P. / Langer, H. / Trunk, C. (Eds.), Operator Theory in Inner Product Spaces (2007).
ISBN 978-3-7643-8269-8

Subseries *Linear Operators and Linear Systems*

BIRKHÄUSER

Edited by

Daniel Alpay, Ben Gurion University of the Negev, Beer Sheeva, Israel
Birgit Jacob, Bergische Universität Wuppertal, Germany
André C.M. Ran, Vrije Universiteit, Amsterdam, The Netherlands

The theory of linear operators has had an important influence on the development of mathematical systems theory. On the other hand, mathematical systems theory serves as a direct source of motivation and new techniques for the theory of linear operators and its applications. The subseries "Linear Operators and Linear Systems" (LOLS) is dedicated to these connections between the theory of linear operators and the mathematical theory of linear systems. LOLS will continue in the tradition of the series Operator Theory: Advances and Applications and maintain the high quality of the volumes. Books published in LOLS will be either monographs or consist of essays presenting the state of the art and new results. The volumes will be addressed to a wide range of mathematicians from beginners to experts in these fields.

OT 208: Elin, M. / Shoikhet, D.,
Linearization Models for Complex Dynamical Systems (2010). ISBN 978-3-0346-0508-3

OT 204: Frazho, A. / Bhosri, W.,
An Operator Perspective on Signals and Systems (2009). ISBN 978-3-0346-0291-4

This book provides an operator perspective to solve problems arising in systems theory and signal processing. It is divided into four parts. The first part is an introduction to basic operator techniques used throughout the monograph. The second part uses finite section methods with interpolation theory to develop algorithms to compute inner-outer factorizations and solve sinusoid estimation problems. The third part develops the Riccati method to solve factorization problems arising in systems and control. We also included the Kalman filter and its connection to Wiener filtering. The final part uses isometric extensions to present an introduction to positive real and contractive interpolation theory.

OT 200: Bart, H. / Gohberg, I. / Kaashoek, M.A., Ran, A.C.M.,
A State Space Approach to Canonical Factorization with Applications (2010).
ISBN 978-3-7643-8752-5

OT 184: van der Mee, C.V.M.,
Exponentially Dichotomous Operators and Applications (2008).
ISBN 978-3-7643-8731-0

In this monograph the natural evolution operators of autonomous first-order differential equations with exponential dichotomy on an arbitrary Banach space are studied in detail. Characterizations of these so-called exponentially dichotomous operators in terms of their resolvents and additive and multiplicative perturbation results are given. The general theory of the first three chapters is then followed by applications to Wiener-Hopf factorization and Riccati equations, transport equations, diffusion equations of indefinite Sturm-Liouville type, noncausal infinite-dimensional linear continuous-time systems, and functional differential equations of mixed type.

OT 178: Bart, H. / Gohberg, I. / Kaashoek, M.A. / Ran, A.C.M.,
Factorization of Matrix and Operator Functions: The State Space Method (2007).
ISBN 978-3-7643-8267-4

OT 165: Alpay, D. / Gohberg, I. (Eds.),
Interpolation, Schur Functions and Moment Problems (2006). ISBN 978-3-7643-7546-1

OT 161: Alpay, D. / Gohberg, I. (Eds.),
The State Space Method. Generalizations and Applications (2005).
ISBN 978-3-7643-7370-2

GPSR Compliance

The European Union's (EU) General Product Safety Regulation (GPSR) is a set of rules that requires consumer products to be safe and our obligations to ensure this.

If you have any concerns about our products, you can contact us on ProductSafety@springernature.com

In case Publisher is established outside the EU, the EU authorized representative is:

Springer Nature Customer Service Center GmbH
Europaplatz 3
69115 Heidelberg, Germany

Batch number: 10732744

Printed by Printforce, the Netherlands